NEURAL NETWORK APPLICATIONS IN CONTROL

Edited by G. W. Irwin, K. Warwick and K. J. Hunt

The Institution of Electrical Engineers

Published by: The Institution of Electrical Engineers, London, United Kingdom

The Institution of Electrical Engineers,
Michael Faraday House,
Six Hills Way, Stevenage,
Herts. SG1 2AY, United Kingdom

British Library Cataloguing in Publication Data

A CIP catalogue record for this book
is available from the British Library

ISBN 0 85296 852 3

Printed in England by Short Run Press Ltd., Exeter

IEE CONTROL ENGINEERING SERIES 53

Series Editors: Professor D. P. Atherton
Professor G. I. Irwin

NEURAL NETWORK APPLICATIONS IN CONTROL

Other volumes in this series:

Contents

Page

Preface

The field of neural networks for control continues to flourish following the enormous worldwide resurgence of activity in the late 1980s. Encouraged by the success of our text earlier in the series, 'Neural networks for control and systems', and realising that the subject had moved on significantly with the applications potential becoming clearer, it was decided to produce the present book. The aim is to present an introduction to, and an overview of, the present state of neural network research and development, with an emphasis on control systems application studies. The book is useful to a range of levels of reader. The earlier chapters introduce the more popular networks and the fundamental control principles, these are followed by a series of application studies, most of which are industrially based, and the book concludes with a consideration of some recent research.

The level of the book makes it suitable for practising control engineers, final year students and both new and advanced researchers in the field. The contents have been chosen to present control applications of neural networks together with a sound theoretical base with a minimum of overly mathematical abstract ideas.

The first four chapters together constitiute a good introductory perspective to the subject, with Chapter 1 in particular giving an easy guide to the more popular neural network paradigms. Chapter 2 describes digital networks which have been used in electronic neural computing hardware like WISARD and MAGNUS. This is followed by a concise overview of the fundamentals of neurocontrol in Chapter 3 which concentrates on the general theoretical framework. Chapter 4 deals with nonlinear modelling by neural networks from a function approximation viewpoint, a key concept in the use of this new technology for control.

The next set of six contributions form the backbone of the book dealing as they do with a variety of different control applications. Chapter 5 reports results from two industrial projects: one on feedforward networks for the nonlinear modelling of a 200 MW oil-fired power station boiler, the other concerned with viscosity prediction for a chemical polymerisation reactor using both feedforward and B-spline networks. Chapter 6 links neural networks for identification, Gaussian feedforward networks and learning algorithms with structures for neurocontrol and concludes with an example of pH control. Further process control applications are reported in Chapter 7, including inferential estimation of polymer quality and nonlinear predictive control of a distillation tower. Here dynamic feedforward networks are studied. A variety of applications, including vision systems, colour recipe predictions and speech processing are the subject of the next chapter which uses a number of different networks. Model predictive control, based on a radial basis function network, is discussed in Chapter 9. Significantly this has been applied to the control of an AC servo drive system and implemented in real-time on parallel processing hardware. By contrast the subject of Chapter 10 is neurofuzzy control in intensive-care blood pressure management.

The book concludes with chapters of a more theoretical nature. Chapter 11 looks more deeply into learning algorithms, radial basis functions and system identification, thereby providing rigour. The last chapter, a long one, takes a comprehensive look at some of the more recent research ideas in neurofuzzy modelling and highlights emerging ideas such as local model networks and constructive learning which will have an impact on applications in the future.

In total the book provides a strong coverage of the state-of-the-art in control applications of neural networks together with the present position in research progress. It is possible for the book to be used as a complete text, comprehensively covering the field or, in terms of individual chapters, detailing areas of special interest. It is our intention that the book will be of use to readers wishing to find out about this important new technology and its potential advantages for control engineering applications.

The Editors would like to express their sincere thanks to all the authors for their contributions and co-operation in the production of the finalised text. Thanks are also due to Robin Mellors-Bourne and Fiona MacDonald of the IEE for their unfailing encouragement and assistance, and to Hilary Morrow and Rafal Żbikowski who helped overcome some of the difficulties with incompatible word-processing formats.

Sadly, Patrick Parks one of the authors and an internationally well known figure in the control research community for many years died earlier in the year. More recently he was an enthusiastic participant in meetings on neural networks for control and his technical contribution and friendship will be missed.

George Irwin, Belfast
Kevin Warwick, Reading
Ken Hunt, Berlin

October 1995

Contributors

K Warwick
Department of Cybernetics
The University of Reading
Whiteknights
PO Box 225
Reading
RG6 2AY

e-mail: kw@cyber.rdg.ac.uk

A Redgers and I Aleksander
Department of Electrical Engineering
Imperial College
Exhibition Road
London
SW7 2BT

e-mail: i.aleksander@ic.ac.uk

R Żbikowski and P J Gawthrop
Department of Mechanical Engineering
University of Glasgow
Glasgow
G12 8QQ

e-mail: rafal@mech.gla.ac.uk

J C Mason[1] and P C Parks[2]
[1]School of Computing and Mathematics
University of Huddersfield
Queensgate
Huddersfield
West Yorkshire
HD1 3DH

e-mail: j.c.mason@hud.ac.uk

[2]Formerly with the University of Oxford

G Irwin, P O'Reilly, G Lightbody, M Brown and E Swidenbank
Department of Electrical and Electronic Engineering
Queen's University of Belfast
Stranmills Road
Belfast
BT9 5AH

e-mail: g.irwin@ee.qub.ac.uk
Internet: http://www-cerc.ee.qub.ac.uk

K J Hunt and D Sbarbaro
Daimler-Benz AG
Alt-Moabit 91b
D-10559 Berlin
Germany

e-mail: hunt@dbresearch-berlin.de

P Turner, J Morris and G A Montague
Department of Chemical and Process Engineering
Merz Court
University of Newcastle upon Tyne
Newcastle upon Tyne
NE1 7RU

e-mail: julian.morris@newcastle.ac.uk

D A Linkens
Department of Automatic Control and Systems Engineering
University of Sheffield
PO Box 600
Mappin Street
Sheffield
S1 4DU

e-mail: d.linkens@sheffield.ac.uk

R J Mitchell and J M Bishop
Department of Cybernetics
The University of Reading
Whiteknights
PO Box 225
Reading
RG6 2AY

e-mail: r.j.mitchell@reading.ac.uk

K M Bossley, M Brown and C J Harris
Department of Electronic and Computer Science
University of Southampton
Highfield
Southampton
SO17 1BJ

e-mail: mqb@ecs.soton.ac.uk

D Neumerkel, J Franz and L Kruger
Systems Technology Research
Daimler-Benz AG
Alt-Moabit 91b
D-10559 Berlin
Germany

e-mail: neumerkel@dbresearch-berlin.de

Chapter 1

Neural networks: an introduction

K. Warwick

1.1 Introduction

The study of neural networks is an attempt to understand the functionality of the brain. In particular it is of interest to define an alternative 'artificial' computational form that attempts to mimic the brain's operation in one or a number of ways, and it is this area of artificial neural networks which is principally addressed here.

Essentially artificial neural networks is a 'bottom up' approach to artificial intelligence, in that a network of processing elements is designed, these elements being based on the physiology and individual processing elements of the human brain. Further, mathematical algorithms carry out information processing for problems whose solutions require knowledge which is difficult to describe.

In the last few years interest in the field of neural networks has increased considerably, due partly to a number of significant break-throughs in research on network types and operational characteristics, but also because of some distinct advances in the power of computer hardware which is readily available for net implementation. It is worth adding that much of the recent drive has, however, arisen because of numerous successes achieved in demonstrating the ability of neural networks to deliver simple and powerful problem solutions, particularly in the fields of learning and pattern recognition, both of which have proved to be difficult areas for conventional computing.

In themselves, digital computers provide a medium for well-defined, numerical algorithm processing in a high performance environment. This is in direct contrast to many of the properties exhibited by biological neural systems, such as creativity, generalisation and understanding. However, computer-based neural networks, both of hardware and software forms, at the present time provide a considerable move forward from digital computing in the direction of biological systems; indeed several biological neural system properties can be found in certain neural network types. This move is supported by a number of novel practical examples, even though these tend to be in fairly scientific areas, e.g. communication processing and pattern recognition.

Because of its inter-disciplinary basis, encompassing computing, electronics, biology and neuropsychology, the field of neural networks attracts a variety of interested researchers and implementers from a broad range of backgrounds. This makes the field very exciting, with a flood of new ideas still to be tried and tested. In this chapter a brief view is given of some of the various artificial neural network techniques presently under consideration, a few application areas are discussed and indications are given as to possible future trends.

1.2 Neural network principles

The present field of neural networks links a number of closely related areas, such as parallel distributed processing, connectionism and neural computing, these being brought together with the common theme of attempting to exhibit the method of computing which is witnessed in the study of biological neural systems. An interesting and succinct definition of neural computing [1] is that it is a study of networks of adaptable nodes which, through a process of learning from task examples, store experiential knowledge and make it available for use. Taking this definition alone, although representing a significant step forward from standard digital computing, it does indicate a rather limited, practically oriented state at the present time, leaving much ground still to be covered towards the artificial reconstruction of complex biological systems.

A fundamental aspect of artificial neural networks is the use of simple processing elements which are essentially models of neurons in the brain. These elements are then connected together in a well-structured fashion, although the strength and nature of each of the connecting links dictates the overall operational characteristics for the total network. By selecting and modifying the link strengths in an adaptive fashion, so the basis of network learning is formed along the lines of the previous definition [1].

One important property of a neural network is its potential to infer and induce from what might be incomplete or non-specific information. This is, however, also coupled with an improvement in performance due to the network learning appropriate modes of behaviour in response to problems presented, particularly where real-world data are concerned. The network can, therefore, be taught particular patterns of data presented, such that it can subsequently not only recognise such patterns when they occur again, but also recognise similar patterns by generalisation. It is these special features of learning and generalisation that make neural networks distinctly different from conventional algorithm processing computers, along with their potential property of faster operational speeds realised through inherent parallel operation.

1.3 Neural network elements

It is not the intention here to provide a detailed history of artificial neural networks, however, earlier stages of development can be found in [2–4], whereas a good guide to more recent events can be found in [5,6]. The discussion here focuses on some of the different approaches taken in the design of neural networks and their basic principles of operation.

In general, neural networks consist of a number of simple node elements, which are connected together to form either a single layer or multiple layers. The relative strengths of the input connections and also of the connections between layers are then decided as the network learns its specific task(s). This learning procedure can make use of an individual data set, after which the strengths are fixed, or learning can continue throughout the network's lifetime, possibly in terms of a faster rate in the initial period with a limited amount of 'open mind' allowance in the steady-state. In certain networks a supervisor is used to direct the learning procedure, and this method is referred to as 'supervised learning', whereas it is also possible to realise self-organising networks which learn in an unsupervised fashion.

The basic node elements employed in neural networks differ in terms of the type of network considered. However, one commonly encountered model is a form of the McCulloch and Pitts neuron [7], as shown in Figure 1.1. In this, the inputs to the node take the form of data items either from the real world or from other network elements, possibly from the outputs of nodes in a previous layer. The output of the node element is found as a function of the summed weighted strength inputs.

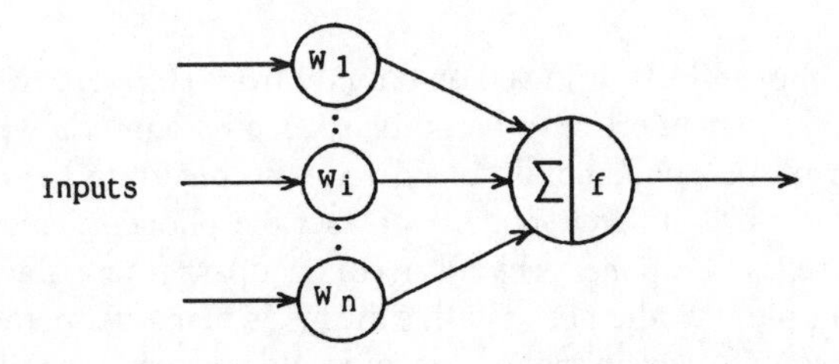

Figure 1.1 *Basic neuron model*

This output signal can then be employed directly, or it can be further processed by an appropriate thresholding or filtering action, a popular form being a sigmoid function. For some networks the node output signal is therefore used directly, whereas in other cases the important feature is simply whether or not the node has fired due to the weighted sum exceeding a previously defined threshold value for that particular node.

1.4 Hopfield networks

In the network model proposed in [8], binary input signals are introduced such that each of the network node elements has an output which acquires one of two possible states, either +1 or −1. Further, the number of node elements is assumed to be extremely large, with each element connected, in some way, to all others, i.e. the output of a node is connected as a weighted input to all other nodes. Each node does, however, have its own separate input which is unique, and this is an effective input interface with the real world.

A signal/data pattern can be applied to the real world network inputs, and the weights adjusted, by such means as Hebbian learning, to produce an output pattern which is, in some way, associated with that particular input pattern. A good description, in slightly more detail, of this operation can be found in [9], in which it is shown how weights are adjusted to arrive at a local minimum of a defined global energy function. This energy function is given by

$$E = -\frac{1}{2}\sum_{ij=1}^{N} b_{ij} S_i S_j \qquad (1.1)$$

where E is the energy function, N is the total number of nodes, S_i and S_j are the output values, either +1 or −1, of the ith and jth node elements and b_{ij} is the weighting applied to the link from the jth node output to the ith node input.

1.5 Kohonen networks

In many ways Kohonen [10] followed on directly from Hopfield by using a single layer analogue neural network as a basis for speech recognition. Here, a speech signal, in the form of a phoneme, is considered with regard to the energy content of the phoneme at different frequencies. In fact the phoneme energy is split up into several frequency components by means of bandpass filters, the output of each filter acting as an input to the net. By this means, when a phoneme is uttered by a particular person, a certain pattern is presented as an overall input to the network.

The weightings on each node input are initially selected randomly. Then for a pattern presented as input to the network, a clustering effect on the network output can be enforced by adjustment of the weightings between nodes. This can be achieved by selecting a node output as that which resembles a specific phoneme. The nodes which are adjacent then have their weights adjusted, the amount of adjustment being dependent on their distance from the selected node. If the selected node weighting is adjusted by an amount x, then the adjacent nodes' weightings are adjusted by an amount Q, where Q is given by the Mexican hat function shown in Figure 1.2. In this way a tailored network response, to a presented input pattern, is enforced, i.e. the network is taught how to respond to a certain input pattern.

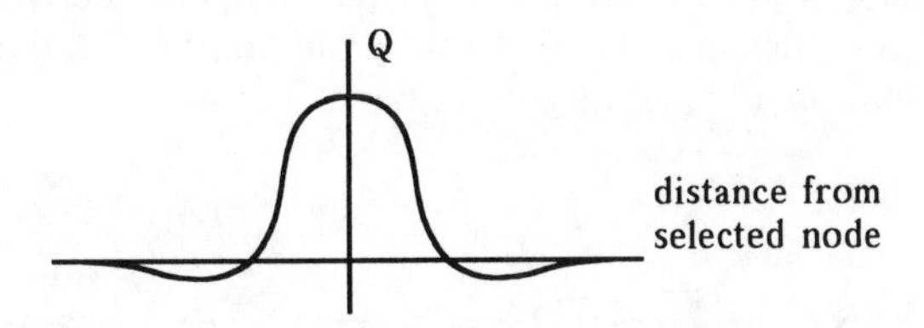

Figure 1.2 *Mexican hat function*

In the Kohonen network, the nodes can be laid out in a two-dimensional matrix pattern, such that when a phoneme pattern is input to the network, a small area of the matrix, centred around one node output, should exhibit active outputs; the actual area being selected due to the training/tailoring process. In this way, having been trained on numerous phoneme energy inputs, when a phoneme is uttered again, this should be recognisable on the matrix. Problems arise due to (*a*) different speakers, therefore different energy spectra, of the same phoneme, (*b*) the same speaker uttering a phoneme in a different way and (*c*) phonemes that are new to the matrix. It is, however, possible [11] to form a hierarchical system where prior knowledge is available of the likely phoneme to be uttered, such cases arising in natural language.

The description of Kohonen networks has been given here principally in terms of speech recognition, in fact this is just one example of their use, which happened to be the area of research originally investigated by Kohonen. Such networks can be used in general as classifier systems, i.e. given a set of data the network can indicate particular characteristics of the data by classifying them as being similar to previously learnt data sets, e.g. for fault indication and diagnosis.

1.6 Multi-layer perceptrons

Multi-layer perceptrons (MLPs) have very quickly become the most widely encountered artificial neural networks, particularly within the area of systems and control [12]. In these networks it is assumed that a number of node element layers exist and further, that no connections exist between the node elements in a particular layer. One layer of nodes then forms the input layer whilst a second forms the output layer, with a number of intermediate or hidden layers existing between them. It is often the case, however, that only one, two or even no hidden layers are employed.

A multi-layer perceptron requires a pattern, or set of data, to be presented as inputs to the input node element layer. The outputs from this layer are then fed, as weighted inputs, to the first hidden layer, and subsequently the outputs from the first layer are fed, as weighted inputs, to the second hidden layer. This construction process continues until the output layer is reached. The structure is shown in a basic three-layer form in Figure 1.3, although it should be stressed that this arrangement only includes one hidden layer and further, no input weightings are immediately indicated.

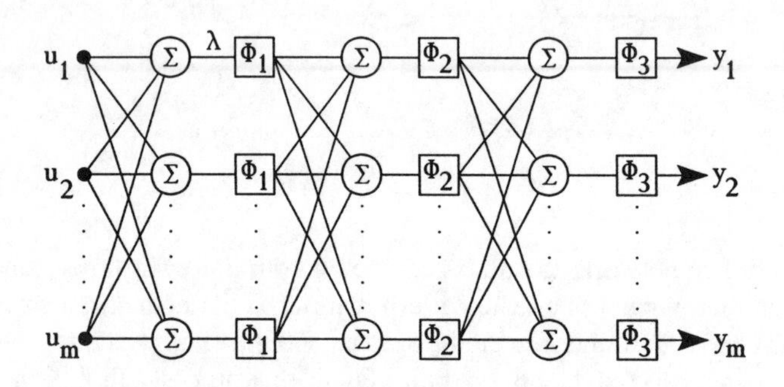

Figure 1.3 *Three-layer multi-layer perceptron*

The three-layer network shown in Figure 1.3 depicts m inputs and m outputs, though it is not, in general, necessary for these values to be equal. With the network connected thus, no fixed feedback exists, other than is necessary for weight training, and the network is termed a static network in that network input signals are fed directly from input to output. Where such feedback exists from neuron states or outputs to inputs, then this is called a recurrent or dynamic network, often with a delay in the feedback path [13].

An individual (ith) node in the network can be described by

$$y_i = \Phi(s_i + \omega_i) = \Phi(\lambda) \tag{1.2}$$

in which y_i is the node output, ω_i is a bias term and Φ is the nonlinear function applied. The term s_i is given as

$$s_i = \sum_{j=1}^{N} \omega_{ij} x_j \tag{1.3}$$

where x_j are the ith neuron's N input signals and ω_{ij} the adjustable weightings associated with each input.

The nonlinear function, Φ, is often common throughout the network, i.e. $\Phi = \Phi_1 = \Phi_2 = \Phi_3$, and can take on one of a number of forms, the most common being sigmoidal, such that

$$\Phi(\lambda) = 1/(1 + \exp(-\lambda)) \tag{1.4}$$

or

$$\Phi(\lambda) = (1 - \exp(-\lambda))/(1 + \exp(+\lambda)) \tag{1.5}$$

Other nonlinear functions encountered are (*a*) Gaussian relationships (*b*) threshold functions, and (*c*) hyperbolic tangents.

In reality it is known that real neurons, located in different parts of the brain, behave in different ways [14], ranging from Gaussian-like for visual needs to sigmoidal for ocular motor needs. It is usually the case for artificial neural networks, however, that only one type of nonlinearity is used in each network, thus linking it closely with the fact that each network is usually suitable for one specific task only.

The multi-layer perceptron of Figure 1.3 is a fully-connected network in that all of the neuron outputs from one layer are connected as inputs to the next layer, and this is normal practice. It is quite possible, however, for only part-connectivity to be realised by simply connecting a group of outputs to only a subset of the next layer's inputs. In this way sub-models can be formed within the network, a feature which is useful when system data can be classified into distinct type groups.

For multi-layer perceptrons, important structural questions are paramount and need to be answered at a relatively early stage; namely how many layers of neurons should there be for a particular problem and how many neurons should there be in each layer? Once the structural selections have been made, the network weights can be chosen such that the network can perform as desired. Although some algorithms for network structure selection do exist [15], in most cases the structural specification of MLPs is carried out in a fairly heuristic way, so for a certain problem a reasonable number of layers and neurons in each layer are initially selected, based on experience. If the numbers selected appear to be too large or too small then adjustments can be made on a trial-and-error basis.

Normal use of an MLP in practice involves training the network on a set of data, mostly obtained from the plant to be controlled. This necessitates training the network weights once the network structure and hence the resultant number and location of the weights, have been specified. The weights are then adjusted so that the network input-output relationship best approximates the plant data. The weights can then be fixed at the values trained and the network subsequently used as a plant model.

Dynamic networks, which have a much greater capacity for modelling complex dynamical plant, can be connected up in a number of ways. One possibility is to employ only a single-layered arrangement, with each individual network node, being described by:

$$T_i y_i(k+1) = \Phi(s_i(k)) + \omega_i(k) \tag{1.6}$$

in which $y_i(k+1)$ indicates the ith neuron's output at time instant $k+1$, with $\omega_i(k)$ showing the bias term at instant k. The term T_i is an optional, adjustable parameter thereby achieving a more general functional description.

For multi-layer perceptrons, weight learning is most commonly carried out by the method of backpropagation [5]. In this approach the network outputs are first compared with a set of the desired values for those outputs namely $d_i (i = 1,..., m)$. An appropriate error function is then given by:

$$E(W) = \frac{1}{2}\sum_{i=1}^{m}(d_i - y_i)^T(d_i - y_i) \tag{1.7}$$

and this error, on the output layer only, must first be minimised by a best selection of output layer weights. Once the output layer weights have been selected the weights in the hidden layer next to the output can be adjusted by employing a linear backpropagation of the error term from the output layer. This procedure is followed until the weights in the input layer are adjusted. A full description of backpropagation for both static and dynamic multi-layer perceptrons can be found in Cichocki and Unbehauen [16].

On the negative side, backpropagation is a nonlinear steepest descent type algorithm and can either converge on local minima or be extremely slow to converge. On the positive side, however [17], a multi-layer perceptron with only one hidden layer is sufficient to approximate any continuous function.

Multi-layer perceptrons can be used in a number of ways in the systems and control area [18] one example being in the formation of process models, where reasonably cost effective and reliable results can be achieved. Such networks are especially useful in this respect, because of their inherent nonlinear mapping capabilities, which can deal effectively with a wide range of process features. Another immediate use of these networks is in the real-time estimation of process parameters within an adaptive model of the plant under control.

The employment of multi-layer perceptron models of a process plant can offer distinct advantages when compared with more conventional techniques. As an example, a network can be used to obtain frequent measurement information of a form which is better than that which can be obtained by hardware methods alone, thus overcoming some problems incurred with delayed signals. Such networks can also be used to provide inferred values for signals which are difficult to measure in a practical situation, something which is a common occurrence in many process control problems.

The use of artificial networks within model-based control schemes has been considered also, such as those employing a model reference strategy [12], a useful mode of implementation being in the form of a multi-layered network processor, something which is particularly beneficial when applied to nonlinear systems.

Multi-layer perceptrons are also well suited for real-time, on-line, computer control systems because of their abilities in terms of rapid processing of collected plant input-output data. One example of this is in the control of a robot manipulator [19], where the network mapping adapts, in terms of its weightings, as it learns with regard to characteristic changes in robot manipulator joints, such changes occurring as the manipulator moves within its field of operation. The network rôle is to produce an output vector which includes signals such as servo motor input power and desired trajectories. Network weightings are adapted on-line as manipulator operating conditions change, such that once an initial learning phase has been completed the network is allowed to continue an element of learning as it is in operation, in order to track changes in motor joint characteristics. This is shown schematically in Figure 1.4.

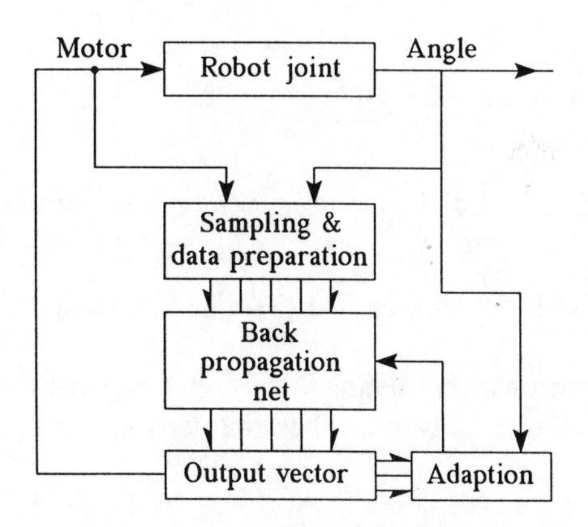

Figure 1.4 *Neural net for robot control*

1.7 Radial basis function networks

Multi-layer perceptron networks, as described in the previous section, have been found, in practice, to perform poorly in a number of ways, e.g. slow convergence of weights in the nonlinear updating procedure and difficulty in modelling differential responses. Recently, popularity has therefore moved towards radial basis function networks which are conceptually simpler and have the ability to model any nonlinear function in a relatively straightforward way [20]. A number of other methods, e.g. CMAC (cerebellar model articulation controller) [21], cellular neural networks [22], and orthogonal basis functions, exhibit several similarities with radial basis functions (RBFs), and hence much of the discussion in this section applies equally to a number of different network types.

A key feature of RBF networks is that, as shown in Figure 1.5, the output layer is merely a linear combination of the hidden layer signals, there being only one hidden layer. RBF networks therefore allow for a much simpler weight updating procedure and subsequently open up greater possibilities for stability proofs and network robustness in that the network can be described readily by a set of nonlinear equations.

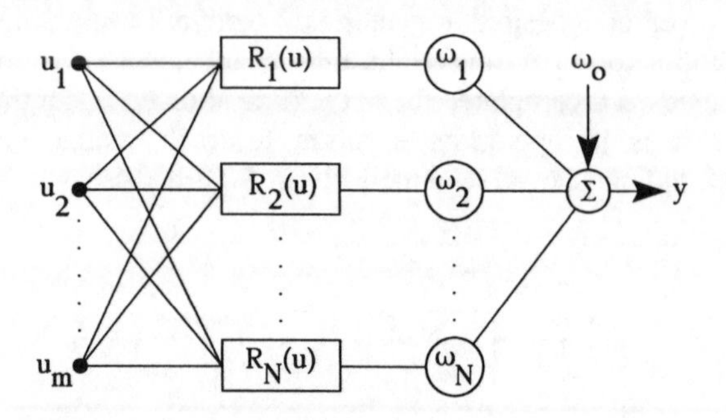

Figure 1.5 *A radial basis function network*

Operation of an RBF network can be described as follows:

Given an input vector u, consisting of m input signals, and a set of weights $\omega_i (i = 1, ..., N)$, with ω_0 a bias term, then the network output is found from:

$$y = \sum_{i=1}^{N} \omega_i R_i(u) + \omega_0 \tag{1.8}$$

where N is the number of hidden layer neurons and R_i are activation functions of the form

$$R_i(u) = \Phi(\| u - c_i \|) \tag{1.9}$$

where $c_i (i = 1, ..., N)$ are radial basis function centres and $\Phi(.)$ is the radial basis function which can be selected in one of a number of ways, examples being:

(*a*) Gaussian,

$$\Phi(r) = \exp(-r^2/\sigma^2) \tag{1.10}$$

(*b*) Piecewise linear approximation,

$$\Phi(r) = r \tag{1.11}$$

(*c*) Cubic approximation

$$\Phi(r) = r^3 \tag{1.12}$$

(*d*) Multiquadratic function

$$\Phi(r) = (r^2 + \sigma^2)^{1/2} \quad (1.13)$$

(*e*) Inverse-multiquadratic function

$$\Phi(r) = 1/(r^2 + \sigma^2)^{1/2} \quad (1.14)$$

(*f*) Thin plate spline

$$\Phi(r) = r^2 \log(r) \quad (1.15)$$

where σ is simply a scaling parameter and

$$r = \| u - c_i \| \quad (1.16)$$

i.e. u is the vector of m input signals, u_1, ..., u_m, and each c_i is a function centre vector. The value r therefore indicates the Euclidean distance from the input vector u to the centre c_i.

Of these possible choices for basis function, (*a*), the Gaussian, is perhaps the most intuitive in that a maximum occurs when the input vector is exactly equal to the function centre, the value trailing off exponentially the further the input vector is away from the centre. The rate of trailing off is in fact directed by the scaling parameter σ, thus allowing for either sharp, specific radial basis functions or broad, generalising types of functions. It is worth pointing out however that, although it is not so intuitively appealing, the thin plate spline (*f*) works very well in practice.

System dynamics can easily be taken account of in an RBF network through the input vector u, in that this vector can take the form

$$u(k) = (y(k-1), y(k-2), \ldots; u(k-1), u(k-2), \ldots) \quad (1.17)$$

i.e. the vector, at time instant k, consists of sampled past values of output and input, where $u(k-1)$ indicates the value of the input signal one sample period ago, etc. The network output then models the plant output at instant k, i.e. it becomes $y(k)$. The RBF network can thus be used either to model a nonlinear plant response, in a sampled-data fashion, or to form the basis of a desired sampled-data controller.

Although only one output is shown in Figure 1.5 it is easily possible for multiple outputs to be modelled with RBF networks, in that either parallel network arrangements can be employed with a common input vector or, more likely, only a subset of the radial basis function outputs feeds each of the network outputs.

In operating an RBF network, in the first instance, the basis function centres are fixed, as is the choice of basis function $\Phi(r)$. The network weights ω_i can then be adjusted so that the network follows a desired response pattern. an error minimisation criterion being by far the simplest way. As a straightforward, linear relationship exists between the weights and the network output, so a standard linear parameter estimation scheme can be employed to train the weights with ordinary linear least squares being perhaps the simplest, although partitioned least squares [23] is more suitable for multivariable problems. This linear learning procedure for RBF networks is in fact a big advantage when comparing such networks with multi-layer perceptrons which require problematic nonlinear learning algorithms.

As was the case for MLP networks, a structural problem exists for RBF networks in terms of the number and position of basis function centres required for a particular problem. The choice of centres is critical in that it directly affects the quality of function approximation achievable by the network, a minimum requirement being that the number of centres is at least sufficient to span the entire input domain [24]. Unfortunately, the number of centres required tends to increase exponentially with respect to the input space dimension and hence RBF networks, in the form described, may not be suitable for modelling some highly complex plant.

The number of RBF centres selected for a particular problem, initiates a direct trade-off with the quality of function approximation achieved. For a plant of low complexity, only a relatively small number of basis function centres are required, and as the plant's complexity increases so the number of centres must also increase if the accuracy of approximation is to be maintained. However, for a plant of low complexity, selecting more centres than are 'necessary' will result in over-parameterising the solution, resulting in poor weight convergence and noise modelling [25]. Conversely, when the plant is of high complexity, a reasonable solution can often be obtained by the selection of a fairly low number of centres, it being remembered that the overall approximation provided by the network is going to be, at best, reasonable.

A number of methods are possible for the selection of basis function centre positions, with either a uniform or random distribution of the centres across the input space being simplest, but certainly not the best. If certain volumes in the input space are more usually active, then these areas will not be given any special treatment through either the uniform or random methods, in fact they would be treated in exactly the same way as volumes which may never be active. A much more sensible procedure to employ for centre selection is the use of a mean-tracking clustering algorithm [26], which delivers a statistically best selection of centre quantity and position, with a relatively low computational requirement, it being remembered that the centre selection procedure is usually a one-off event.

For adaptive control or plant characteristic tracking, it is quite possible to adjust the weights ω_i adaptively on-line, by employing linear tracking/estimation algorithms such as recursive least squares. However, the adjustment of basis function centres, also in an on-line fashion, although conceptually possible, is a much more difficult process to quantify and analyse in terms of stability and convergence.

1.8 *N*-tuple networks

N-tuple networks, also known as digital nets, are based on a slightly different neuron model and also a different learning strategy from those methods already discussed [7, 27]. Each node element makes use of binary data and the node is actually designed in terms of digital memory. Essentially the node inputs are memory address lines whilst the output is the value of the data stored at that address, as shown in Figure 1.6.

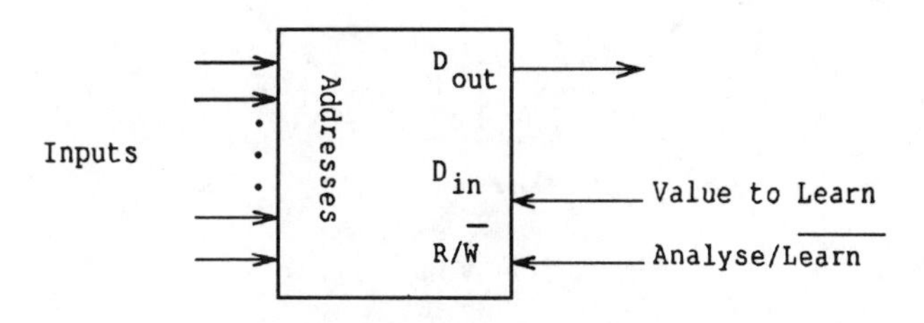

Figure 1.6 *RAM neuron*

When the node element is in its learning mode, the pattern/data being taught are entered on the memory address lines, and the appropriate value stored. When in analysis mode, the node is addressed and the stored value thus appears at the node output.

Figure 1.7 shows schematically how input patterns are fed to the *n*-tuple neural network, such that for each node element, *n* bits of the input pattern are sampled (as an *n*-tuple). The first *n*-tuple is used to address the first node element, such that a '1' is stored at that address. A further sampled *n* bits are learnt by the second node element, and so on until the whole input pattern has been dealt with in a random fashion.

When the *n*-tuple network is required to analyse a pattern, once a particular input pattern has been presented, the number of node elements outputting a '1' is counted. This means that if an 'exactly' identical input pattern is presented, when compared with the learnt pattern, the count will be 100%, whereas if the pattern is similar the count will be very high. In practice the neural network can be considered to 'recognise' an object if the count is greater than *x*%, where *x* is some previously defined confidence value. If a new input pattern needs to be learnt however, the network can be cleared by setting all node elements to zero and starting the procedure again.

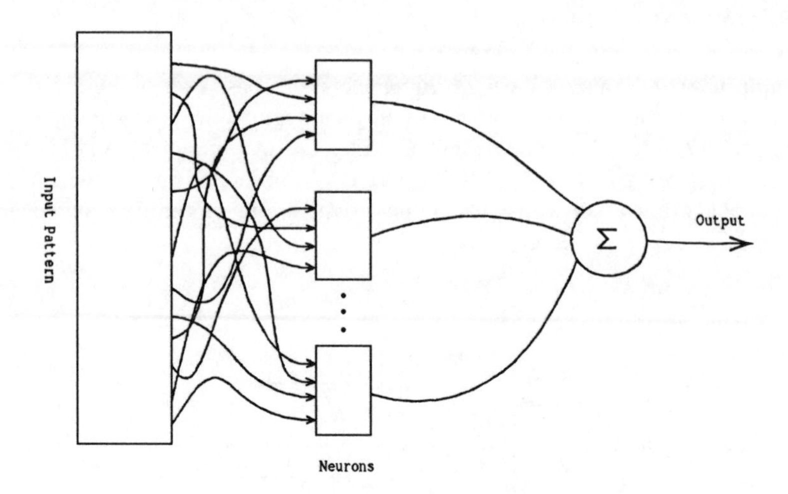

Figure 1.7 *N-tuple network*

1.9 Conclusions

To regard the presently available neural net techniques as being a solution to the search for true artificial intelligence would be a falsehood, but rather they are techniques which exhibit certain intelligent properties and are useful for solving some specific tasks. Indeed many 'intelligent' systems combine features of other artificial intelligent tools such as expert systems or genetic algorithms along with neural nets.

An overall intelligent system, however, requires much more than a central processing unit, which is the part played by a neural net, such that sensory systems, genetic features and problem understanding are required along with their appropriate interface to the central unit.

At present neural nets are extremely powerful, rapidly operating devices which exhibit many advantages when compared with more conventional computing methods. They can be constructed in a number of forms, some of which have been described here, and are already in use in a wide variety of applications.

The number of possible neuron models is, however, very large with more complex models representing the operation of human neurons more closely. The artificial neural networks considered here, and those which are generally in use for systems and control, tend to be much simpler and easier to realise physically.

A number of different types of neural networks have been considered in this chapter, and each has a field of use for which it is best suited; for *n*-tuple nets it is image processing and vision, for Kohonen nets it is classification and fault diagnosis, whereas for both multi-layer perceptrons and radial basis functions it is signal processing and control. All of the networks though are nonlinear mapping devices and have a rôle to play as such. They are not suitable as simple controllers to replace PID controllers or to replace linear industrial control schemes [25], as in these cases they are in general over-parameterising such that a control action which is far worse than a PID controller or a straightforward linear digital controller will most likely be obtained.

1.10 References

[1] Aleksander, I. and Morton, H., 1990, *An introduction to neural computing*, Chapman and Hall.

[2] Hebb, D.E., 1949, *Organisation of behaviour*, Wiley.

[3] Rosenblatt, F., 1962, *Principles of neurodynamics*, Spartan.

[4] Minsky, M. and Papert, S., 1969, *Perceptrons*, MIT Press.

[5] Rumelhart, D.E. and McClelland, J.L., 1986, *Parallel distributed processing*, MIT Press, Vols. 1 and 2.

[6] Pao, Y-H., 1989, *Adaptive pattern recognition and neural networks*, Addison Wesley.

[7] Aleksander, I., 1991, Introduction to neural nets, in *Applied Artificial Intelligence*, K. Warwick (Ed.), Peter Peregrinus Ltd.

[8] Hopfield, J.T., 1982, Neural networks and physical systems with emergent collective computational abilities, *Proc. Natl. Acad. Sci. USA 79*, pp. 2554-2558.

[9] Benes, J., 1990, On neural networks, *Kybernetica*, Vol. 26, No. 3, pp. 232-247.

[10] Kohonen, T., 1984, *Self-organisation and associative memory*, Springer-Verlag.

[11] Wu, P. and Warwick, K., 1991, A structure for neural networks with artificial intelligence, in *Research and Development in Expert Systems VIII*, I.M. Graham and R.W. Milne (Eds.), Cambridge, Cambridge University Press, pp. 104-111.

[12] Narendra, K.S. and Parthasarathy, K., 1990, Identification and control of dynamical systems using neural networks, *IEEE Trans. on Neural Networks*, Vol. NN-1, No. 1, pp. 4-27.

[13] Narendra, K.S., 1994, Adaptive control of dynamical systems using neural networks, in *Handbook of Intelligent Control*, Van Nostrand.

[14] Ballard, D.H., 1988, Cortisal connections and parallel processing: structure and function, in *Vision, Brain and Cooperative Computation*, Arbib and Hamson (Eds.), MIT Press, pp. 563-621.

[15] Psichogios, D.C. and Ungar, L.H., 1994, SVD-NET: An algorithm that automatically selects network structure, *IEEE Trans. on Neural Networks*, Vol. 5, pp. 514-515.

[16] Cichocki, A. and Unbehauen, R., 1993, *Neural networks for optimization and signal processing*, John Wiley & Sons.

[17] Cybenko, G., 1989, Approximations by superpositions of a sigmoidal function, *Mathematics of Control, Signals and Systems*, Vol. 2, pp. 303-314.

[18] Willis, M.T., Di Massimo, C., Montague, G.A., Tham, M.T. and Morris, A.J., 1991, Artificial neural networks in process engineering, *Proc. IEE*, Part D, Vol. 138, No. 3, pp. 256-266.

[19] Warwick, K., 1991, Neural net system for adaptive robot control, in *Expert Systems and Robotics*, T. Jordanides and B. Torbey (Eds.), Springer-Verlag, Computer Systems Science Series, Vol. 71, pp. 601-608.

[20] Girosi, F. and Poggio, T., 1990, Neural networks and the best approximation property, *Biol. Cybernetics*, Vol. 63, pp. 169-176.

[21] Albus, J.S., 1975, Data storage in the cerebellar model articulation controller (CMAC), *Trans. of the ASME Journal of Dynamic Systems, Measurement and Control*, Vol. 97, pp. 228-233.

[22] Chua, L.O. and Yang, L., 1988, Cellular neural networks, *IEEE Trans. on Circuits and Systems: Theory*, pp. 1257-1272.

[23] Karny, M. and Warwick, K., 1994, Partitioned least squares, *Proc. Int. Conference Control 94*, Warwick University, Vol. 1, pp. 827-832.

[24] Chen, S., Billings, S.A., Cowan, C.F. and Grant, P.M., 1990, Practical identification of NARMAX models using radial basis functions, *Int. Journal of Control*, Vol. 52, pp. 1327-1350.

[25] Warwick, K., 1995, A critique of discrete-time linear neural networks, *Int. Journal of Control*, Vol. 61, No. 6, pp. 1253-1264.

[26] Warwick, K., Mason, J.D. and Sutanto, E., 1995, Centre selection for radial basis function networks, *Proc. ICANNAGA*, Ales, France, pp.309-312.

[27] Aleksander, I. (Ed.), 1989, *Neural computing architectures*, North Oxford Academic Publishers.

Chapter 2

Digital neural networks

A. Redgers and I. Aleksander

2.1 Classification of artificial neural networks

2.1.1 McCulloch and Pitts versus Boolean nodes

When McCulloch and Pitts first studied artificial neural networks (ANNs), their neuron model consisted of binary signals contributing to a sum which was then thresholded to produce the output of the neuron. This model quickly evolved to the well known 'function of weighted sum of inputs' model, the function usually being a sigmoidal squashing function such as *output* = $1/(1+\exp(-\lambda(\text{sum}-\text{bias})))$ or *output* = $\text{erf}(\lambda(\text{sum}-\text{bias}))$.

The binary threshold function *output* = $\Theta(\text{sum, bias})$ can be thought of as the limiting case of these when the parameter λ becomes large. Function-of-weighted-sum-of-inputs neuron models are collectively termed McCulloch and Pitts (McCP) models to distinguish them from other models.

The weights and bias can be considered to be other parameters of McCP nodes which can be set, and trained, so that a network of McCP nodes performs the desired function of its inputs. However, there are functions of their inputs which McCP nodes cannot perform. For example, there is no setting of weights and threshold (bias) which will allow a 2-input McCP neuron to perform the XOR function: 'fire if and only if a single input is set to one'.

Boolean nodes are based on random access memories, RAMs, which are look-up tables. The inputs form the address of a memory location and the RAM outputs the contents of that location during operation or changes it during training. RAMs can perform any binary function of their inputs, but they do not scale well as the number of addressable memory locations increases as 2^{inputs}. In practice they are decomposed, split into large logic neurons (LLNs), consisting of smaller Boolean nodes.

Conventionally the contents of memory locations in RAMs are binary valued, and recent research has focused on Boolean models with three or more valued contents. The output of the node is then a *function* of the contents of the addressed memory location. In the case of RAMs this function is just the identity function; in the case of three-valued probabilistic logic nodes (PLNs), the function outputs 1 and 0 for memory contents 1 and 0, but outputs 1 or 0 randomly for memory contents *U*, the third value. Multi-PLNs (MPLNs) have even more allowed values, activation states, of memory contents.

2.1.2 Definition of ANNs

The definition 'an interconnected system of parameterised functions' covers many types of ANNs and neuron models. For example, in McCP and Boolean nodes the parameters are, respectively: the weights (and bias), and the memory contents. The functions are respectively, 'output a function of the weighted sum of inputs', and 'output a function of the contents of the addressed memory location'.

It is useful to be able to classify ANNs in order to compare them and place them relative to each other in our understanding. Attempts have been made to do this such as that in [1]. One simple classification scheme which achieves this requires five pieces of information for the description of an ANN as defined above:

- *Topology* — how the functions are interconnected
- *Architecture* — type/use of network (four possibilities)
- *Neuron model* — what the functions are
- *Training algorithm* — how the parameters are set
- *Operation schedule* — timing of function interactions.

The topology of an ANN is a set of connections between inputs, outputs and nodes. All topologies are of course subsets of the fully connected topology — any topology can be considered as just the fully connected topology with the node functions independent of certain inputs. Architecture is more vague — it is the user's intention of how the ANN is to work. Simple architectures come in four main types: feedforward, external feedback, internal feedback and unsupervised (Figure 2.1).

With external feedback the previous output provides the new input. There is only one external stimulus, at the very start, and any new stimulus elicits the same response no matter what other stimuli have previously been shown to the network.

With internal feedback the net sees both an external stimulus and its own response to the previous stimulus. Its response depends not only on the current stimulus but also on the previous stimuli and the order in which they were presented to the net. More complex architectures are formed of these simple components. It may be that there are still other architectures, that cannot be classified using any of the four existing possibilities, which should be added to this list.

The McCulloch and Pitts and Boolean models already mentioned are examples of frequently studied neuron models. The training algorithm conducts a search through the space of parameter values for a set of values with which the network will perform the required function. Networks of McCP nodes are frequently

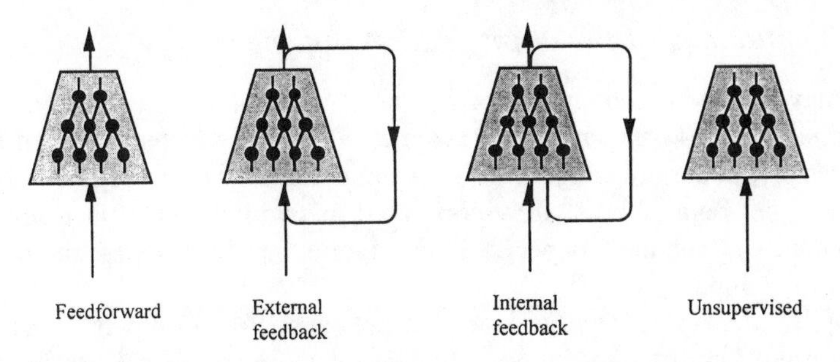

Figure 2.1 *Four possible architectures of ANNs*

trained using variations of Rumelhart's error backpropagation. Several algorithms have been developed for supervised training of a tree or pyramid (see Figure 2.2) network of PLNs [2-4]. All have similar characteristics, Myers' algorithm being the simplest. It is:

1 Present a new training example to the input nodes at the base of the pyramid.

tries := 1

2 Allow signals from the training example to propagate through the network until they reach the output node at the top of the pyramid.

3 If the output is correct then in each node if the addressed location contains a *U* then set it to the (random) value that the node has just output, **gel**.

Else

If **tries** < maxTries then

tries := **tries** + 1

Goto step 2

Else

Set contents of addressed locations to U, **punish**.

4 Goto step 1

In the training algorithms in [3,4] a flag is set for each node to say whether its output was probabilistic. If the output is incorrect the search for improving node contents is conducted on those nodes whose output was probabilistic.

The operation schedule is a list of times and neurons which respond at those times. In a layered feedforward network the neurons closest to the retina respond first, their responses feed into the neurons of the next layer, which then respond, and so on. In networks with feedback the set of neurons responding at a given time may be chosen randomly. In a synchronous feedback network all nodes respond at the same time. In an asynchronous network only one node responds at a time. But for any number of nodes, synchronicity between these two extremes is possible. A synchronicity of 1 means asynchronous; a synchronicity of *N*, the number of nodes in the network, means synchronous.

2.1.3 Limitations of the classification scheme

There may be ANNs which do not fit into this classification scheme, or whose inclusion is strained or trivial. For example, in networks such as those of Binstead and Jones [5], nodes are added during training, that is, the topology changes. But these can be regarded as fully-connected networks with functions at first independent of, but later dependent on, certain inputs as extra parameters are allowed to change.

Compiled networks, such as Hopfield networks [6], do not have an iterative training procedure, instead the node function parameters are calculated explicitly from the data examples. Either such off-line procedures can be considered to be training algorithms of a sort, or else one can regard the network as not having a training algorithm.

One normally expects that a function called twice with the same arguments will give the same answer both times - not so if the function has an internal state which changes every time the function is called. Gorse and Taylor [7,8] have proposed neuron models whose functions change continuously over time, for example modelling refractory periods in brain neurons. Although these can be covered by considering internal states of nodes to be parameters of the functions there is a case for adding a sixth category to the classification scheme above.

2.2 Functions of Boolean nodes

2.2.1 MPLN activation functions

PLNs are three-valued RAMs with the third, U, activation state representing a 0.5 probability of outputting 1. MPLNs have more activation states, each with their own probability of outputting 1. The contents of memory locations are changed to neighbouring activation states during training, providing an ordering on the set of activation states. Making the activation function, mapping activation states to probabilities of outputting 1, sigmoidal rather than linear reduces the search space. Myers [9] has obtained results indicating that for MPLNs the threshold function is an efficient function of the activation state s:

$$f(s)=\begin{cases} s<U: f(s)=0 \\ s=U: f(s)=0.5 \\ s>U: f(s)=1 \end{cases}$$

This looks like a PLN with extensions to either side of the U value, and during training the extensions serve to keep a count of how many times a value is used correctly.

2.2.2 *The address decoder function*

Martland [10] gives the probability, r, of a two-input PLN outputting a 1 when addressed as:

$$r = a_{00}(1-x_1)(1-x_2) + a_{01}(1-x_1)x_2 + a_{10}x_1(1-x_2) + a_{11}x_1x_2 \qquad (2.1)$$

where a_{00} is the contents of address location 00, i.e. the probability of the PLN firing if it is addressed with 00; similarly for a_{01}, a_{10} and a_{11} and x_1, x_2 are the (binary) values on the two input lines.

The inputs x_1 and x_2 need not be just binary but can take any real value between 0 and 1; in Taylor and Gorse's pRAM these represent probabilities of the inputs being 1. The resulting probability r might not even be run through a generator to obtain 1 or 0, but instead passed on directly. Martland himself used the address decoder formula eqn. 2.1 to implement the error backpropagation algorithm on pyramids of these 'continuous RAMs' (training took far longer than Myers' algorithm for three-valued PLNs).

The address decoder function for an n-input Boolean node is:

$$r = \sum_{\omega \in B^n} a_\omega \prod_{i=1}^{n} v(\omega_i, x_i) \qquad (2.2)$$

where $v(p, q)$ is a function such that

$$\begin{cases} v(1,q) = q \\ v(0,q) = 1-q \end{cases}$$

e.g.

$$v(p, q) = 2pq - p - q + 1 \qquad (2.3)$$

B^n is the set of Boolean strings of n 1s and 0s (e.g. $0101 \varepsilon B^4$)
ω is one such string (the sum is taken over all strings)
ω_i is the ith component of ω
a_ω is the contents of the memory location whose address is ω
x_i is the ith input to the node.

Equation 2.1 can be rewritten as a power series in x_1 and x_2:

$$r = a_{00} + x_1(a_{10}-a_{00}) + x_2(a_{01}-a_{00}) + x_1x_2(a_{11}+a_{00}-a_{10}-a_{01}) \qquad (2.4)$$

From the n-input address decoder function eqn. 2.2, eqn. 2.3 can be similarly rewritten as a power series:

$$r = \sum_{\omega \in B^n} g_\omega \prod_{i=1}^{n} w(\omega_i, x_i) \tag{2.5}$$

where $w(p, q)$ is a function such that

$$\begin{cases} w(1,q) = q \\ w(0,q) = 1 \end{cases}$$

e.g.

$$w(p, q) = pq - p + 1 \tag{2.6}$$

Given either the coefficients a_ω or g_ω one can obtain the coefficients for the other form:

$$a_\omega = \sum_{\phi \in B^n} g_\phi \prod_{i=1}^{n} y(\omega_i, \phi_i) \tag{2.7}$$

where:

$$y(p, q) = pq - q + 1 \tag{2.8}$$

In the other direction:

$$g_\omega = \sum_{\phi \in B^n} a_\phi \prod_{i=1}^{n} z(\omega_i, \phi_i) \tag{2.9}$$

$$z(p, q) = pq - q + 1 \tag{2.10}$$

In a McCP node only the constant term $g_{000...}$ and the first degree coefficients $g_{100...}$, $g_{010...}$, $g_{001...}$, the *weights*, are non-zero. In McCP nodes with $\prod$-units [11] second ($g_{110...}$, $g_{101...}$, $g_{011...}$) or higher degree coefficients are non-zero. To discover whether the function performed by a Boolean node can be performed by a McCP node, the a_ω is transformed into g_ω and checked to see which coefficients are non-zero. The reverse transformation will implement a function performed by a McCP node on a Boolean node.

2.2.3 Kernel functions and transforming output functions

Equations 2.3, 2.6, 2.8 and 2.10 are examples of kernel functions of the form

$$v(p, q) = hpq + kp + lq + m \tag{2.11}$$

Given any kernel function eqn. 2.11 for an output function

$$r = \sum_{\omega \in B^n} h_\omega \prod_{i=1}^{n} v(\omega_i, x_i) \tag{2.12}$$

the kernel function for the transformation to the address decoder function eqn. 2.2 and eqn. 2.3 can be obtained:

$$a_\omega = \sum_{\phi \in B^{0n}} h_\phi \prod_{i=1}^{n} u(\omega_i, \phi_i) \tag{2.13}$$

where
$$u(p, q) = hpq + lp + kq + m \tag{2.14}$$

The kernel function for the transformation to eqns. 2.5 and 2.6 is:

$$u(p, q) = (h-k)pq + (l-m)p + kq + m \tag{2.15}$$

The kernel function for the identity transformation from h_ω to h_ϕ is:

$$h_\omega = \sum_{\phi \in B^n} h_\phi \prod_{i=1}^{n} i(\omega_i, \phi_i) \tag{2.16}$$

with

$$i(p, q) = 2pq - p - q + 1 \qquad (2.17) = (2.3)$$

If linear transformations are made of the variables x_i : $y_i = ax_i + b$, $x_i = (y_i - b)/a$ (e.g. binary $x_i y$ bipolar y_i : $y_i = 2x_i - 1$, $x_i = (y_i + 1)/2$)), then in the kernel function eqn. 2.11 the variable q, representing x_i, is transformed in the same way: $q := (q - b)/a$.

2.3 Large logic neurons

2.3.1 Decomposition of Boolean nodes

The number of weights in McCP nodes increases as the number of inputs to the node; with current computers it is possible to simulate or build McCP nodes with thousands or even millions of inputs. However, the number of memory locations in Boolean nodes (and weights in Π-unit models) grows exponentially with the number of inputs: to implement a node with 240 inputs would require 2^{240} bits of artificial storage, which is of the order of the number of particles in the universe! Hence when scaling up Boolean nodes it is necessary to decompose them into devices called large logic neurons, LLNs, consisting of many small Boolean nodes linked up so as to produce a single output.

There are two main types of LLNs: discriminators, in which the large Boolean node is replaced by a single layer of small nodes whose outputs are summed (the output of the LLN is a function of that sum); and pyramids, in which the single large Boolean node is replaced by small nodes arranged in a tree-like structure feeding finally into a single node with a single output.

When nodes are decomposed in this way they are no longer universal, i.e. they cannot perform every possible (binary valued) function of their (binary) inputs. As part of an attempt to find *which* functions they can perform we have begun by calculating *how many* functions they can perform, functionality, and functionality formulae are presented for discriminators and pyramids — the pyramidal functionality formula presented here being a slight improvement on that independently derived elsewhere [3,12].

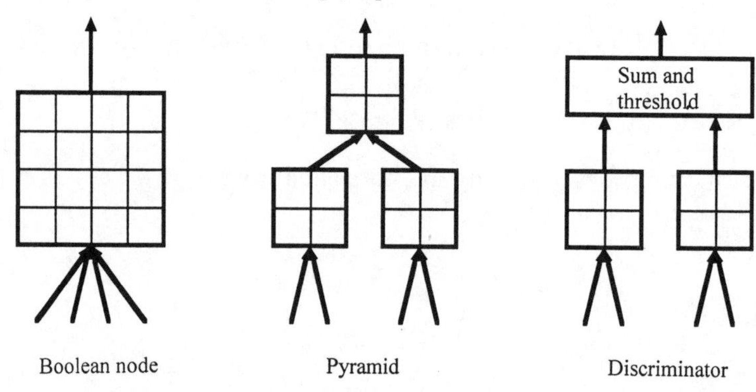

Figure 2.2 *Decomposing a 4-input Boolean node*

2.3.2 Generalisation and generalising RAMs

Generalisation is the complement of functionality. It refers the way that training an ANN on one example affects its response to other examples; hence generalisation also depends on the training algorithm used. McCP nodes are not universal and so have generalisation built in. Boolean nodes *are* universal and so generalisation in a network of Boolean nodes is a result of the topology of the network, not the nodes themselves. Thus, generalisation is seen to be an emergent property of the network; the generalisation of the whole is greater than the sum of the generalisations of the parts. At a different level, LLNs can be thought of as neuron models with built-in generalisation, and the distinction between nodal and topological generalisation in a network of LLNs is again blurred.

A generalising RAM, GRAM, is a PLN with an extra training step, spreading, which distributes already learnt information within the PLN [13]. Spreading is similar to training with noise, in which a neural network is trained on an input/output pair and then trained to give the same output for randomly distorted versions of the input. A typical training and spreading algorithm for a PLN is

```
For each memory location ω in the PLN                    (initialise)
        Set contents of ω to U
        Unset hiFlag for ω
Train on input / output pairs using, say, Myers' algorithm.
For each memory location ω in the PLN                    (spread)
        If contents of ω is U
                sum := 0
For each bit in ω make address φ from ω by reversing that bit.
If not (contents of φ is U or hiFlag for φ set)
                sum := sum + 2 × contents - 1
        If sum > 0
                Set contents of ω to 1
                Set hiFlag for ω
        If sum < 0
                Set contents of ω to 0
                Set hiFlag for ω
For each memory location ω in the PLN                    (clean up)
        Unset hiFlag for ω
```

The spreading algorithm needs a distance measure, metric, on the memory locations: in this case the Hamming distance is the number of bits different in two addresses, and spreading is to Hamming distance 1. Spreading can be thought of as an explicit training step or else as part of the node function of a GRAM with generalisation built in. In the latter case a PLN can be thought of as a GRAM generalising to Hamming distance 0. In other words a PLN is a special case of a GRAM.

2.3.3 Discriminators

WISARD [14] is a set of discriminators consisting of RAMs whose outputs are summed. Each RAM is mapped to a group of pixels, n-tuple, in the retina. During training the patterns seen on the n-tuples are stored in the corresponding RAMs. When a test image is put to the retina the discriminators report the number of n-tuple patterns that were seen during training. Variations of WISARD can improve discrimination: for example, allowing a count of the number of times an n-tuple pattern is seen rather than just logging it seen or unseen; or weighting the contribution of each RAM to the sum according to its individual discriminative ability, assessed in a second pass of the training set.

The functionality of a thresholded WISARD discriminator is:

$$F = 2 + R.2^{R.2^n} - 2\sum_{m=1}^{R}\sum_{p=m}^{R} C_p^R\,(2^{2^n} - 1)^{R-p} \tag{2.18}$$

where:

$C^p_q = p!/((p-q)!\,q!)$ ways of choosing q items from a set of p items.
R is the number of RAMs in the discriminator.
n is the number of inputs to each RAM, n-tuple size.

It is assumed that the R n-input nodes are connected to nR independent pixels in the retina, i.e. the retina is covered once.

Discriminators are typically used as classifiers, with one discriminator for each class. Trained on single images they act similarly to Hamming comparators, which are simpler and faster. Trained on multiple images, discriminators can out-perform Hamming comparators trying to find the average distance from an image to members of a class.

2.3.4 *Reversing discriminators*

Having trained a discriminator it is possible to reverse it to get a class example or set of class examples of what it has learnt. This is particularly useful in unsupervised nets and feature detection/description from noisy training examples. The aim is to find values of the inputs which cause the discriminator, and therefore its component Boolean nodes, to output 1. Partially differentiating eqn. 2.3 with respect to the x_i and equating to zero, yields a set of simultaneous equations:

$$\frac{\partial r}{dx_m} = \sum_{\omega\in B^n} a_\omega \frac{v(\omega_m, x_m) - v(1,0)}{x_m\, v(\omega_m, x_m)} \prod_{i=1}^{n} v(\omega_i, x_i) = 0 \tag{2.19}$$

For $n=2$ this gives solutions:

$$x_1 = \frac{a_{01}}{a_{01} + a_{10} - a_{11} + a_{00}}, \quad x_2 = \frac{a_{10}}{a_{01} + a_{10} - a_{11} + a_{00}}$$

which may be out of the range [0,1], e.g. if $a_{00}+a_{11} \geq a_{01}+a_{10}$. In other situations there may be infinitely many solutions or none. Instead solutions are based on:

$$x_m = 2^{-n}\sum_{\omega\in B^n} a_\omega\,(2\omega_m - 1) + \frac{1}{2} \tag{2.20}$$

which for $n=2$ looks like:

$$x_1 = (\tfrac{1}{4})\,(a_{10} - a_{01} + a_{11} - a_{00}) + (\tfrac{1}{2})$$

$$x_2 = (\tfrac{1}{4})\,(a_{01} - a_{10} + a_{11} - a_{00}) + (\tfrac{1}{2})$$

Values of x_i close to ($\tfrac{1}{2}$) indicate contradictory features or features depending on more than n variables. Constraint contrast enhancement is a technique whereby the contrast is enhanced by sytematically setting such inputs to 1 or 0 and using them in turn to constrain in other low contrast inputs. It is used in conjunction with a class of complex, high coverage mappings between the discriminator and the retina, and returns a list of high contrast images rather than a single low contrast image.

2.3.5 Pyramids

Boolean nodes are also commonly connected together in tree or pyramid structures. In a regular pyramid, with a fan-in of n, each bottom-layer node receives its input from the retina and propagates its response to one node in the next layer. Each node in a successive layer is fed by n nodes from the previous layer until one top-level node outputs a binary response for the whole net. Higher bandwidth responses can be made by having several pyramids, each of which sees the whole input and responds independently.

For a regular pyramid the total number of functions is:

$$F = \sum_{k=0}^{n} C_k^n \, S_{n-k} \, p^{(n-k)} \tag{2.21}$$

where

$$S_m = \sum_{j=0}^{m} (-1)^j \, C_j^m \, 2^{2^{(m-j)}} \tag{2.22}$$

$$p = \begin{cases} (F_s - 2)/2 \text{ if } F_s > 1 \\ 1 \text{ if } F_s = 1 \end{cases}$$

the number of non-trivial functions, F_s is the total number of functions that sub-pyramid s can perform. S_{n-k} is the number of functions, given a choice of k inputs, of the top node which are independent of just those k inputs.

The functionality formula is recursive, working down the sub-pyramids until it comes to one of the 'base' cases: a single line, direct to the retina, or a Boolean node with n input lines. In the case of a single line there is one non-trivial function: the value on the line itself, so $F_s = 1$ and $p = 1$ (not $(F_s - 2)/2$). For a

Boolean node with n input lines $F_s = 2^{2^n}$)) the number of states that the node can be in. For example, in a 2-input node $n=2$, $F_s=16$ and $p=7$.

Because the functionality of each sub-pyramid is calculated, the formula for the functionality as it stands can cope with changes of fan-in between levels. To deal with changes of fan-in within a level (i.e. nodes on the same level have different fan-ins) eqn. 2.22 is changed so that, instead of multiplying by the combinatorial coefficient $C^k{}_n$, a sum is made of all choices of k sub-pyramids and the term $p^{(n-k)}$ is replaced by the product of the p values for the chosen sub-pyramids:

$$F = \sum_{k=0}^{n} S_{n-k} \sum_{\text{choices } \lambda \text{ of } n-k \text{ sub-pyramids}} \prod_{j=1}^{k} p_{\lambda j} \qquad (2.23)$$

where

$$p_{\lambda j} = \begin{cases} (F_{\lambda j}-2)/2 \text{ if } F_{\lambda j} > 1 \\ 1 \text{ if } F_{\lambda j} = 1 \end{cases}$$

$F_{\lambda j}$ is the functionality of the jth sub-pyramid in choice λ. In the case of a single line, direct to the retina, $F_{\lambda j} = 1$ and $p = 1$ (not $(F_{\lambda j}-2)/2$).

2.3.6 *Networks of LLNs*

All McCP networks have their Boolean or LLN analogues. Discriminators are comparable with Rosenblatt's perceptron. As mentioned before, Martland has demonstrated error backpropagation on pyramids of Boolean nodes. Analogues of Hopfield nets are sparsely connected networks of PLNs [15] and the fully connected WIS auto-associator. The WIS is a set of WISARD discriminators whose outputs are thresholded. There are as many discriminators as retinal pixels so their thresholded outputs form a new binary image which can be fed back as a new retinal image (external feedback). WIS has also demonstrated internal feedback and performed some classification of simple formal grammars. Kohonen networks of c-discriminators, containing continuous RAMs, have been developed [16].

Functionality of networks increases explosively with size, and the specific neuron model, and local topology, matter much less than the broad flow of information within a network. For example, the mapping from the retina to the inputs of an LLN is usually taken to be random, and feedback a fraction of the input bandwidth. At large scales it becomes likely that the network will be able to perform the desired function whatever type of node is used, so long as dependent signals are at least connected. For real applications the speed of training and operation, and the ability to create large networks are important. Networks of Boolean nodes compare favourably in these respects with networks of McCP nodes.

2.4 Implementing large ANNs using MAGNUS

2.4.1 General neural units

A general neural unit, GNU, is an internal feedback ANN made of Boolean nodes or LLNs. It can be represented as in Figure 2.3.

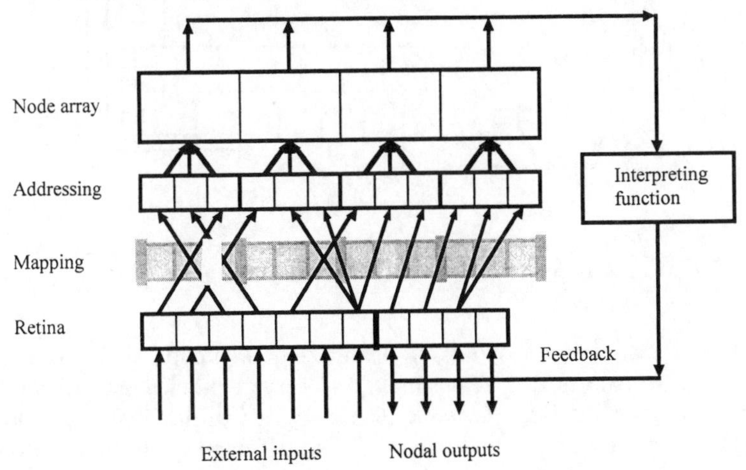

Figure 2.3 *Representation of a GNU*

This representation is known as MAGNUS (Mapping Array GNU Scheme) and it can be used to regularise a variety of Boolean ANNs. The topology of the network is a list of offsets to the retina array; it is held in the mapping array. The nodes are taken to be Boolean and their outputs are passed through the interpreting function before being fed back into the retina and also output to the user.

Using offsets to a retina containing both externally applied and fed-back signals allows nodes to receive signals of either type or both. Figure 2.4 shows how this scheme could implement a PLN pyramid.

The mapping array is created off-line, although it can be dynamically altered if required. MAGNUS software uses defaults, such as random mappings, and queries the user for any extra information it requires. Thus users have few decisions to make (they are not, however, compelled to use the defaults). With GNUs of GRAMs the only decisions that need be taken are the number of external inputs, the *n*-tuple size of the nodes, and the degree of feedback.

2.4.2 MAGNUS and LLNs

Implementing LLNs or other groupings of nodes requires extra information e.g. number of nodes in the group, *n*-tuple size of nodes within a group, output

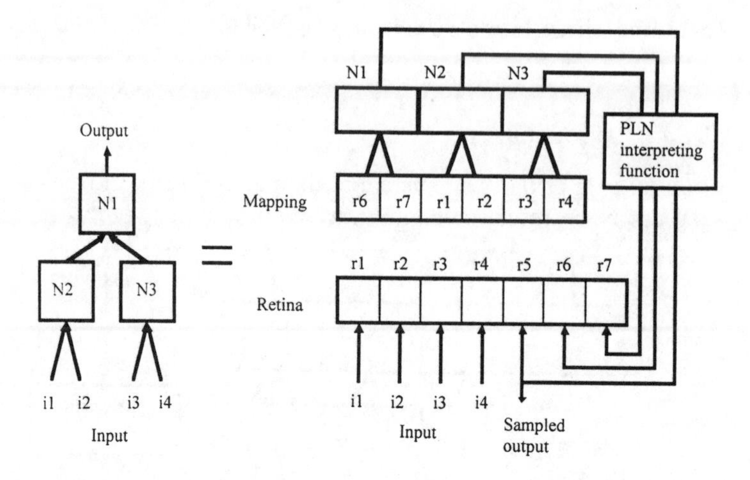

Figure 2.4 *Implementing a PLN pyramid using MAGNUS*

function (e.g. sum and threshold) of members of a group. Even with these additions the scheme is still very simple and easy to implement efficiently in software, meaning that large networks can be run with minimal overhead and bringing real applications (retina sizes ~ 10^4 to 10^5 inputs, 10^3 to 10^4 Boolean nodes) within the range of moderately powerful computers.

MAGNUS is a neural state machine which allows its user to develop up to 100 000 neural state variables. It is presently being used in language and vision understanding tasks [17].

2.5 Conclusion

Function-of-weighted-sum-and-threshold model neurons are so common that definitions of neural networks, and software implementations, may even tacitly assume them. However, Boolean model neurons, based upon random access memories, are very different. The simple definition of a neural network as 'an interconnected system of parameterised functions' assumes very little, but is useful for building a classification system for neural networks based upon: topology, 'architecture', neuron model, training algorithm and operation schedule.

The original digital neural networks made of RAMs have evolved in two ways: they have become probabilistic, multi-valued and finally, using the address decoder function, analogue Boolean nodes, and thus equivalent to McCulloch and Pitts nodes with Π-units; and they have been decomposed into large logic neurons - discriminators and pyramids, with limited functionality. Boolean nodes are universal, and in digital neural networks generalisation is a property of the network topology not of the nodes. GRAMs are nodes for which the generalisation may be explicitly defined.

All the major architectures — perceptrons, Hopfield nets, error backpropagation, Kohonen unsupervised learning, even ART1 — have their Boolean analogues. Digital neural networks are faster to train and easier to implement in hardware than networks of McCulloch and Pitts nodes, and the MAGNUS scheme for implementing large networks of Boolean nodes is being developed so as to reduce processing overheads and facilitate large networks suitable for 'real' applications. It is intended that MAGNUS be implemented on a PCB, and then perhaps on a single chip.

2.6 References

[1] Rumelhart, Hinton and McClelland, 1986, A general framework for parallel distributed processing, *Parallel Distributed Processing*, MIT, Vol. 1, ch. 2, pp. 45-76.

[2] Myers and Aleksander, 1988, Learning algorithms for probabilistic logic nodes, *Abstracts of INNS 1st Ann. Meeting*, Boston, p. 205.

[3] Al-Alawi and Stonham, 1989b, *A training strategy and functionality analysis of multi-layer Boolean neural networks*, Brunel University.

[4] Filho, Bisset and Fairhurst, 1990, A goal-seeking neuron for Boolean neural networks, *Proc. INNC-90-PARIS*, Paris, Vol. 2, pp. 894-897.

[5] Binstead and Jones, 1987, Design techniques for dynamically evolving n-tuple nets, *Proc. IEEE*, Vol. 134 E, No. 6, November.

[6] Tank and Hopfield, 1987, Collective computation in neuron-like circuits, *Sci. Am.*, December 1987, pp. 62-70.

[7] Gorse and Taylor, 1988, On the equivalence and properties of noisy neural and probabilistic RAM nets, *Physics Letters A*, Vol. 131, No. 6.

[8] Gorse and Taylor, 1990, Training strategies for probabilistic RAMs, in *Parallel Processing in Neural Systems and Computers*, Eckmiller, Hartmann and Hauske, (Eds.), North Holland, Elsevier Science, pp. 161-164.

[9] Myers, 1989, Output functions for probabilistic logic nodes, *Proc. 1st IEE Int. Conf. on ANNs*, London, pp. 163-185.

[10] Martland, 1989, Adaptation of Boolean networks using back-error propagation, *Abstracts IJCNN-89*, Washington D.C., p. 627.

[11] Williams, 1986, The logic of activation functions, *Parallel Distributed Processing*, MIT, Vol. 1, ch. 8, pp. 318-362.

[12] Al-Alawi and Stonham, 1989a, The functionality of multi-layer Boolean neural networks, *Electronics Letters*, Vol 25, No. 10, pp. 657-658.

[13] Aleksander, 1990, Ideal neurons for neural computers, in *Parallel processing in neural systems and computers*, Eckmiller, Hartmann and Hauske (Eds.), North Holland, Elsevier Science, pp. 225-232.

[14] Aleksander, Thomas and Bowden, 1984, WISARD - a radical step forward in image processing, *Sensor Review*, Vol. 4, No. 3, pp. 29-40, July.

[15] Wong and Sherrington, 1989, Theory of associative memory in randomly connected Boolean neural networks, *J. Physics A*, Math Gen 22, pp. 2233-2263.
[16] Ntourntoufis, 1990, Self-organisation properties of a discriminator-based neural network, *Proc. IJCNN-90*, San Diego, Vol. 2, pp. 319-324.
[17] Aleksander, I., 1994, Developments in artificial neural systems: towards intentional machines, *Science Progress*, Vol. 77, No. 1/2, pp. 43-55.
[18] Kan and Aleksander, 1987, A probabilistic logic neuron network for associative learning, *Proc. IEEE 1st Ann. Int. Conf. on Neural Networks*, San Diego.

Chapter 3

Fundamentals of neurocontrol: a survey

R. Żbikowski and P. J. Gawthrop

3.1 Introduction

This chapter gives a concise overview of fundamentals of neurocontrol. Since a detailed and comprehensive survey [24] and its recent update [71] are available, we concentrate on the general theoretical framework only. Specific control structures are not covered, as their discussion can be found in the aforementioned surveys.

Both feedforward and recurrent networks are considered and the foundations of approximation of nonlinear dynamics with both structures are briefly presented. A continuous-time, state-space approach to recurrent neural networks is presented; it gives valuable insight into the dynamic behaviour of the networks and may be fast in analogue implementations [36]. Recent learning algorithms for recurrent networks are surveyed with emphasis on the ones relevant to identification of nonlinear plants. The generalisation question is formulated for dynamic systems and discussed from the control viewpoint. A comparative study of stability is made discussing the Cohen-Grossberg and Hopfield approaches.

The chapter is organised as follows. First, basic notions of neural networks are introduced in Section 3.2. Then feedforward and recurrent networks are described in Sections 3.3 and 3.4, respectively.

Section 3.3 on feedforward networks is structured in the following way. Basic definitions are given in Section 3.3.1. Approximation issues arising in the neural context are described in Section 3.3.2. Sections 3.3.3–3.3.5 focus on theoretical foundations of modelling with feedforward neural networks for the purposes of nonlinear control. The major approaches are briefly described and discussed with emphasis on relevance and applicability. Section 3.3.2 defines the approximation problem to be tackled with FNNs. Section 3.3.3 describes the ideas related to the Stone-Weierstrass theorem. Section 3.3.4 analyses the approach based on Kolmogorov's solution to Hilbert's 13th problem. Section 3.3.5 gives an account of the application of multi-dimensional sampling. Additionally, the celebrated backpropagation learning algorithm is presented in Section 3.3.6.

Section 3.4 on recurrent networks begins with their mathematical description in Section 3.4.1. This is followed by a brief account of the approximation problems in Section 3.4.2. Sections 3.4.3–3.4.4 present learning algorithms for recurrent networks and make use of Section 3.3.6. The following Section, Section 3.4.5, comments on certain control aspects of recurrent structures. Finally, Section 3.4.6 briefly addresses the important issue of the networks' stability.

The chapter ends with conclusions.

3.2 Basic notions

Neural networks or connectionist models have been intensively investigated recently [65, 33]. Originally, the inspiration came from the nervous systems of higher organisms, most notably from the human brain [18]. The philosophical, psychological and biological issues which arise are complex and ambiguous and will not be addressed here. From now on a *neural network* or a *connectionist model* will be considered as a computing machine or a dynamic system without any reference to living matter. It is assumed that a neural network is characterised by:

- parallel architecture; it is composed of many self-contained, parallel interconnected processing elements or *neurons*;

- similarity of neurons; each basic processor is described by a standard nonlinear algebraic or differential equation;

- adjustable weights; there are multiplicative parameters each associated with a single interconnection and they are adaptive.

The above structure can be configured in different ways from the signal flow point of view. When the input and intermediate signals are always propagated forwards, the system is called a static or *feedforward* network. The flow of information is directed towards the output and no returning paths are allowed (see Figure 3.2). If either states or outputs are fed back then this is a dynamic or *recurrent* network (see Figure 3.5). The signals are re-used, thus their current values are influenced by the past ones, which is an important characteristic absent in feedforward models.

The key features of neural networks are: nonlinearity, adaptation abilities and parallelism. If we specify a network structure and a learning rule, the network is defined.

As is common in control theory [55] and dynamic systems theory [2], the following general description of finite-dimensional, deterministic dynamic systems will be assumed:

$$\dot{x} = f(x, u) \tag{3.1}$$

$$y = g(x, u) \tag{3.2}$$

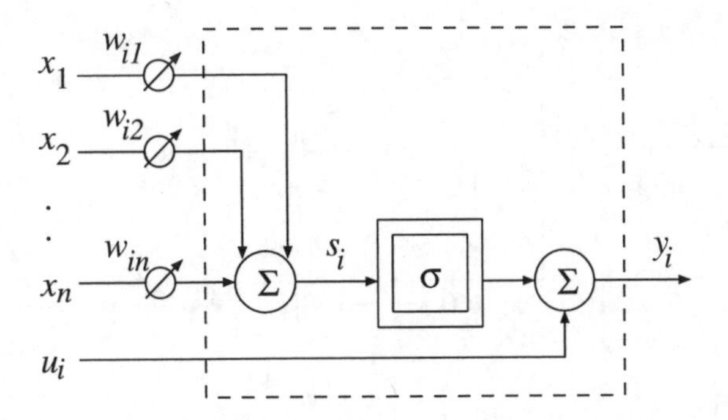

Figure 3.1 *Static neuron. The frame stands for a double square in Figure 3.2*

with $x(t_0) = x^0$. Here, $u = u(t)$, $x = x(t)$, $y = y(t)$ are time-dependent vectors and $f(\cdot, \cdot)$, $g(\cdot, \cdot)$ are vector functions of appropriate dimensions. The function $f(\cdot, \cdot)$ satisfies the Lipschitz condition with respect to x.

3.3 Feedforward networks

As mentioned above, there are two general structures of neural networks: feedforward and recurrent. This Section is devoted to a discussion of the former.

Feedforward networks are widely used for pattern recognition/classification [39]. Another relevant area is approximation theory [54, 20], where the problem may be stated as follows: given two signals $x(t)$ and $y(t)$ find a 'suitable' approximation of the functional relation between the two. This aspect is also of interest for control.

3.3.1 Feedforward networks' description

The most common approach to the approximation problem is to consider a non-dynamic relation

$$y = F(x) \tag{3.3}$$

where y and x are vectors and the parameter t is eliminated. This equation may be approximated with a network having the structure of Figure 3.2 composed of neurons (as in Figure 3.1) described by

$$y_i = \sigma(s_i) + u_i, \quad s_i = \sum_j w_{ij} x_j \tag{3.4}$$

A common choice of $\sigma(\cdot)$ is a *sigmoid* function.

Definition 1 *A C^k-sigmoid function $\sigma : \mathbf{R} \to \mathbf{R}$ is a nonconstant, bounded, and monotone increasing function of class C^k (continuously differentiable up to order k).*

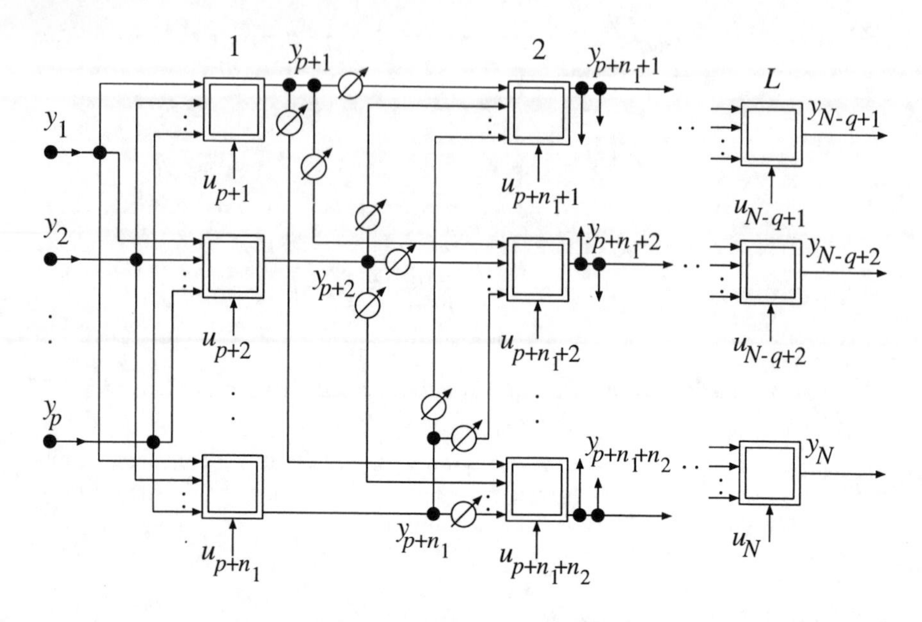

Figure 3.2 *Feedforward neural network*

Thus, the sigmoid is, roughly speaking, a (smooth) nonlinearity with saturation.

Notice that in Figure 3.2 all signals are denoted as y_i and numbered from top to bottom. Equation 3.4 and Figure 3.1 use x_j as inputs to emphasise that the signals fed into static neurons are *not* feedback ones, so not equal to outputs. The reason for introducing the homogeneous notation $y_1, y_2, \ldots, y_N$ will become clear when deriving learning algorithms (see Sections 3.3.6 and 3.4.3–3.4.4).

A usual way of arranging the architecture of static networks is to group neurons into *layers*,[1] or to design connections in such a way that each of them sends its signal directly only to a certain group of other neurons. As seen in Figure 3.2, the input layer composed of p distribution nodes (not neurons) feeds the first hidden layer, containing n_1 neurons with outputs $y_{p+1}, \ldots, y_{p+n_1}$. This layer, in turn, is connected through weights to the second hidden layer of n_2 units and so on until the last layer of q neurons, called the output layer. Notice that each layer l has a direct connection only to the subsequent one, i.e., $l + 1$, making the total of L layers. This results in the description

$$y = \sigma(s) + u, \qquad s = Wy \tag{3.5}$$

[1] This is motivated by the biological knowledge and greater transparency of analysis.

where the matrix W is sparse

$$W = \begin{bmatrix} 0 & 0 & \dots & 0 & 0 \\ W^1 & 0 & \dots & 0 & 0 \\ 0 & W^2 & \dots & 0 & 0 \\ \vdots & \vdots & \ddots & \vdots & 0 \\ 0 & 0 & \dots & W^L & 0 \end{bmatrix} \tag{3.6}$$

W^i are matrices of interlayer connection weights. The vectors y and u in eqn. 3.5 have the form $y = [y^0\ y^1 \dots y^L]^T$ and $u = [u^0\ u^1 \dots u^L]^T$, where y^i, u^i are vectors corresponding to the ith layer; u is the constant bias (offset) vector.

Recently, radial basis function feedforward networks have become increasingly important (*cf.* Section 3.3.5).

Definition 2 *A C^k-radial basis function (RBF) $g_{c,m}\colon \mathbf{R}^n \to \mathbf{R}$, with $c \in \mathbf{R}^n$ and $m \in \mathbf{R}_+$, is a C^k function constant on spheres $\{x \in \mathbf{R}^n \mid \|x - c\|/m = r\}$, centre c, radius rm, $r \in \mathbf{R}_+$, where $\|\cdot\|$ is the Euclidean norm on $\mathbf{R}^n$.*

The RBF has spherical (radial) symmetry with the centre c and the positive number m controlling the size of the spherical neighbourhood of c. It can be alternatively defined as the function $g\colon \mathbf{R}_+ \cup \{0\} \to \mathbf{R}$, where the argument is taken to be the radius of spheres centered at c.

The structure of the RBF network is:

$$\begin{aligned} y_i &= \sum_{j=1}^{N} w_{ij} g_{c_{ij},m_{ij}}(U) \\ &= \sum_{j=1}^{N} w_{ij} g(\|U - c_{ij}\|/m_{ij}) \quad i = 1, \dots, q \end{aligned} \tag{3.7}$$

Here U is the input; see footnote 2.

3.3.2 *Approximation: feedforward networks*

The use of feedforward neural networks (FNNs) for nonlinear control is based on the input-output discrete-time description of systems

$$y(t+1) = f\Big(y(t), \dots, y(t-n+1); u(t), \dots, u(t-m+1)\Big) \tag{3.8}$$

where the minimal assumption is that f is continuous. By limiting the bandwidth of y via antialiasing filters and then time sampling, a convenient use of digital computers is possible. Since the current and delayed samples of y and u are then readily available, the issue is (in the SISO case) to approximate the continuous mapping $f\colon \mathbf{R}^{m+n} \to \mathbf{R}$. The essence of the neural approach is to use an FNN model for this purpose (it may also apply to approximation of continuous feedback control laws).

The problem of feedforward neural modelling can be formulated as follows. Consider a continuous mapping $f: K \rightarrow \mathbf{R}^q$, where K is an uncountable compact subset of $\mathbf{R}^{pm+qn}$ (here p is the number of inputs, q the number of outputs of eqn. 3.8). Compactness of K means that it is closed and bounded. The mapping f is not given explicitly, but by a *finite* number of pairs $(U_k, Y_k) \in K \times \mathbf{R}^q$, $k = 1, \ldots, s$ (here s is the number of observed input-output pairs).[2] The problem is: find a representation of f by means of known functions and a finite number of real parameters, such that the representation yields uniform approximation of f over K.

Mathematically, this is an approximation problem: a *possibility* of representing f by some standard functions to within an arbitrary accuracy. If successfully solved, it will provide us with an *existence* result, but not necessarily a constructive one. It will then give meaning to our efforts, but may not offer practical clues. This is the case with the Stone-Weierstrass theorem (Section 3.3.3) and Kolmogorov's results (Section 3.3.4).

However, the essential part of the formulation of our problem is that f is given in the form of the samples (U_k, Y_k). From the practical viewpoint this means that we have to *interpolate* the continuum $f(K)$ from the samples (U_k, Y_k) only. The problem is thus intrinsically orientated towards construction (see Section 3.3.5). Possibility of its solution is ensured by the (Stone-Weierstrass' and Kolmogorov's) existence results.

3.3.3 Stone-Weierstrass theorem

This Section presents the Stone-Weierstrass theorem and its use for neural approximation with feedforward networks.

The classical theorem due to Weierstrass [4] says that an arbitrary continuous function $f: [a, b] \rightarrow \mathbf{R}$ can be uniformly approximated by a sequence of polynomials $\{p_n(x)\}$ to within a prescribed accuracy. In other words, for any $\varepsilon > 0$, an $N \in \mathbf{N}$ can be found, such that for any $n > N \quad |f(x) - p_n(x)| < \varepsilon$ uniformly on $[a, b]$. The approximation requires, for a given ε, a finite number of parameters: coefficients of the polynomial p_{N+1}.

The Weierstrass approach was analysed by M. H. Stone [58], who found universal properties of approximating functions, not intrinsic to polynomials. To explain his rather abstract approach, we begin with a motivating example.

Real numbers $\mathbf{R}$ are composed of rational and irrational numbers, $\mathbf{Q}$ and $\mathbf{R} - \mathbf{Q}$, respectively. All practical computations are performed on rationals, because any real can be approximated to a desired accuracy by a sequence in $\mathbf{Q}$. This is formalised by saying that $\mathbf{Q}$ is *dense* in $\mathbf{R}$, or, equivalently, that $\mathbf{R}$ is the *closure* of $\mathbf{Q}$. In other

[2] For example, if $t + 1$ is replaced with t in eqn. 3.8, then $U_k = [y(t-1), \ldots, y(t-n); u(t-1), \ldots, u(t-m)]^T$ is a $q \times (pm + qn)$ vector and $Y_k = y(t)$ a $q \times 1$ vector.

words, **R** is the smallest set in which all rational Cauchy sequences have limits.

Stone considered the converse problem [7, 47]. Given the set B (playing the role of **R** above) of *all* continuous functions from a compact K to **R**, find a *proper* subset $A \subset B$, such that B is the closure of A (here A is like **Q** above). Thus, if $\{p_n\}$ is a sequence of functions from A, such that $p_n \to f$, its limit f is required to be in B. Since we deal with *function* approximation (as opposed to *number* approximation), it is desirable to perform algebraic operations on A, e.g., taking linear combinations. We also insist that sequences with terms in A converge uniformly on K. This reasoning motivates the following definitions.

Definition 3 *A set A of functions from $K \subset \mathbf{R}^{pm+qn}$ to* **R** *is called an* algebra of functions *iff $\forall f, g \in A$ and $\forall \alpha \in \mathbf{R}$*

(i) $f + g \in A$;

(ii) $fg \in A$;

(iii) $\alpha f \in A$.

Definition 4 *Let B be the set of all functions which are limits of uniformly convergent sequences with terms in A, a set of functions from $K \subset \mathbf{R}^{pm+qn}$ to* **R**. *Then B is called the* uniform closure *of A.*

Two more conditions of nondegeneracy are needed; see Reference [34, page 142].

Definition 5 *A set A of functions from $K \subset \mathbf{R}^{pm+qn}$ to* **R** *is said to* separate points on K *iff* $\forall x_1, x_2 \in K \quad x_1 \neq x_2 \Rightarrow \exists f \in A,\ f(x_1) \neq f(x_2)$.

Definition 6 *Let A be a set of functions from $K \subset \mathbf{R}^{pm+qn}$ to* **R**. *We say that A* vanishes at no point of K *iff $\forall x \in K\ \exists f \in A$, such that $f(x) \neq 0$.*

The main result is as follows:

Theorem 1 (Stone-Weierstrass) *Let A be an algebra of* some *continuous functions from a compact $K \subset \mathbf{R}^{pm+qn}$ to* **R**, *such that A separates points on K and vanishes at no point of K. Then the uniform closure B of A consists of* all *continuous functions from K to* **R**.

The original formulation [47, 7] is for $f: \mathbf{R}^{pm+qn} \to \mathbf{R}$ due to condition (ii) of Definition 3. But the codomain of a vector-valued function is the Cartesian product of its components, so the result remains valid for $f: \mathbf{R}^{pm+qn} \to \mathbf{R}^q$.

The theorem is a relatively simple criterion which given functions have to satisfy in order to uniformly approximate arbitrary continuous functions on compacts. This approach is the essence of a series of publications demonstrating approximation capabilities of multi-layer FNNs, e.g. References [23] and [6] to name two.

For both sigmoidal and RBF networks it suffices to show that the set of all finite linear combinations of sigmoids and RBFs, respectively, is a nonvanishing algebra separating points on a compact $K \subset \mathbf{R}^{pm+qn}$, as specified in Definitions 3, 5, 6. This can readily be done with the additional condition of convexity of K for Gaussians [52].

Thus, according to the Stone-Weierstrass theorem, both sigmoids and RBFs are suitable for uniform approximation of an arbitrary continuous mapping. However, their interpolation properties are different, as RBFs are, in a sense, designed for interpolation, unlike sigmoids (see Section 3.3.5).

3.3.4 Kolmogorov's theorem

This Section describes Kolmogorov's theorem and its neural network applicability.

Kolmogorov's theorem arose from his solution to the so-called 13th Hilbert's problem, which was the following:

Is every analytic function of three variables a superposition of continuous functions of two variables? Is the root $x(a, b, c)$ of the equation

$$x^7 + ax^3 + bx^2 + cx + 1 = 0$$

a superposition of continuous functions of two variables?

One has to consider here the class of functions of which the superposition is constituted. Any function of three variables can be represented as a superposition of functions of two variables by a proper choice of the latter functions [3]. This may not be possible within a given class of smoothness.

Theorem 2 *There is an analytic function of three variables which is not a superposition of infinitely differentiable functions of two variables.*

This motivates the problem of the possibility of reduction to superpositions of continuous functions. Hilbert's hypothesis was that a reduction of this kind would not, in general, be possible.

Kolmogorov disproved Hilbert's hypothesis [71]. As a result of the research he proved the representation theorem [27].

Theorem 3 (Kolmogorov) *Any function continuous on the n-dimensional cube E^n, $E = [0, 1]$, can be represented in the form*

$$f(x_1, \ldots, x_n) = \sum_{i=1}^{2n+1} \chi_i\left(\sum_{j=1}^{n} \phi_{ij}(x_j)\right) \tag{3.9}$$

where χ_i and ϕ_{ij} are real continuous functions of one variable.

Thus, we may *exactly* represent every continuous function as a superposition of a *finite* number of continuous functions of one variable and of a single particular function of two variables, *viz.* addition. The functions ϕ_{ij} are standard and independent of $f(x_1, \ldots, x_n)$. Only the functions χ_i are specific for the given function f.

Kolmogorov's representation theorem was improved by several other authors. Lorentz [35] showed that the functions χ_i may be replaced by only one function χ. Sprecher [57] replaced the functions ϕ_{ij} by $\alpha^{ij}\phi_j$, where α^{ij} are constants and ϕ_j are monotonic increasing functions. The latter has been reformulated by Hecht-Nielsen [19] to motivate the use of neural networks. Thus Sprecher's version of Kolmogorov's representation theorem in terms of neural networks is as follows.

Theorem 4 *Any continuous function defined on the n-dimensional cube E^n can be implemented* exactly *by a three-layered network having $2n + 1$ units in the hidden layer with transfer functions $\alpha^{ij}\phi_j$ from the input to the hidden layer and χ from all of the hidden units to the output layer.*

Girosi and Poggio [16] criticised Hecht-Nielsen's approach, pointing out that:

1. the functions ϕ_{ij} are highly nonsmooth,
2. the functions χ_i depend on the specific function f and are not representable in a parameterised form.

The difficulties were overcome by Kúrková [29, 30] by the use of staircase-like functions of a sigmoidal type in a sigmoidal feedforward neural network. This type of function has the property that it can approximate any continuous function on any closed interval with an arbitrary accuracy. Using this fact, Kolmogorov's representation theorem modified by Kúrková is:

Theorem 5 *Let $n \in \mathbf{N}$ with $n \geq 2, \sigma: \mathbf{R} \to E$ be a sigmoidal function, $f \in C^0(E^n)$, and ε be a positive real number. Then there exist $k \in \mathbf{N}$ and staircase-like functions $\chi_i, \phi_{ij} \in S(\sigma)$ such that for every $(x_1, \ldots, x_n) \in E^n$*

$$\left| f(x_1, \ldots, x_n) - \sum_{i=1}^{k} \chi_i \Big(\sum_{j=1}^{n} \phi_{ij}(x_j) \Big) \right| < \varepsilon \tag{3.10}$$

where $S(\sigma)$ is the set of all staircase-like functions of the form $\sum_{i=1}^{k} a_i \sigma(b_i x + c_i)$.

The theorem implies that any continuous function can be approximated arbitrarily well by a four-layer sigmoidal feedforward neural network. However, comparing eqn. 3.10 with eqn. 3.9 we see that the original *exact* representation in Kolmogorov's result (Theorem 3) is replaced by an *approximate* one (Theorem 5). It is worth noting that it has been established, not necessarily by Kolmogorov's argument, that even three layers are sufficient for approximation of general continuous functions [6, 14, 23].

In practice one has to consider the question of whether an arbitrary given multivariate function f can be represented by an approximate realisation of the corresponding functions χ_i of one variable. The answer [32] is that an *approximate* implementation of χ_i does not, in general, guarantee an approximate implementation of the original function f, i.e. χ_i must be *exactly* realised.

3.3.5 *Multi-dimensional sampling*

This Section describes the fundamental results on multi-dimensional sampling and their applicability for neural modelling with RBF networks.

We have seen in Section 3.3.3 that both sigmoids and RBFs are suitable for approximation of the right-hand side of eqn. 3.8. It is interesting to see how they perform in the more practical problem of interpolation; see Section 3.3.2.

We first recall the basic result of the classical Shannon theory [53].

Theorem 6 (Sampling) *Let $f: \mathbf{R} \to \mathbf{R}$ be such that both its direct, F, and inverse Fourier transforms are well-defined. If the spectrum $F(\omega)$ vanishes for $|\omega| > 2\pi\nu_N$, then f can be exactly reconstructed from samples $\{f(t_k)\}_{k \in \mathbf{Z}}$, $t_k = k/2\nu_N$.*

The essential assumption is that f is bandlimited, i.e. its spectrum is zero outside the interval $[-2\pi\nu_N, 2\pi\nu_N]$. For the one-dimensional case this is ensured by the use of lowpass antialiasing filters. Then, with $t_k = k/2\nu_N$,

$$f(t) = \sum_{k=-\infty}^{\infty} f(t_k) \frac{\sin \pi(2\nu_N t - k)}{\pi(2\nu_N t - k)} = \sum_{k=-\infty}^{\infty} f(t_k) g(t - t_k) \qquad (3.11)$$

where $g(t - t_k) = \sin[2\pi\nu_N(t - t_k)]/2\pi\nu_N(t - t_k)$. This is the canonical interpolation *exact* between all samples. If the sampling period is T, so that $t_k = kT$, then

$$F(\omega) = \frac{G(\omega)}{T} \sum_{k=-\infty}^{\infty} F\left(\omega + \frac{2\pi k}{T}\right) \qquad (3.12)$$

Thus F is periodic and the repetitive portions of the spectrum will not overlap iff $T \leq 1/2\nu_N$, as stipulated in Theorem 6.

The ideal spectrum of G is

$$G(\omega) = \begin{cases} T, & |\omega| < 2\pi\nu_N \\ 0, & \text{elsewhere} \end{cases} \qquad (3.13)$$

which corresponds to $g(t - t_k)$ in eqn. 3.11. Approximate interpolation occurs when the spectrum $G(\omega)$ of eqn. 3.13 is realised approximately, e.g. by a suitably scaled Gaussian.

This reasoning can, in principle, be carried over to n dimensions [43], i.e. when $f: \mathbf{R}^n \to \mathbf{R}$ (it remains true if f is defined on a compact subset K of $\mathbf{R}^n$). A fact worth noting is the necessity of multi-dimensional bandlimiting of the spectrum of $f: \mathbf{R}^n \to \mathbf{R}$.

Thus, there are essentially two steps in the approximate interpolation of f via multi-dimensional sampling:

1. multi-dimensional bandlimiting of the spectrum of f;
2. design of an interpolating filter with characteristic approximating eqn. 3.13.

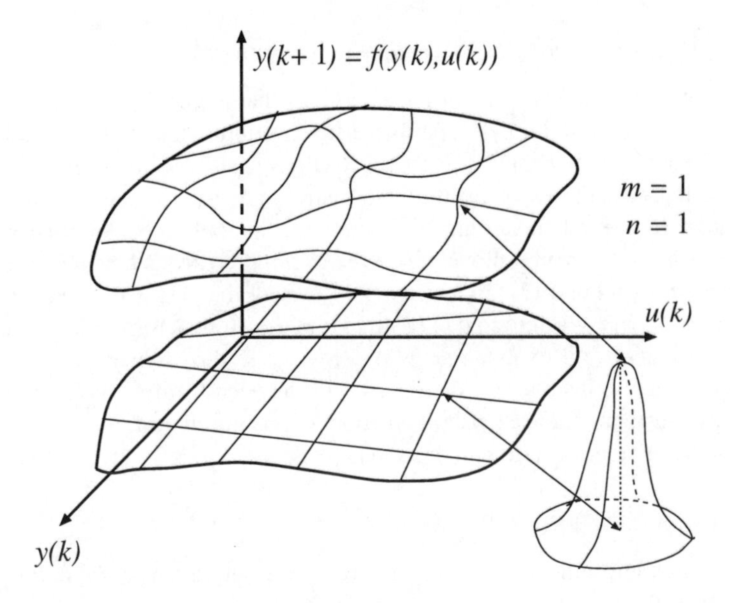

Figure 3.3 *Approximation with feedforward networks. See formula (3.8)*

The possibility of 1 is assumed to be available either due to properties of f or by explicit design. The essence of the neural approach is step 2. It is based on the observation [49] that the Fourier transform of a Gaussian is also Gaussian and may approximate eqn. 3.13 with quantifiable accuracy.

Thus, after a realistic bandlimiting of f, it can be approximately interpolated by linear combinations of Gaussians (see Figure 3.3) with series of the form eqn. 3.11, where $g(t - t_k)$ will correspond to the approximation of eqn. 3.13. Since the formulation of our problem (Section 3.3.2) allows only a *finite* number of samples, the series has to be truncated. It is important that an upper bound for the truncation error is available [49].

However, it should be pointed out that the method bears similarity (on the conceptual level) to the direct methods of optimisation [28]. No *a priori* knowledge about f (except for its continuity and Fourier transformability) is incorporated and the problem scales exponentially with the dimension of K. This phenomenon, known in optimisation as the 'curse of dimensionality' [12], led to the abandonment of direct methods in favour of the ones using some *a priori* knowledge about f.

By assuming little about f the approach gains generality, but the price is the problem of dimensionality. In most control applications some non-trivial *a priori* information is usually available. On the other hand, the computational (real-time) constraints may be severe.

Finally, the whole set-up of uniform (regular) sampling is an idealisation and thus the irregular sampling context seems more appropriate.

3.3.6 *Backpropagation for feedforward networks*

Neural networks are adaptive nonlinear systems that adjust their parameters automatically in order to minimise a performance criterion. This obviously links them to adaptive and optimal control. Previous Sections dealt with approximation and/or interpolation capabilities of feedforward neural networks. They give theoretical foundation for neural modelling of discrete-time, input-output nonlinear plants. Identification of the neural models involves learning, which is covered extensively in References [24] and [71]. Here we only present the classical backpropagation (BP) [48, 62, 20] algorithm to demonstrate it in the Werbos form [63] and to show later its extensions for dynamic networks. Moreover, the essence of the algorithm gives an immense insight into the differences between feedforward and recurrent networks. Last, but not least, useful notation will be introduced.

Define the ordered system of equations (see eqns. 3.5–3.6 and Figure 3.2)

$$y_i = f_i(y_{i-1}, \ldots, y_1), \quad i = 1, \ldots, N+1 \tag{3.14}$$

Then a systematic and formally proper way of calculating the partial derivatives of y_{N+1} is as follows

$$\frac{\partial^+ y_{N+1}}{\partial y_i} = \sum_{j>i}^{N+1} \frac{\partial^+ y_{N+1}}{\partial y_j} \frac{\partial f_j}{\partial y_i} \tag{3.15}$$

which is a recursive definition of the ordered derivative $\partial^+ y_{N+1}/\partial y_i$, valid only for the systems of eqns. 3.14.[3]

The network in eqn. 3.5 is given a reference signal d only for the output layer, i.e. $d_i \neq 0$ for $i = N - q + 1, \ldots, N$ (see Figure 3.2), and it is required to minimise

$$E = \frac{1}{2} \sum_{t=1}^{P} (d^t - y^t)^T (d^t - y^t) \tag{3.16}$$

which is the nonlinear least-squares fitting problem [10] for P patterns. The gradient algorithm for adjusting weights yields

$$w_{ij}^{\text{new}} = w_{ij}^{\text{old}} - \alpha \frac{\partial E}{\partial w_{ij}}, \quad \alpha > 0 \tag{3.17}$$

[3] The need for some new notation is apparent from the following example (Exercise 3.14/p. 89 in Reference [8]). Let $w = f(x, y, z)$ and $z = g(x, y)$, so that x, y are the independent variables. Then, mechanically applying the standard notation, one obtains

$$\begin{aligned} \partial w/\partial x &= (\partial w/\partial x)(\partial x/\partial x) + (\partial w/\partial y)(\partial y/\partial x) + (\partial w/\partial z)(\partial z/\partial x) \\ &= \partial w/\partial x + (\partial w/\partial z)(\partial z/\partial x) \end{aligned}$$

as $\partial x/\partial x = 1$ and $\partial y/\partial x = 0$. But now $\partial w/\partial x$ can be cancelled on both sides resulting in the nonsensical $0 = (\partial w/\partial z)(\partial z/\partial x)$ for *any* functions w and z. See also the notational difficulties in Reference [13], pp. 12–13, and the way round them in Reference [9], p. 49.

It should be emphasised that this is solely a *notational* issue of no theoretical meaning.

where ($P = 1$ for simplicity; see eqn. 3.16)

$$\frac{\partial E}{\partial w_{ij}} = \frac{\partial E}{\partial y_i}\frac{dy_i}{ds_i}\frac{\partial s_i}{\partial w_{ij}} = \frac{\partial E}{\partial y_i}\sigma'(s_i)y_j \tag{3.18}$$

For the output layer $\partial E/\partial y_i = y_i - d_i$, but it is not straightforward to find this expression for the hidden layers. Backpropagation computes the error derivative as

$$\frac{\partial E}{\partial y_i} = \sum_j \frac{\partial E}{\partial y_j}\frac{dy_j}{ds_j}\frac{\partial s_j}{\partial y_i} = \sum_j \frac{\partial E}{\partial y_j}\sigma'(s_j)w_{ji} \tag{3.19}$$

where y_i belongs to the layer l, and y_j to $l + 1$. Starting from the output layer this can be recursively solved. This formulation requires care with neuron indexing and raises doubts about notation; see footnote 3. If, however, the necessary care is exercised, then eqn. 3.19 is simply a result of the Chain Rule.

On the other hand, the network structure is ordered (compare eqns. 3.5–3.6 and 3.14). Thus, eqn. 3.19 can be expressed in Werbos' notation as

$$\frac{\partial^+ E}{\partial y_i} = \frac{\partial E}{\partial y_i} + \sum_{j>i}^{N} \frac{\partial^+ E}{\partial y_j}\frac{\partial y_j}{\partial y_i} \tag{3.20}$$

with E treated as y_{N+1} in eqn. 3.15. Then the BP algorithm for feedforward networks is

$$\begin{aligned} z_i &= \sum_{j>i} w_{ji}\sigma'(s_j)z_j + \epsilon_i \\ \frac{\partial E}{\partial w_{ij}} &= \sigma'(s_i)z_i y_j \\ w_{ij}^{\text{new}} &= w_{ij}^{\text{old}} - \alpha\frac{\partial E}{\partial w_{ij}} \end{aligned} \tag{3.21}$$

where $z_i = \partial^+ E/\partial y_i$ and $\epsilon_i = \partial E/\partial y_i$, or in vector-matrix form:

$$\begin{aligned} S &= \text{diag}[\sigma'(s_1), \ldots, \sigma'(s_N)] \\ z &= W^T S z + \epsilon \\ W^{\text{new}} &= W^{\text{old}} - \alpha S z y^T \end{aligned} \tag{3.22}$$

This is inherently an off-line technique as the forward pass must be computed for all t (see eqn. 3.16) before eqn. 3.22 is applied (backward pass).

Both methods lead to the same results for networks of the structure shown in Figure 3.2 and require only *local* computations. However, Werbos' method corresponds to a fully connected network and thus is more general. Moreover, it is very systematic [63] and straightforward in implementations requiring network architecture modification [64].

It should be mentioned that learning algorithms for RBF networks are linear in the parameters [52]. This is because only w_{ij} are adjusted in eqn. 3.7, while c_{ij} and m_{ij} are preset.

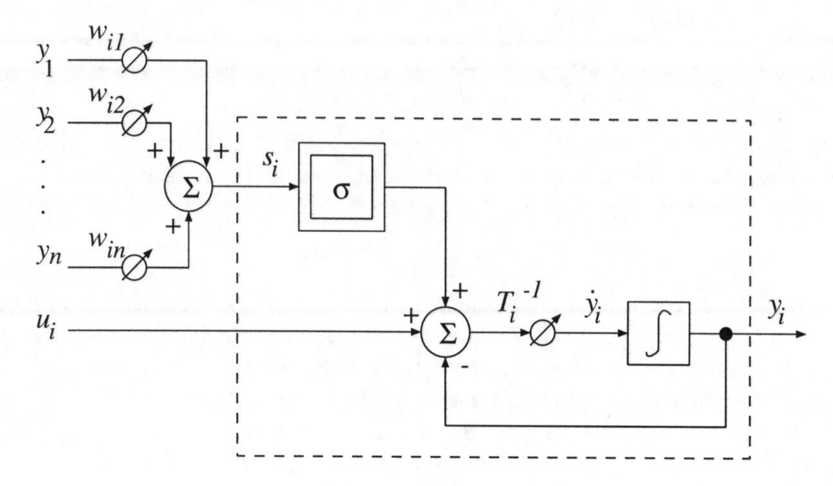

Figure 3.4 *Dynamic neuron. The frame stands for a double square in Figure 3.5*

3.4 Recurrent networks

Dynamic or recurrent networks are qualitatively different from feedforward ones, because their structure incorporates feedback. In general, the output of every neuron is fed back with varying gains (weights) to the inputs of all neurons. A scheme of such a neuron is shown in Figure 3.4. A neural system composed of these elements, with all weights non-zero in general, is called a *fully connected network*. The network architecture is inherently dynamic and usually one-layered, since its complex dynamics gives it powerful representation capabilities; see Section 3.4.2. Figure 3.5 shows a fully connected network configured for control purposes, as explained later. Other characteristics of recurrent networks will be discussed below. Convergence and stability will be analysed after the presentation of learning algorithms.

3.4.1 Mathematical description

To exhibit the dynamic behaviour of eqn. 3.1 the network must have a dynamic structure. These kinds of networks [21, 22, 42] are built from dynamic neurons (see Figure 3.4)

$$T_i \dot{y}_i = -y_i + \sigma(s_i) + u_i, \quad s_i = \sum_{j=1}^{N} w_{ij} y_j \tag{3.23}$$

with $y_i(t_0) = y_i^0$. Here y_j are feedback signals from all neurons and N is the number of neurons in the network.

There are two basic recurrent network descriptions. The first [22] assumes separate state and output, while the other gives linear output equal to state, i.e. in vector matrix form

$$\begin{aligned} s &= Wy, \\ T\dot{y} &= -y + \sigma(s) + u. \end{aligned} \tag{3.24}$$

The matrix W is dense, contrary to that in eqns. 3.5–3.6, since the network is fully connected. The discrete time version is:

$$\begin{aligned} s(t) &= Wy(t), \\ Ty(t+1) &= \sigma(s(t)) + u(t). \end{aligned} \tag{3.25}$$

We shall use eqns. 3.24–3.25, which result in less cumbersome formulae.

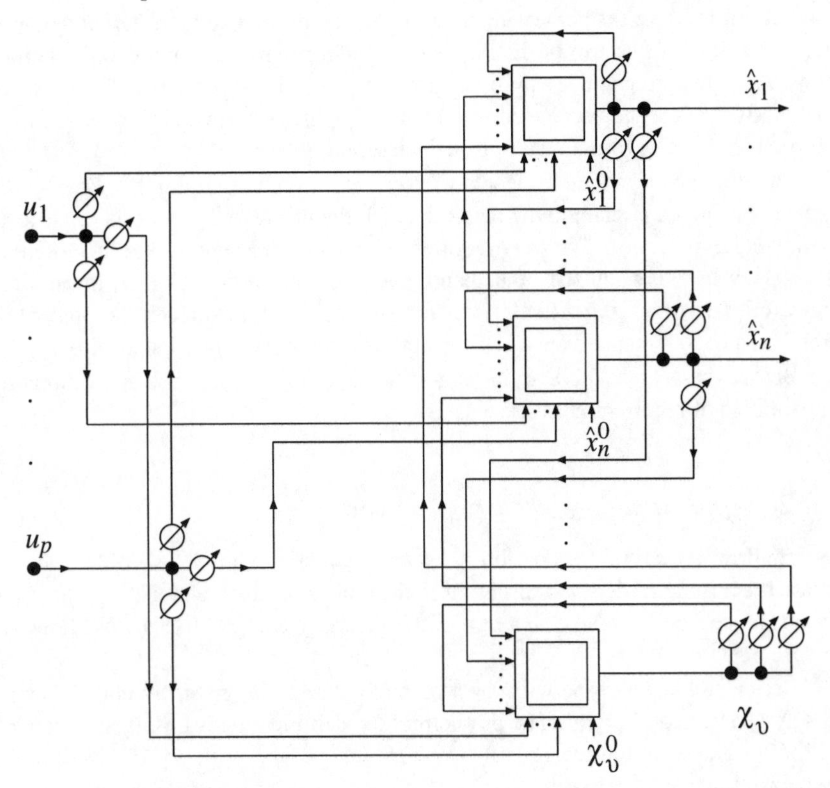

Figure 3.5 *Recurrent neural network configured for control purposes*

It should be emphasised that eqns. 3.24 and 3.25 describe a *general* model of recurrent networks. The homogeneous notation (all signals denoted as y_i) is used for uniformity and compactness of matrix notation. However, this model may be *configured for control purposes*, which means that some elements of the matrices T and W in eqn. 3.24 will be set to 0. The aim is to set up the network in such a way that it approximates the system of eqn. 3.1. This corresponds to the following design steps (with $\sigma(0) = 0$); see Figure 3.5. Input neurons are obtained from eqns. 3.23–3.24 by setting $T_i = 0$ and $w_{ij} = 0$ for $i = 1, \ldots, p$ and $j = 1, \ldots, N$. For all other neurons ($i = p + 1, \ldots, N$) the signals u_i are set to zero. Amongst the neurons different from the input ones, n neurons are picked and their outputs are treated as the *network outputs*. The remaining $\nu > 0$ neurons represent the network's internal *hyperstate*. Thus the network has p inputs u_i, ν hyperstate variables χ_i and n outputs $\hat{x}_i$ making $p + \nu + n = N$ neurons. Note that eqns. 3.23–3.24 describe *uniform* neurons with their outputs denoted y_i for $i = 1, \ldots, N$. This homogeneity is replaced in Figure 3.5 by the specialised configuration with neuron outputs denoted according to their function from the control viewpoint.

Recurrent networks possessing the same structure can exhibit different dynamic behaviour, due to the use of distinct learning algorithms. The network is defined when its architecture and learning rule are given. In other words, it is a composition of two dynamic systems: transmission and adjusting systems. The overall input-output behaviour is thus a result of the interaction of both.

There are two general concepts of recurrent structures training. *Fixed-point learning* is aimed at making the network reach the prescribed equilibria or perform steady-state matching. The only requirement on the transients is that they die out. *Trajectory learning*, on the other hand, trains the network to follow the desired trajectories in time. In particular, when $t \to \infty$, it will also reach the prescribed steady-state, so it can be viewed as a generalisation of fixed point algorithms.

Before we present the learning algorithms, we briefly describe approximating capabilities of recurrent networks.

3.4.2 *Approximation: recurrent networks*

We shall now analyse the possibility of approximating eqn. 3.1 with 3.27 and/or related recurrent models and discuss it in the context of differential approximation, as introduced below. Note that eqn. 3.27 is simply eqn. 3.24 recast as shown in Figure 3.5.

Let an unknown plant with an accessible state x be given by eqn. 3.1 with $x \in X \subseteq \mathbf{R}^n$, $u \in U \subseteq \mathbf{R}^p$. The problem of its identification is unsolved in general [17, 69]. The neural approach offers two parametric approaches [70]. One [60] is to approximate eqn. 3.1 with

$$\dot{\hat{x}} = \sum_{i=1}^{N} w_i g\left(\frac{\|\bar{x} - c^i\|^2}{m_i}\right) \tag{3.26}$$

where $g(\cdot)$ is a radial basis function, typically Gaussian, w_i, m_i are scalar and c^i vector parameters. Here c^i and m_i define the grid and w_i is adaptive; see also eqn. 3.7 and Figure 3.3. Also, $\bar{x} \in \mathbf{R}^{n+p}$ is concatenation of $\hat{x}$ and u with $\hat{x} \in X \subseteq \mathbf{R}^n$ being

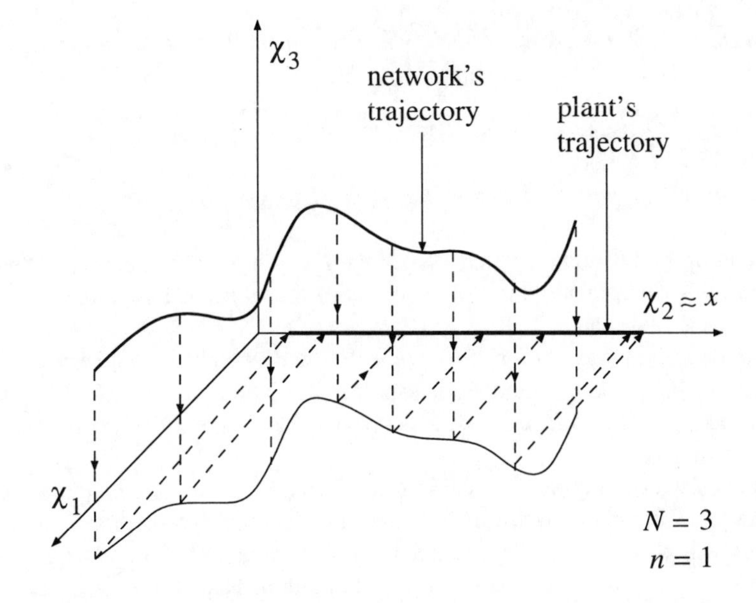

Figure 3.6 *Approximation with recurrent networks. See formula (3.27)*

the approximation of the plant's state $x \in X \subseteq \mathbf{R}^n$. Roughly speaking, the right-hand side of eqn. 3.1 is expanded into a finite weighted sum (linear combination) of nonlinear functions retaining the dimension of the original state, i.e. both x and $\hat{x}$ are n-vectors.

The other method [70] postulates a network (see Figure 3.5)

$$\begin{aligned} \dot{\chi} &= -\chi + \vec{\sigma}(W\chi) + u, \\ \hat{x} &= C\chi, \end{aligned} \tag{3.27}$$

where $\chi \in H \subseteq \mathbf{R}^N, \hat{x} \in X \subseteq \mathbf{R}^n, W \in V \subseteq \mathbf{R}^{N^2}$ with $N \geq n$, so that $\hat{x}$, the approximation of x from eqn. 3.1, is a projection of χ to n dimensions. Roughly speaking, eqn. 3.27 fixes its right-hand side and expands its order. The form of eqn. 3.27 is, however, arbitrary, in the sense that it is not motivated by any mathematical reasoning. Thus, there naturally arises a question about the sense of investigating eqn. 3.27 and not some other dynamic, parametric model. The following definition [69] attempts to address the problem (see Figure 3.6).

Definition 7 (Differential approximation) *Given a nonlinear dynamic system*

$$\Sigma\colon\ \dot{x} = f(x, u)$$

where $x(t_0) = x^0$, $x \in X \subseteq \mathbf{R}^n$, $u \in U \subseteq \mathbf{R}^p$, and $f(\cdot,\cdot)$ Lipschitz in $X \times U$, find a dynamic parametric system

$$\Sigma_W\colon\ \dot{\chi} = \phi(\chi, u, W)$$

with $\chi(t_0) = \chi^0$, $\chi \in H \subseteq \mathbf{R}^N$, $N \geq n$, *parameters* $W \in V \subseteq \mathbf{R}^r$, *and* $\phi(\cdot, \cdot, \cdot)$ *Lipschitz in* $H \times U \times V$, *such that*

$$\forall \varepsilon > 0 \quad \exists r \in \mathbf{N} \quad \exists W \in V \quad \forall t \geq 0 \quad \|x - \hat{x}\| < \varepsilon$$

where $\hat{x} = \Pi(\chi)$, *with* Π *the projection from* H *to* X.

Definition 7 postulates, strictly speaking, *local* differential approximation (for a given x^0). The global definition would have the additional quantifier $\forall x^0 \in X$ (between $\exists W \in V$ and $\forall t \geq 0$).

The definition leaves room for manoeuvre with the choice of the norm $\|\cdot\|$, the sets X, H, V and the numbers N and r, but obviously $X \subseteq H$. Finally, the projection $\Pi(\cdot)$ may be replaced by a less trivial mapping if this would lead to more useful results.

The insistence of parameterisation is motivated by the possibility of employing algebraic tools along with the analytic ones it offers for investigation of nonlinear dynamics. Fixing the model's structure allows focusing on W, simplifying identification and design, as we then deal with the finite-dimensional space V. Apart from the existence and uniqueness questions, the issue of interest is that the parameterisation of Σ_W leads to generic nonlinear control [69]. This means that if a nonlinear system Σ can be replaced by its parametric equivalent Σ_W, then given the special, yet generic, form of $\phi(\cdot, \cdot, \cdot)$ of Σ_W the control problem should be soluble with existing techniques or a new one. An obvious suggestion is to look for a parametric version of the affine model [25]; see eqn. 3.27.

In the setting of differential approximation the problem is still unsolved, but there are some existence results available [56, 15]. They ensure a theoretical possibility of using eqn. 3.27 for approximation of eqn. 3.1, but in a context different from differential approximation; for details see Reference [70].

3.4.3 Fixed-point learning

For the network in eqn. 3.24 the error is defined as

$$E = \frac{1}{2} \sum_{k=1}^{P} (y_\infty^k - y)^T (y_\infty^k - y) \tag{3.28}$$

where y_∞^k are vectors of the desired equilibria, being solutions of eqn. 3.24 with the left-hand side set to 0:

$$0 = -y_\infty^k + \sigma(s) + u, \quad k = 1, \ldots, P \tag{3.29}$$

During learning the network does not receive any external inputs. It is excited by initial conditions corresponding to the expected workspace and evolves with y^∞ as a constant reference signal, which is called *relaxation* [45]. Practically, this is done using *recurrent backpropagation* [44, 1] (with $P = 1$):

$$\begin{aligned} e_i &= \begin{cases} y_i^{\infty} - y_i, & \text{if } y_i \text{ is a network output} \\ 0, & \text{otherwise} \end{cases} \\ \dot{z}_i &= -z_i + \sum_j w_{ji}\sigma'(s_j)z_j - e_i \\ \dot{w}_{ij} &= -\alpha\sigma'(s_i(\infty))z_i(\infty)y_j(\infty) \end{aligned} \tag{3.30}$$

or, in vector-matrix form

$$\begin{aligned} e &= [0 \vdots y^{\infty}_{\cdot} - y]^T \\ S &= \text{diag}[\sigma'(s_1), \ldots, \sigma'(s_N)] \\ \dot{z} &= -z + W^T S z - e \\ \dot{W} &= -\alpha S(\infty)z(\infty)y^T(\infty) \end{aligned} \tag{3.31}$$

The calculation of z corresponds to the backpropagation rule for feedforward networks of eqn. 3.21 with $\epsilon = -e$. The equation for W can be solved after the network has settled, since $y(\infty)$ and $z(\infty)$ are needed. Notice that the right-hand side of the differential equation for W is *constant*, which means that the solution is just the product of the constant and time. Adjustment is thus a one-shot procedure, once the equation for $z(t)$ is solved.

This algorithm differs from Hopfield nets [21, 22], since there the learning was performed by the Hebbian rule [54] or the weights were established by the designer, not the network adaptive algorithm [59]. Thus, there the relaxation is done only during the recall phase (the Hopfield net is a content-addressable memory). Recurrent backpropagation creates basins of attraction around the given equilibria by adjusting the weights, learning them through relaxation eqn. 3.31. The Hopfield net generates its attractors and basins by applying fixed weights and thus produces equilibria on its own, representing input patterns or an optimisation problem solution.

If the forward pass in recurrent backpropagation is stable, so is the backward one [1]. On the other hand, the Hopfield net is always stable, because defining it requires guessing a Lyapunov function appropriate to an application [59, 40].

3.4.4 Trajectory learning

In this case, the error for the network of eqn. 3.24 is given by

$$E = \frac{1}{2}\int_{t_0}^{t_1} \Big[(d(\tau) - y(\tau))^T (d(\tau) - y(\tau))\Big] d\tau \tag{3.32}$$

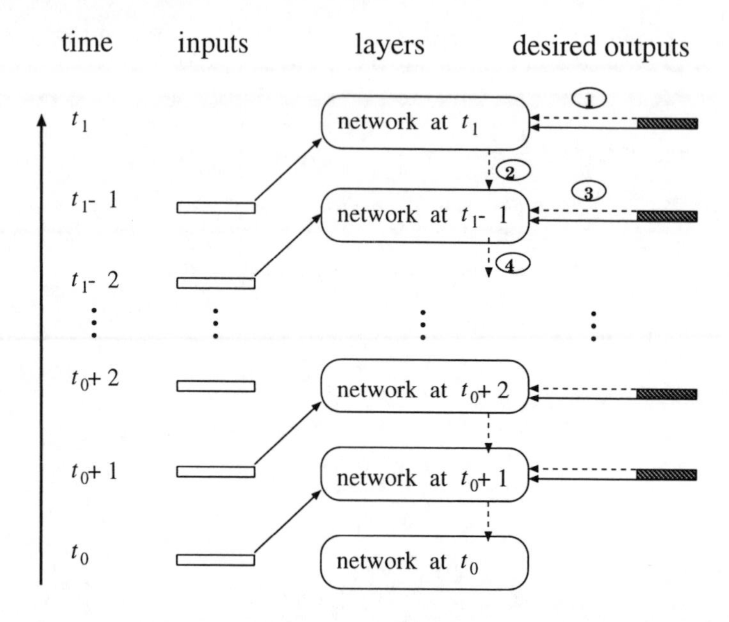

Figure 3.7 *Backpropagation through time. Dashed arrows, together with the numbers in ovals, show the order of backwards computations [68]*

where $d(\tau)$ is a vector of the desired trajectories and t_1 may be a constant (off-line techniques) or a variable (on-line algorithms). The discrete-time version (see eqn. 3.25) yields:

$$E = \frac{1}{2}\sum_{\tau=t_0}^{t_1}\Big[(d(\tau) - y(\tau))^T(d(\tau) - y(\tau))\Big] \tag{3.33}$$

with the previous remarks valid. The case of multiple reference trajectories $d^k(\tau)$, $k = 1, \ldots, P$ is omitted for simplicity.

3.4.4.1 Backpropagation through time (BPTT)

The simplest method, already mentioned in Reference [48], is to unfold the network through time, i.e. replace a one-layer recurrent network with a feedforward one with t_1 layers. The equivalent static network has as many layers as time instants, which is easily understood in the discrete time context and therefore we start with this. Standard backpropagation is applied, and since each layer contains the same weights 'seen' in different moments, their final value is the sum of intermediate values [64]; see Figure 3.7. Thus, we have

$$E = \frac{1}{2}\sum_{\tau=t_0}^{t_1}\sum_{i=1}^{N}(d_i(\tau) - y_i(\tau))^2, \quad t_1 = \text{const}$$

$$
\begin{aligned}
e_i(\tau) &= \begin{cases} d_i(\tau) - y_i(\tau), & \text{if } y_i(\tau) \text{ is a network output} \\ 0, & \text{otherwise} \end{cases} \\
z_i(\tau) &= \sum_{j=1}^{N} w_{ji}(t_0)\sigma'(s_j(\tau+1))z_j(\tau+1) - e_i(\tau), \quad z_i(t_1) = 0 \\
\frac{\partial E}{\partial w_{ij}} &= \sum_{\tau=t_0}^{t_1} \sigma'(s_i(\tau))z_i(\tau)y_j(\tau) \\
w_{ij}^{t_1} &= w_{ij}^{t_0} - \alpha \frac{\partial E}{\partial w_{ij}}
\end{aligned}
\tag{3.34}
$$

where $z_i(\tau) = \partial^+ E/\partial y_i(\tau)$, or, in vector-matrix form,

$$
\begin{aligned}
E &= \frac{1}{2}\sum_{\tau=t_0}^{t_1}\Big[(d(\tau) - y(\tau))^T(d(\tau) - y(\tau))\Big], \quad t_1 = \text{const} \\
e(\tau) &= [0 \,\vdots\, d(\tau) - y(\tau)]^T \\
S(\tau) &= \text{diag}[\sigma'(s_1(\tau)), \ldots, \sigma'(s_N(\tau))] \\
z(\tau) &= W^T(t_0)S(\tau+1)z(\tau+1) - e(\tau), \quad z(t_1) = 0 \\
\nabla_W E &= \sum_{\tau=t_0}^{t_1} S(\tau)z(\tau)y^T(\tau) \\
W^{t_1} &= W^{t_0} - \alpha \nabla_W E
\end{aligned}
\tag{3.35}
$$

The definition of $e_i = e_i(\tau)$ in eqn. 3.34 and/or $e = e(\tau)$ in eqn. 3.35 will be further used without additional explanation. Theoretically, there exists the possibility [68] of an on-line algorithm, but the above calculations would have to be performed every step, requiring unlimited memory and computational power.

Comparison of eqns. 3.35 and 3.31 shows that BPTT is a generalisation of recurrent backpropagation. The trajectory to be learned in the fixed-point learning is constant, i.e. $d(\tau) \equiv y^{\infty}$. The relationship between W^{t_1} and W^{t_0} corresponds to the one-shot solution of eqn. 3.31.

The continuous-time version was introduced heuristically by Pearlmutter [41], and then rediscovered by Sato [51], whose derivation, based on the calculus of variations, is mathematically rigorous and will be briefly presented here. Introduce the Lagrangian

$$
L = \int_{t_0}^{t_1} \Big\{ \frac{1}{2}\sum_{N-n}^{N}(d_i - y_i)^2 - \sum_{i=1}^{N} z_i[T_i\dot{y}_i + y_i - \sigma(s_i) - u_i] \Big\} d\tau \tag{3.36}
$$

where z_i are Lagrange multipliers (for explanation of index n see Figure 3.5). The first variation yields

$$\delta L = \int_{t_0}^{t_1} \Big\{ \sum_{N-n}^{N} [(y_i - d_i) - z_i]\delta y_i + \sum_{i=1}^{N} z_i \sigma'(s_i) \sum_{j=1}^{N} w_{ij}\delta y_j - \sum_{i=1}^{N} z_i T_i \delta \dot{y}_i + \sum_{i=1}^{N} z_i \sigma'(s_i) \sum_{j=1}^{N} y_j \delta w_{ij} \Big\} d\tau \tag{3.37}$$

which can be reduced by defining the auxiliary equation

$$T_i \dot{z}_i = z_i - \sum_{j=1}^{N} w_{ji} \sigma'(s_j) z_j + e_i \tag{3.38}$$

Multiplying both sides of eqn. 3.38 by δy_i and summing with respect to i gives

$$\delta L = \sum_{i=1}^{N} [T_i z_i(t_0) \delta y_i(t_0) - T_i z_i(t_1) \delta y_i(t_1)] + \int_{t_0}^{t_1} \Big[\sum_{i=1}^{N} z_i \sigma'(s_i) \sum_{j=1}^{N} y_j \delta w_{ij} \Big] d\tau \tag{3.39}$$

Since the initial conditions $y_i(t_0)$ do not depend on the weights, $\delta y_i(t_0) = 0$. If additionally the boundary values

$$z_i(t_1) = 0 \tag{3.40}$$

are imposed then

$$\delta L = \int_{t_0}^{t_1} \Big[\sum_{i=1}^{N} z_i \sigma'(s_i) \sum_{j=1}^{N} y_j \delta w_{ij} \Big] d\tau \tag{3.41}$$

and hence (compare with eqn. 3.34)

$$\frac{\partial E}{\partial w_{ij}} = \int_{t_0}^{t_1} \sigma'(s_i) z_i y_j d\tau \tag{3.42}$$

$$\dot{w}_{ij} = -\alpha \frac{\partial E}{\partial w_{ij}} \tag{3.43}$$

which together with eqn. 3.38 completes the algorithm. The vector-matrix form looks as follows:

$$\begin{aligned} E &= \frac{1}{2} \int_{t_0}^{t_1} \Big[(d(\tau) - y(\tau))^T (d(\tau) - y(\tau)) \Big] d\tau, \quad t_1 = \text{const} \\ T\dot{z} &= z - W^T S z + e \\ z(t_1) &= 0 \\ \nabla_W E &= \int_{t_0}^{t_1} S(\tau) z(\tau) y^T(\tau) d\tau \end{aligned}$$

$$\dot{W} = -\alpha \nabla_W E \tag{3.44}$$

Notice that the auxiliary equation 3.38 is defined on $[t_0, t_1]$ with $z_i(t_1) = 0$ (see eqn. 3.40), which means that it must be integrated backwards in time. The method for adjusting time constants and delays can be found in Reference [42]. A speeding up modification of the algorithm was proposed in Reference [11].

The method is essentially non-recurrent, because it uses a feedforward equivalent for learning, so there is no explicit use of the network's feedback. It requires recording of the whole trajectories and playing them back to calculate the weights, which are again adjusted according to a one-shot procedure. However, once the weights are established the network runs as a recurrent one and this differentiates it from static backpropagation as described in Section 3.3.6.

3.4.4.2 Forward propagation

Backpropagation through time is inherently an off-line technique. The following algorithm [46, 67] overcomes the difficulty. For the recurrent network (T_k suppressed for clarity; compare with eqn. 3.23)

$$\dot{y}_k = -y_k + \sigma(s_k) + u_k, \quad k = 1, \ldots, N \tag{3.45}$$

with

$$\frac{\partial \dot{y}_k}{\partial w_{ij}} = -\frac{\partial y_k}{\partial w_{ij}} + \sigma'(s_k)\Big[\delta_{ik} y_j + \sum_{l=1}^{N} w_{kl} \frac{\partial y_l}{\partial w_{ij}}\Big] \tag{3.46}$$

where δ_{ik} is the Kronecker symbol, define error

$$E = \int_{t_0}^{t} \Big[\frac{1}{2}\sum_{k=1}^{N} (d_k(\tau) - y_k(\tau))^2\Big] d\tau, \quad t = \text{var} \tag{3.47}$$

i.e. with free termination time (t is a variable). The rationale for introducing the subscript k in eqn. 3.45 is that each neuron is fully connected, so it is influenced by all weights w_{ij} (compare with eqn. 3.46).

Consider the first variation of the error

$$\delta E = -\sum_{k=1}^{N}\Big[\int_{t_0}^{t} [e_k(\tau)\delta y_k] d\tau\Big] \tag{3.48}$$

where $e_k(\tau)$ is defined as in eqn. 3.34. Since the w_{ij} are independent variables, the following holds

$$\delta y_k = \sum_{i=1}^{N}\sum_{j=1}^{N} \frac{\partial y_k}{\partial w_{ij}} \delta w_{ij} \tag{3.49}$$

Taking into account constraints, eqn. 3.45 gives, with $p_{ij}^k = \partial y_k / \partial w_{ij}$ substituted to eqn. 3.46,

$$\frac{\partial E}{\partial w_{ij}} = -\sum_{k=1}^{N}\Big[\int_{t_0}^{t} [e_k(\tau) p_{ij}^k(\tau)] d\tau\Big]$$

$$\begin{aligned}\dot{p}_{ij}^{k} &= -p_{ij}^{k} + \sigma'(s_k)\Big[\delta_{ik} y_j + \sum_{l=1}^{N} w_{kl} p_{ij}^{l}\Big]\\ \dot{w}_{ij} &= -\alpha \frac{\partial E}{\partial w_{ij}}\end{aligned} \tag{3.50}$$

and

$$p_{ij}^{k}(t_0) = 0 \tag{3.51}$$

because the initial conditions $y_k(t_0)$ do not depend on weights. The replacement of the boundary conditions of eqn. 3.40 with the initial conditions of eqn. 3.51 allows on-line calculations in eqn. 3.50. It is not straightforward to write eqns. 3.50–3.51 in the vector-matrix form, because variables p_{ij}^{k} require a three-dimensional array. One possibility is:

$$\begin{aligned}E &= \frac{1}{2}\int_{t_0}^{t}\Big[(d(\tau) - y(\tau))^T (d(\tau) - y(\tau))\Big]d\tau, \qquad t = \text{var}\\ \nabla_W E &= -\sum_{k=1}^{N}\Big[\int_{t_0}^{t}[e_k(\tau) P_k(\tau)]d\tau\Big]\\ \dot{P}_k &= -P_k + \sigma'(s_k)[\lambda^k y^T + \Pi_k], \quad k = 1, \ldots, N\\ P_k(t_0) &= 0, \quad k = 1, \ldots, N\\ \dot{W} &= -\alpha \nabla_W E\end{aligned} \tag{3.52}$$

where $P_k = [p_{ij}^{k}]$ and $\Pi_k = [\pi_{ij}]$, $\pi_{ij} = \sum_{l=1}^{N} w_{kl} p_{ij}^{l}$ are square matrices of dimension N and $\lambda^k = [0, \ldots, \underbrace{1}_{\lambda_k}, \ldots, 0]^T$ is a vector. A modification of the method can be found in Reference [72].

Forward propagation, called also *real-time recurrent learning (RTRL)* [66], requires non-local computations, since to calculate $\partial E/\partial w_{ij}$ knowledge of all $p_{ij}^{k}, k = 1, \ldots, N$ is needed (see eqn. 3.50). Real-time solving of the auxiliary equation on p_{ij}^{k} contributes to computing complexity, but the algorithm is truly on-line.

It is worth noting that dynamic networks allow an elegant solution of the key problem of multi-layer static networks, *viz.* the reference signal for hidden units. The answer is straightforward (see eqn. 3.34) by setting the error to 0. It does not make impossible calculation of weights for hidden units (compare with eqn. 3.50), since through *feedback* they get information about the system performance. It may also be noted that eqn. 3.46 is the network's sensitivity model [13], p_{ij}^{k} being the sensitivity functions.

3.4.5 *Some control aspects of recurrent networks*

The most promising area of neural networks applications is nonlinear control. If the plant is time-invariant and its description is well-known then feedback linearisation [38, 25] gives a systematic method of controller design. It is, however, sensitive to noise [55] and assumes the canonical model

$$\dot{x} = f(x) + h(x)u \tag{3.53}$$

which, although fairly general, is not universal. Sliding control [61] offers some robustness, but requires great skill in guessing an appropriate Lyapunov function. Adaptive nonlinear control [50, 55, 26] is still in an early stage of development and the describing function method is very limited. Thus, when faced with model uncertainty and real-world noise/disturbances, it is not trivial to design a working controller for a nonlinear plant.

In this context, recurrent neural networks offer an interesting alternative to the conventional approach. They give a generic model [69] for a very broad class of systems considered in control theory. This is an important feature, since the power of *linear* control lies, among other reasons, in the universal method of description. Comparison of the plant state equation and the dynamic network model

$$\begin{aligned} \dot{x} &= f(x, u) && \text{for plant} \\ T\dot{y} &= -y + \sigma(s) + u && \text{for general dynamic network} \end{aligned} \tag{3.54}$$

where $x \in \mathbf{R}^n$, $u \in \mathbf{R}^p$ and $y \in \mathbf{R}^N$, may give the false impression that the network is not more general than 3.53. However, if p neurons (eqn. 3.23) are chosen with $T_i = 0, w_{ij} = 0$ for $i = 1, \ldots, p,\ j = 1, \ldots, N$ then their outputs are equal to the inputs u_i and are distributed *through weights* w_{ji} to the remaining, unreduced neurons (see Figure 3.5). Massive parallelism of the network in eqn. 3.24 is not only introduced to speed it up in a hardware implementation. It also plays the role of plant representation through the network's internal *hyperstate* (compare Figure 3.5). To avoid the false impression that plant's state $x \in \mathbf{R}^n$ and network's hyperstate $\chi \in \mathbf{R}^\nu$ are equivalent ($n \neq \nu$), x and χ were introduced, respectively. The homogeneous notation y (see eqn. 3.54) as against the structure specific $u, \chi, \hat{x}$ in Figure 3.5 was discussed in Section 3.4.1. The network output is denoted as $\hat{x}$ to emphasise that the network is supposed to output an estimate of the plant's state x. Notice that the network has the same number of inputs, p, as the system and the dimension of its output space (in the network's control configuration in Figure 3.5) is equal to the dimension of the original state space (i.e. n) or, in other words, $\hat{x} \in \mathbf{R}^n$. It is obvious that the right-hand side of eqn. 3.24 cannot be of the same dimension as $f(\cdot, \cdot)$ in eqn. 3.1, because it has an entirely different structure; see also Section 3.4.2.

3.4.6 Stability of recurrent networks

It is characteristic of the enthusiastic approach to neural networks that the central issue of recurrent networks stability did not receive as much attention as simulations and experiments. The most powerful result, the Cohen-Grossberg Theorem [5], was obtained under conservative assumptions, often violated by practitioners without much harm [44]. Stability of the most popular recurrent architecture (which was used throughout this paper), the Hopfield network [22], is obtained (as was the Cohen-Grossberg Theorem) via Lyapunov's direct method. However, as discussed below, the derivation is not complete.

3.4.6.1 Cohen-Grossberg Theorem

This result [5] is the most fundamental one in recurrent networks stability theory. It applies to the general model

$$\dot{y}_i = a_i(y_i)[b_i(y_i) - \sum_{j=1}^{N} w_{ij}\sigma_j(y_j)], \quad i = 1, \ldots, N \tag{3.55}$$

The following assumptions are made for $i, j = 1, \ldots, N$:
1° $w_{ij} \geq 0$ and $w_{ij} = w_{ji}$;
2° $a_i(\xi)$ is continuous for $\xi \geq 0$ and positive for $\xi > 0$;
3° $b_i(\xi)$ is continuous;
4° $\sigma_i(\xi) \geq 0$ for $\xi \in (-\infty, +\infty)$, differentiable and monotone non-decreasing for $\xi \geq 0$;
5° $\limsup_{\xi\to\infty}[b_i(\xi) - w_{ii}\sigma_i(\xi)] < 0$;
6° either
(a) $\lim_{\xi\to 0^+} b_i(\xi) = \infty$
or
(b) $\lim_{\xi\to 0^+} b_i(\xi) < \infty$ and $\int_0^{\varepsilon} \frac{d\xi}{a_i(\xi)} = \infty$ for some $\varepsilon > 0$.
The assumption of weights symmetry, 1°, is the most common in recurrent networks theory, but here it is augmented with the requirement of non-negativity of w_{ij}. The continuity assumptions 2°–3° and positivity of $a_i(\cdot)$ for positive argument are needed together to prove that if the system of eqn. 3.55 starts from *positive* initial conditions then its trajectories remain positive. Therefore all considerations are limited to the positive orthant $\mathbf{R}_+^N$, which together with the non-negativity constraint on w_{ij} is fairly stringent. However, the assumptions about network nonlinearity, 4°, are mild from the practical point of view, unless the hardlimiter is used.

The Lyapunov function (consider $1^o, 4^o$ and positivity of trajectories) for the system of eqn. 3.55 is

$$V(y) = -\sum_{i=1}^{N} \int_0^{y_i} b_i(\xi)\sigma_i'(\xi)d\xi + \frac{1}{2}\sum_{i=1}^{N}\sum_{j=1}^{N} w_{ij}\sigma_i(y_i)\sigma_j(y_j) \tag{3.56}$$

and has the time derivative $\dot{V}(y) \leq 0$ of the form

$$\dot{V}(y) = -\sum_{i=1}^{N} a_i(y_i)\sigma_i'(y_i)[b_i(y_i) - \sum_{j=1}^{N} w_{ij}\sigma_j(y_j)]^2 \tag{3.57}$$

Notice that in eqn. 3.57 assumptions 2° and 4° were used.

The generality of the theorem is achieved at the cost of a variety of technical (5°–6°) and strong (1°) conditions, which are often violated in practice. The violations do not always lead to instability, because the theorem gives only a sufficient condition.

3.4.6.2 *Hopfield nets stability*

Hopfield published two fundamental papers [21, 22] on certain network architectures, later called Hopfield nets. However, the structure in Reference [21] differs substantially from that of Reference [22]. We shall concentrate here on the continuous Hopfield net [22].

The name Hopfield net is widely used for the continuous model described by

$$\begin{aligned} T_i\dot{x}_i &= -x_i + \sum_{j=1}^{N} w_{ij}y_j + u_i \\ y_i &= \sigma_i(x_i), \qquad i = 1, \ldots, N \end{aligned} \tag{3.58}$$

where $\sigma(\cdot)$ is a sigmoid function. These equations were mentioned when eqn. 3.24 was introduced. The Lyapunov function candidate proposed by Hopfield is

$$E = -\frac{1}{2}\sum_{i=1}^{N}\sum_{j=1}^{N} w_{ij}y_iy_j + \sum_{i=1}^{N}\rho_i\int_0^{y_i}\sigma_i^{-1}(\xi)d\xi - \sum_{i=1}^{N}u_iy_i \tag{3.59}$$

where $\rho_i > 0$ are constants, $\sigma(\cdot)$ is monotone increasing and $w_{ij} = w_{ji} \;\; \forall i, j$. It is straightforward to calculate that $\dot{E} \leq 0$.

According to Hopfield w_{ij} can be of both signs (or 0), which results in the indefiniteness of W in the general case. Neglected discussion of positivity conditions for eqn. 3.59 in Reference [22] was addressed in Reference [37] and worked out in detail, showing which functions of the type of eqn. 3.59 will lead to stable designs. Although through a simple transformation eqn. 3.58 can be obtained from eqn. 3.55 it does not remove the strong assumptions associated with the Cohen-Grossberg Theorem [31].

As in the discrete model, the continuous Hopfield net does not give a method of weights adjusting (compare Section 3.4.3)—on the contrary it assumes that they are given beforehand and are symmetric, usually with $\mathrm{tr}W = 0$. Lyapunov functions similar to that of eqn. 3.59 are used to obtain such w_{ij} [59].

3.5 Conclusions

Neural networks are nonlinear, adaptive and interconnected large-scale systems having two general classes: feedforward (static) and recurrent (dynamic).

Feedforward networks are described by a set of algebraic equations and cannot directly emulate systems dynamics. However, feedforward networks possess proven capabilities of arbitrary nonlinear function approximation which can be exploited in the context of discrete-time input-output models. Thus, multivariate continuous function approximation and interpolation were identified as fundamental problems for neurocontrol with feedforward networks. The methods stemming from the Stone-Weierstrass and Kolmogorov theorems address only the general approximation problem, i.e. they establish pure existence results. In particular, they imply that the multi-dimensional sampling approach is feasible. However, the latter also gives (unlike the others) a constructive, quantifiable methodology for carrying out the interpolation. Since the existence results are now firmly established, it seems advantageous to do more in-depth research on the practical control aspects of the multi-dimensional sampling approach.

Recurrent networks are universal, parametric, nonlinear dynamic systems of possible use as identifiers or dynamic controllers. The complexity of analysis and synthesis involved is, however, a profound theoretical challenge. Approximation with recurrent neural networks builds on the results for feedforward ones and therefore does not solve the problem of differential approximation. The construction used in the proofs of relevant theorems is not practicable from the engineering point of view. Thus the approach gives pure existence results only. However, recurrent continuous networks offer a universal approach to nonlinear adaptive control as an alternative to the limitations of existing nonlinear control techniques.

3.6 Acknowledgments

Rafał Żbikowski's work was supported by the TEMPUS Grant no. IMG–PLS–0630–90 and ESPRIT III NACT project. Peter J. Gawthrop is Wylie Professor of Control Engineering. The authors would like to thank Dr. Andrzej Dzieliński for his cooperation in preparing the Section on approximation.

3.7 References

[1] Almeida, L. B., 1988, Backpropagation in perceptrons with feedback. In *Neural Computers*, edited by Eckmiller, R. and v. d. Malsburg, C. Springer-Verlag

[2] Anosov, D. V. and Arnold, V. I., 1988, *Dynamical Systems I*. Springer-Verlag, Berlin

[3] Arnold, V. I., 1958, Some questions of approximation and representation of functions. In *Proceedings of the International Congress of Mathematicians*, pp. 339–348. Cambridge University Press, (English translation: American Mathematical Society Translations, Vol. 53)

[4] Burkill, J. C. and Burkill, H., 1970, *A Second Course in Mathematical Analysis*. Cambridge University Press, Cambridge, England

[5] Cohen, M. A. and Grossberg, S., 1983, Stability of global pattern formation and parallel memory storage by competitive neural networks. *IEEE Transactions on Systems, Man, and Cybernetics* 13: 815–826

[6] Cybenko, G., 1989, Approximation by superposition of a sigmoidal function. *Mathematics of Control, Signals, and Systems* 2: 303–314

[7] Dugundji, J., 1966, *Topology*. Allyn and Bacon, Boston

[8] Edwards, C. H., 1973, *Advanced Calculus of Several Variables*. Academic Press, New York

[9] Elsgolc, L. E., 1961, *Calculus of Variations*. Pergamon Press, Oxford

[10] Eykhoff, P., 1974, *System Identification: Parameter and State Estimation*. Wiley, London

[11] Fang, Y. and Sejnowski, T. J., 1990, Faster learning for dynamic recurrent backpropagation. *Neural Computation* 2: 270–273

[12] Fletcher, R., 1987, *Practical Methods of Optimization. 2nd Edition*. Wiley, Chichester

[13] Frank, P. M., 1978, *Introduction to System Sensitivity Theory*. Academic Press, New York

[14] Funahashi, K., 1989, On the approximate realization of continuous mappings by neural networks. *Neural Networks* 2: 183–192

[15] Funahashi, K. and Nakamura, Y., 1993, Approximation of dynamical systems by continuous time recurrent neural networks. *Neural Networks* 6: 801–806

[16] Girosi, F. and Poggio, T., 1989, Representation properties of networks: Kolmogorov's theorem is irrelevant. *Neural Computation* 1: 465–469

[17] Haber, R. and Unbehauen, H., 1990, Structure identification of nonlinear dynamic systems—a survey on input/output approaches. *Automatica* 26: 651–677

[18] Hebb, D. O., 1949, *The Organization of Behavior*. Wiley, New York

[19] Hecht-Nielsen, R., 1987, Kolmogorov's mapping neural network existence theorem. In *Proceedings of the International Joint Conference on Neural Networks*, vol. 3, pp. 11–14

..echt-Nielsen, R., 1989, *Neurocomputing*. Addison-Wesley, Reading, Massachusetts

[21] Hopfield, J. J., 1982, Neural networks and physical systems with emergent collective computational abilities. *Proceedings of the National Academy of Sciences of the United States of America* 79: 2554–2558

[22] Hopfield, J. J., 1984, Neurons with graded response have collective computational properties like those of two-state neurons. *Proceedings of the National Academy of Sciences of the United States of America* 81: 3088–3092

[23] Hornik, K., Stinchcombe, M. and White, H., 1989, Multilayer feedforward networks are universal approximators. *Neural Networks* 2: 359–366

[24] Hunt, K. J., Sbarbaro, D., Żbikowski, R. and Gawthrop, P. J., 1992, Neural networks for control systems: A survey. *Automatica* 28, no. 6: 1083–1112

[25] Isidori, A., 1989, *Nonlinear Control Systems: An Introduction. 2nd Ed.*. Springer-Verlag, New York

[26] Kokotović, P. V., 1991, *Foundations of Adaptive Control*. Springer-Verlag, Berlin

[27] Kolmogorov, A. N., 1957, On the representation of continuous functions of many variables by superposition of continuous functions of one variable and addition. *Dokl. Akad. Nauk SSSR* 114: 953–956, (English translation: American Mathematical Society Translations, Vol. 28)

[28] Kowalik, J. S. and Osborne, M. R., 1968, *Methods for Unconstrained Optimization Problems*. Elsevier, New York

[29] Kůrková, V., 1991, Kolmogorov's theorem is relevant. *Neural Computation* 3: 617–622

[30] Kůrková, V., 1992, Kolmogorov's theorem and multilayer neural networks. *Neural Networks* 5: 501–506

[31] Li, J. H., Michel, A. N. and Porod, W., 1988, Qualitative analysis and synthesis of a class of neural networks. *IEEE Transactions on Circuits and Systems* 35: 976–986

[32] Lin, J. N. and Unbehauen, R., 1993, On the realization of Kolmogorov's network. *Neural Computation* 5: 18–20

[33] Lippmann, R. P., 1987, An introduction to computing with neural nets. *IEEE ASSP Magazine* pp. 4–22

[34] Lipschutz, S., 1965, *General Topology*. McGraw-Hill, New York

[35] Lorentz, G. G., 1966, *Approximation of Functions*. Holt, Reinhart and Winston, New York

[36] Mead, C., 1989, *Analog VLSI and Neural Systems*. Addison-Wesley, Reading, Mass.

[37] Michel, A. N., Farrell, J. A. and Porod, W., 1989, Qualitative analysis of neural networks. *IEEE Transactions on Circuits and Systems* 36: 229–243

[38] Nijmeijer, H. and van der Schaft, A., 1990, *Nonlinear Dynamical Control Systems*. Springer-Verlag, New York

[39] Pao, Y., 1989, *Adaptive Pattern Recognition and Neural Networks*. Addison-Wesley, Reading, Mass.

[40] Park, J. and Lee, S., 1990, Neural computation for collision-free path planning. In *Proceedings of the IEEE International Joint Conference on Neural Networks, IJCNN'90, Washington D.C., USA*

[41] Pearlmutter, B. A., 1989, Learning state space trajectories in recurrent neural networks. *Neural Computation* 1: 263–269

[42] Pearlmutter, B. A., 1990, Dynamic recurrent neural networks. Tech. Rep. CMU-CS-90-196, Carnegie Mellon University, School of Computer Science

[43] Petersen, D. P. and Middleton, D., 1962, Sampling and reconstruction of wave-number-limited functions in n-dimensional euclidean spaces. *Information and Control* 5: 279–323

[44] Pineda, F. J., 1987, Generalization of back-propagation to recurrent neural networks. *Physical Review Letters* 59: 2229–2232

[45] Pineda, F. J., 1989, Recurrent backpropagation and the dynamical approach to adaptive neural computation. *Neural Computation* 1: 161–172

[46] Robinson, A. J. and Fallside, F., 1987, Static and dynamic error propagation networks with application to speech coding. In *Proceedings of Neural Information Processing Systems*, edited by Anderson, D. Z. American Institute of Physics

[47] Rudin, W., 1976, *Principles of Mathematical Analysis, 3rd Edition*. McGraw-Hill, Auckland

[48] Rumelhart, D. E. and McClelland, J. L., 1986, *Parallel Distributed Processing: Explorations in the Microstructures of Cognition, Vol. 1: Foundations*. MIT Press, Cambridge, Mass.

[49] Sanner, R. M. and Slotine, J.-J. E., 1992, Gaussian networks for direct adaptive control. *IEEE Transactions on Neural Networks* 3: 837–863

[50] Sastry, S. and Bodson, M., 1989, *Adaptive Control. Stability, Convergence, and Robustness*. Prentice-Hall, Englewood Cliffs

[51] Sato, M., 1990, A learning algorithm to teach spatiotemporal patterns to recurrent neural networks. *Biological Cybernetics* 62: 259–263

[52] Sbarbaro, D., October 1992, *Connectionist Feedforward Networks for Control of Nonlinear Systems*. Ph.D. Thesis, Department of Mechanical Engineering, Glasgow University, Glasgow, Scotland

[53] Shannon, C. E., 1949, Communication in the presence of noise. *Proceedings of the IRE* 37: 10–21

[54] Simpson, P. K., 1989, *Artificial Neural Systems*. Pergamon Press, New York

[55] Slotine, J. E. and Li, W., 1991, *Applied Nonlinear Control*. Prentice-Hall, Englewood Cliffs

[56] Sontag, E., 1992, Neural nets as systems models and controllers. In *Proc. Seventh Yale Workshop on Adaptive and Learning Systems*, pp. 73–79. Yale University

[57] Sprecher, D. A., 1965, On the structure of continuous functions of several variables. *Transactions of the American Mathematical Society* 115: 340–355

[58] Stone, M. H., 1948, The generalized Weierstrass approximation theorem. *Mathematics Magazine* 21: 167–184, 237–254

[59] Tank, D. W. and Hopfield, J. J., 1986, Simple neural optimization networks: An A/D converter, signal decision circuit, and a linear programming circuit. *IEEE Transactions on Circuits and Systems* 33: 533–541

[60] Tzirkel-Hancock, E., August 1992, *Stable Control of Nonlinear Systems Using Neural Networks*. Ph.D. thesis, Trinity College, Cambridge University, Cambridge, England

[61] Utkin, V. I., 1978, *Sliding Modes and Their Application in Variable Structure Systems*. Mir Publishers, Moscow

[62] Werbos, P. J., 1974, *Beyond Regression: New Tools for Prediction and Analysis in the Behavior Sciences*. Ph.D. thesis, Harvard University, Committee on Applied Mathematics

[63] Werbos, P. J., 1989, Maximizing long-term gas industry profits in two minutes in Lotus using neural network methods. *IEEE Transactions on Systems, Man, and Cybernetics* 19: 315–333

[64] Werbos, P. J., 1990, Backpropagation through time: What it does and how to do it? *Proceedings of IEEE* 78: 1550–1560

[65] Widrow, B. and Lehr, M. A., 1990, 30 years of adaptive neural networks: Perceptron, madaline, and backpropagation. *Proceedings of IEEE* 78: 1415–1442

[66] Williams, R. J. and Zipser, D., 1989, Experimental analysis of the real-time recurrent learning algorithm. *Connection Science* 1: 87–111

[67] Williams, R. J. and Zipser, D., 1989, A learning algorithm for continually running fully recurrent neural networks. *Neural Computation* 1: 270–280

[68] Williams, R. J. and Zipser, D., 1990, Gradient-based learning algorithms for recurrent connectionist networks. Tech. Rep. NU-CCS-90-9, Northeastern University, Boston, College of Computer Science

[69] Żbikowski, R., 1993, The problem of generic nonlinear control. In *Proc. IEEE/SMC International Conference on Systems, Man and Cybernetics, Le Touquet, France*, vol. 4, pp. 74–79

[70] Żbikowski, R., July 1994, *Recurrent Neural Networks: Some Control Problems*. Ph.D. Thesis, Department of Mechanical Engineering, Glasgow University, Glasgow, Scotland, (Available from FTP server `ftp.mech.gla.ac.uk` as `/rafal/zbikowski_phd.ps`)

[71] Żbikowski, R., Hunt, K. J., Dzieliński, A., Murray-Smith, R. and Gawthrop, P. J., 1994, A review of advances in neural adaptive control systems. Technical Report of the ESPRIT NACT Project TP-1, Glasgow University and Daimler-Benz Research, (Available from FTP server `ftp.mech.gla.ac.uk` as `/nact/nact_tp1.ps`)

[72] Zipser, D., 1989, A subgrouping strategy that reduces complexity and speeds up learning in recurrent networks. *Neural Computation* 1: 552–558

Chapter 4

Selection of neural network structures: some approximation theory guidelines

J.C. Mason and P.C. Parks

4.1 Introduction: artificial neural networks (ANNs) in control

Control engineers have not been slow in making use of recent developments in artificial neural networks: a pioneering paper was written by Narendra and Partnasarathy [41] and more recent developments are surveyed in this book.

Neural networks allow many of the ideas of system identification and adaptive control originally applied to linear (or linearised) systems to be generalised, so as to cope with more severe nonlinearities. Such strong nonlinearities occur in a number of applications e.g. in robotics or process control.

Two possible schemes for 'direct' adaptive and 'indirect' adaptive control are shown in Figure 4.1 and other schemes will be found elsewhere in this book, but in this chapter we shall concentrate on the modelling to be carried out by the artificial neural networks continued in the boxes C and IM in Figure 4.1. For example, the box IM in Figure 4.1b seeks to build a nonlinear dynamic model of the plant P in discrete-time as

$$y_t = f(y_{t-1}, y_{t-2}, \ldots y_{t-r}, u_{t-1}, u_{t-2}, \ldots u_{t-s})$$

where $\{y_t\}$ is the output and $\{u_t\}$ the input to the plant P which is to be modelled using the nonlinear function f. An immediate problem which arises here is how large the integers r and s should be. They should be kept as small as possible to reduce the complexity of the neural network topology and the weight learning procedure but large enough to model the significant dynamics of the plant P. Some theoretical and practical knowledge of P will be invaluable here; alternatively r and s may be chosen to be large initially and then subsequently 'pruned' in a process later called 'parsimony'.

However, we shall now concentrate our thoughts on the general problem of modelling a single function of ℓ variables $f(x_1, x_2, \ldots x_\ell)$.

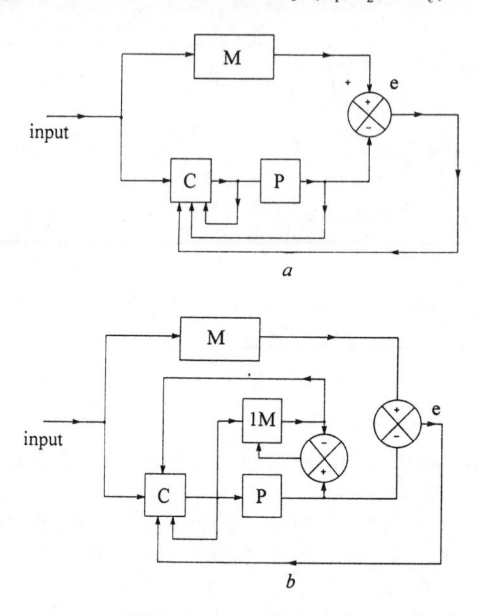

Figure 4.1 *Adaptive control*
(a) Direct
(b) Indirect
M = model; P = plant; C = controller; IM = identification model; e = error signal (ANNs replace C and IM boxes in a neural network implementation)

4.2 Approximation of functions

A neural network effectively creates a function of several variables from a number of sums and products of functions of one variable. It may use several layers of calculations, corresponding to the layers of the network, and it may modify the parameters in the resulting form of approximation by a learning or training procedure which passes through the network repeatedly or which moves forward and backward through the network.

Once the neural network has been designed and trained, it operates as a model function, outputting the value of a specific function F corresponding to the input of a value of each independent variable $x_1, x_2, \ldots, x_\ell$. The function F so created does not coincide in general with the true function f being modelled, but acts as an approximation to it. Typically f is known at a discrete set of points $\underline{x}^{(1)}, \ldots, \underline{x}^m$, where

$$\underline{x}^{(i)} = (x_{1i}, x_{2i}, \ldots, x_{\ell i}), \qquad (i = 1, \ldots, m)$$

corresponding to m sets of values of the vector variable

$$\underline{x} = (x_1, x_2, \ldots x_\ell)$$

and so it is possible to compute the error f–F in the model F at each of the points $\underline{x}^{(i)}$.

The modelling process may thus be viewed in its total effect, as a problem in the 'approximation of functions', and we may appeal to the mathematical principles of 'approximation theory' to guide us in carrying out the process. For this purpose, there is a very substantial and rapidly growing literature available. The bulk of this literature is concerned with nonlinear functions, in the sense that f is not modelled as a hyperplane. Much of the literature is, however, restricted to 'linear approximation' in the sense that the parameters, say

$$\underline{c} = (c_1, c_2, \ldots, c_n)$$

enter linearly into the form of approximation, as each parameter c_j multiplies a nonlinear basis function ϕ_j. Specifically,

$$F(x) = \sum_{j=1}^{n} c_j \phi_j(x) \tag{4.1}$$

An 'approximation problem' may then be posed, typically as one of the following three problems, based on the specification of a norm $\|.\|$ over the points $\underline{x}^{(i)}$:

(i) Good approximation problem: determine an approximation F, of the given form such that $\|f$–$F\|$ is acceptably small.
(ii) Best approximation problem: determine an F, F^{β} say, of the given form such that $\|f$–$F\|$ has its minimum value $\|f$–$F^{\beta}\|$.
(iii) Near-best approximation problem: determine an F of the given form such that

$$\|f - F\| \le (1 + p).\|f - F^{\beta}\|$$

where F^{β} is the best approximation and p is acceptably small. (The norm is then within a relative distance p of its smallest value. The maximum acceptable value for the constant p is up to the user. A value of 9 ensures a loss of accuracy of at most one significant figure, and a value of 0.1 ensures that the norm is minimised correct to about one significant figure.)

In practice, it is common to require that an approximation should be both good and best, or both good and near-best, in the senses of the above definitions.

Obviously there is no great virtue in determining the best approximation of a form which is not capable of representing f accurately. Nor is it necessarily acceptable to obtain a good approximation, when an alternative procedure may yield a much better one or where a much smaller number of parameters may still provide a comparable one. The idea of a 'near-best' approximation is introduced because it is not usually essential to find the actual 'best' approximation, and indeed it may be time-consuming to do so. Although least squares methods readily yield best approximations in the ℓ_2 (or Euclidean) norm (see below), it is typically necessary to adopt iterative procedures to determine best approximations in other norms, and it may be sensible and acceptable to terminate these procedures in advance of their convergence.

The literature in approximation is vast and progressively more specialised. However, general discussions of approximation may be found in textbooks such as [44], and elementary introductions are given in many numerical analysis textbooks, such as in the last chapter of [31]. State-of-the-art discussions may be found in conference proceedings, such as the Texas 'Approximation Theory' series [13] and, on the more practical aspects, in the UK 'Algorithms for Approximation' conference series [15].

In posing a good, best, or near-best approximation problem as defined above, it is clear that a number of choices or specifications are needed, which make up what we call the 'framework' of approximation. Without this framework, it is easy to attempt to solve the wrong problem. The framework consists of the approximation problem, together with three essential components: function, form, and norm.

(i) **Function** The given function $f(\underline{x})$ may be specified on a continuum or on a discrete data set $\{\underline{x}^{(1)},\ldots,\underline{x}^{(m)}\}$. (In neural networks it is likely to be the latter.) Moreover f may be exact or it may have errors in it, and in the latter case it is desirable to specify the nature and overall size of the data error. (For example, it might be assumed that the errors are normally distributed with a prescribed standard deviation.)

In theoretical studies, the function f is often assumed to be from a vector space F of functions, which is infinite-dimensional in the case of a continuum and of finite dimension m (and typically isomorphic to R^m) in the case of discrete data.

(ii) **Form of approximation** The form is an approximation, such as eqn. 4.1, which includes a set of undetermined parameters c_j. It may be linear or nonlinear in the c_j, but in either case it can be identified with a vector $(c_1,\ldots,c_n)$ in R^n. The form can be viewed as a vector space in the case of a linear form such as eqn. 4.1, with ϕ_j as the basis, and typically a subspace A of the function space F.

(iii) **Norm of approximation** The norm is a measure of how well a specific approximation F of the given form matches the given function f. (We refer the reader to [44] for the definition of a norm.) The most common general class of norms, in the case of discrete data, is the ℓ_p (Hölder) norm:

$$\|f - F\|_p = \left(\sum_{i=1}^{m} \left| f(x^{(i)}) - F(x^{(i)}) \right|^p \right)^{1/p} \tag{4.2}$$

4.2.1 *Choices of norm*

Assuming that f is given on a discrete data set $\underline{x}^{(1)},\ldots,\underline{x}^{(m)}$, then the choice of the appropriate norm or measure of approximation is usually restricted to one of four, depending on the nature of the data and, to some extent, of the approximation algorithms available:

(i) $\ell_1 : \|f - F\|_1 = \sum_i \left| f(x^{(i)}) - F(x^{(i)}) \right|$

(ii) $\ell_2 : \|f - F\|_2 = \left(\sum_i [f(x^{(i)}) - F(x^{(i)})]^2 \right)^{1/2}$

(iii) $\ell_\infty : \|f - F\|_\infty = \max_i \left| f(x^{(i)}) - F(x^{(i)}) \right|$ (uniform norm)

(iv) Smoothing (penalised least squares) measure

$$\|f - F\|_2 + \lambda \int_{R^\ell} (PF)^2 dS$$

where P is a constraint operator, typically an appropriate second derivative operator.

Each of these choices has an ideal area of application. The ℓ_1 norm is excellent for data which have a small number of isolated wild points (or outliers), and tends to ignore such data. The ℓ_2 norm is best suited to data with errors in a normal (or statistically related) distribution, where the standard deviation in the error is not large and where approximations of at best comparable accuracy to the data are required. The ℓ_∞ norm is ideal for data which are exact, or very accurate, or have errors in a uniform distribution (such as data which have been rounded). Finally a smoothing measure is aimed at data with significant but statistically distributed noise, or at data with gaps [17]. The inclusion of the integral term in the measure tends to 'regularise' the data, by keeping certain derivatives small, and to fill gaps without undue oscillation.

In the context of neural networks, the choice of the ℓ_2 norm is almost universal, partly on the grounds that errors are present in the data, and partly because algorithms are seen to be simpler. However, excellent algorithms are available in libraries such as NAG for ℓ_∞ approximation [2], and for ℓ_1 approximation [3]. Smoothing measures have been adopted, for example [43], and the reader should refer to discussions such as [51] and [17].

4.2.2 Choice of form

There are a variety of popular choices of forms of approximation, all of which have been used in neural networks.

(i) **Algebraic polynomial** (of degree n–1)

(a) $$F = \sum_{j=1}^{n} c_j x^{j-1}$$

(b) $$F = \sum_{j=1}^{n} c_j \phi_{j-1}(x)$$

where $\{\phi_{j-1}\}$ is an orthogonal polynomial system.

(ii) **Trigonometric polynomial** (of order n)

$$F = a_o + \sum_{j=1}^{n} (a_j \cos jx + b_j \sin jx) \quad \text{(parameters } a_j, b_j\text{)}$$

(iii) **Spline of degree s** (knots $\xi_1, \ldots, \xi_r$)

There are 2 common forms used which, in the case of cubic splines (s = 4), are as follows. (Splines of higher/lower degree may be defined similarly.)

(a) $$F = \sum_{j=1}^{4} c_j x^{j-1} + \sum_{j=5}^{r+4} c_j (x - \xi_{j-4})_+^s \quad \text{(truncated power form)}$$

where $$(x-a)_+^3 = \begin{cases} 0 & \text{for } x < a \\ (x-a)^3 & \text{for } x \geq a \end{cases}$$

(b) $F = \sum_{j=1}^{r+4} c_j B_j(x)$ (B-spline form)

where $B_j(x) = \sum_{k=o}^{4} d_k^{(j)} (x - \xi_{j+k-2})_+^3$ (B-spline centred on ξ_j)

$\xi_{-1}, \xi_o, \xi_{r+1}, \xi_{r+2}$ are added exterior or end point knots, and $d_k^{(j)}$ are so chosen that $B_j = 0$ for $x \geq \xi_{j+2}$. (The latter condition gives 4 equations for $d_k^{(j)}$. It remains to normalise B_j, and there are two standard ways of doing this [5]. Note also that B_j vanishes already for $x \leq \xi_{j-2}$ and hence it has support only on $[\xi_{j-2}, \xi_{j+2}]$.)

(iv) **Radial basis function (multivariate)**

$$F = c_o + \sum_{j=1}^{n} c_j \phi(\|x - y^{(j)}\|) \tag{4.3}$$

where ϕ is a chosen function and $y^{(j)}$ are a set of chosen centres.

(v) **Ridge function**

$$F = \sum_{j=1}^{n} c_j \phi_j (w_j^T x + \theta_j) \tag{4.4}$$

where $\underline{w}_j, \underline{x}$ are ℓ-vectors, θ_j is a scalar.

Here, constant values of

$$z_j = w_j^T x + \theta_j = \sum_{i=1}^{\ell} w_{ij} x_i + \theta_j \tag{4.5}$$

represent hyperplanes on which F is constant, and hence a ridge is formed in F.

Typical choices of ϕ_j are sigmoids of the form $\phi_j = \sigma$, where

$$\text{either} \quad \sigma(t) - \begin{cases} 1 & t \geq 0 \\ 0 & t < 0 \end{cases} \quad \text{or} \quad \sigma(t) - \frac{1}{1 + e^{-t}}$$

These sigmoids satisfy the boundary conditions: $\sigma(-\infty) = 0, \quad \sigma(+\infty) = 1$.

The choices (iv) and (v) are those which appear to be most popular in current work on neural networks. The ridge function has been the traditional choice, but the radial basis function is becoming a popular competitor. A simple structure for the realisation of a ridge function has an input layer, a hidden layer, and an output layer. Here the variables $x_1, x_2, \ldots, x_\ell$ are inputs in the first layer, these are collected in the form $\underline{w}_j^T \underline{x} + \theta_j$ at the node j of the hidden layer, where they are processed by node function $\phi_j(.)$. They are then collected and the output at the 3rd layer is in the form of eqn. 4.4. Suppose there are ℓ, n and p nodes in the 1st, 2nd and 3rd layers, respectively, then there are p output functions

$$F^{(k)} = \sum_{j=1}^{n} c_j^{(k)} \phi_j(z_j), \quad (k - 1, \ldots, p) \tag{4.6}$$

of the form of eqns. 4.4 and 4.5 with connection weights $c_j^{(k)}$ between the 2nd and 3rd layers, and connection weights w_{ij} between the 1st and 2nd layers. This structure is illustrated in Figure 4.2a. Strictly speaking, if $F^{(k)}$ approximates a known output $f^{(k)}$, say, then the fitting problem consists of p simultaneous approximation problems $f^{(k)} \simeq F^{(k)}$ each involving a function of ℓ variables.

Further hidden layers could build more complicated functions than eqn. 4.4, for example one extra layer would form linear combinations of sigmoids whose arguments are formed from eqn. 4.4.

Radial basis functions have been recently adopted and may also be readily realised in a three-layer structure, with an input layer, one hidden layer, and an output layer [8]. Again the input nodes contain the variables $x_1, \ldots, x_\ell$. The hidden layer has n nodes, one for each centre $\underline{y}^j$, and each connection has a scalar y_{ij} assigned to it. Here y_{ij} is the ith component in $\underline{y}^j$, and its value links the ith input node to the jth hidden node. The fan-in to the jth hidden node is

$$z_j = \sum_{i=1}^{\ell} (x_i - w_{ij})^2 = \left\| x - y^{(j)} \right\|^2,$$

and the hidden layer applies the (radial) function ϕ to $(z_j)^{1/2}$. Thus the form of eqn. 4.3 is the output at the 3rd layer. Assuming again that there are p output nodes with output functions $F^{(k)}$ and weights $c_j^{(k)}$ $(k = 1, \ldots, p)$ each of the form of eqn. 4.3, then the connection weights in the network are $c_j^{(k)}$ between the 2nd and 3rd layers and w_{ij} between the 1st and 2nd layers. This structure is

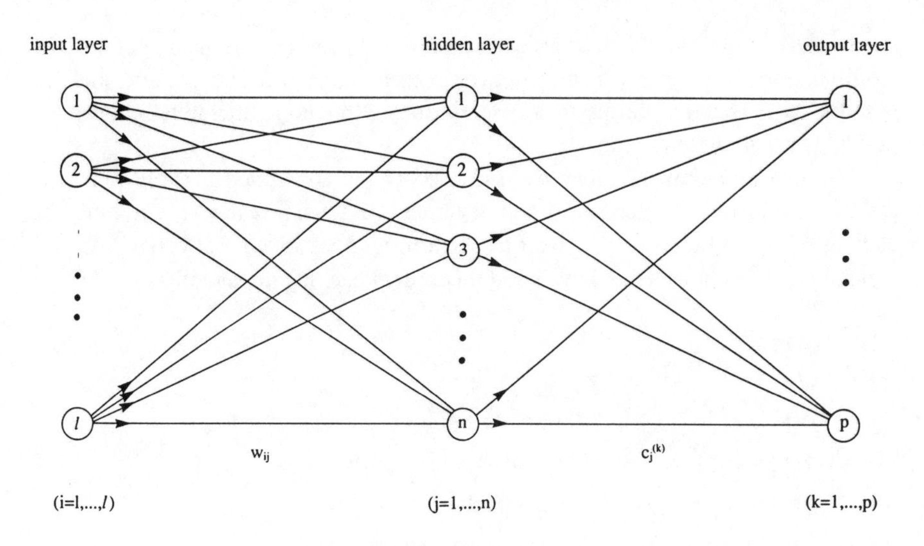

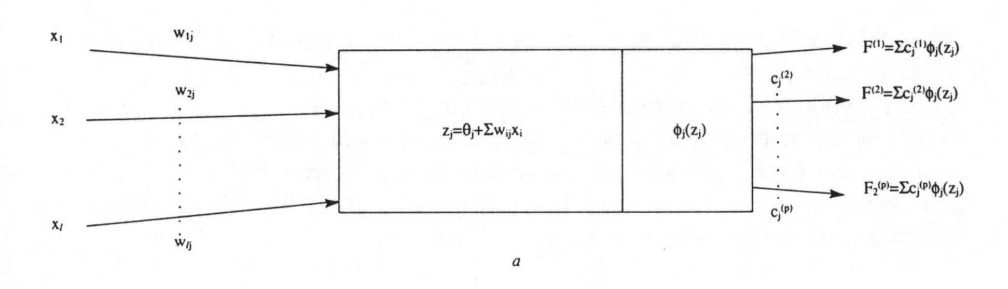

a

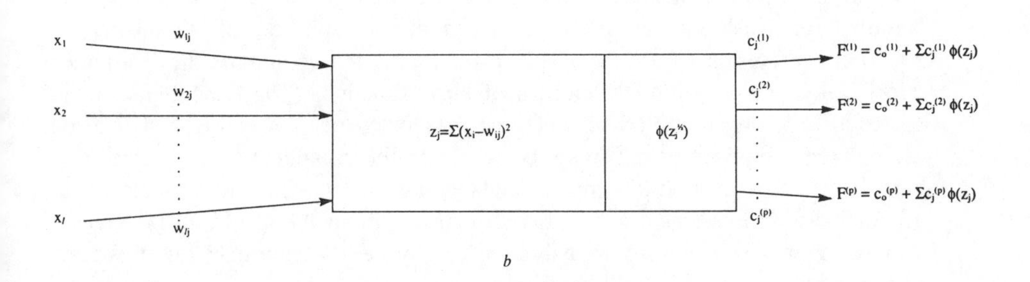

b

Figure 4.2 *Feed forward networks*
(a) Hidden layer: ridge functions
(b) Hidden layer: radial basis functions

illustrated in Figure 4.2b. Radial basis functions have been adopted (possibly in modified form) by several neural network researchers in recent papers, such as [43, 10, 25]. Detailed discussions of radial basis functions, including choices of functions ϕ, are given in [45, 9].

A number of radial functions were also proposed and tested in early practical discussions in [20]. There are good arguments for using each of a number of choices of ϕ, and indeed all appear to perform with reaso: able effectiveness and versatility. The choices that have good theoretical backing are as follows:

(a) $\phi(r) = r$

(b) $\phi(r) = r^3$

(c) $\phi(r) = r^2 \log r$ (thin plate splines)

(d) $\phi(r) = \exp(-r^2/2)$ (Gaussian)

(e) $\phi(r) = (r^2 + c^2)^{\frac{1}{2}}$ (multi-quadric)

(f) $\phi(r) = (r^2 + c^2)^{-\frac{1}{2}}$ (inverse multi-quadric)

For neural networks, thin plate splines have been adopted in [10], and multi-quadrics and Gaussians have been used in [8].

Polynomials and spline functions, which have a long history in approximation theory have also been adopted for neural networks. Incidentally, the forms (i), (ii) and (iii), as we have defined them above, are given as functions of one variable x. However, analogous multivariate forms may be defined, typically by forming tensor products of univariate basis functions, such as

$$B_{j_1}(x_1)B_{j_2}(x_2)\ldots B_{j_\ell}(x_\ell)$$

in the case of multivariate B-splines.

Orthogonal polynomials have two particular advantages. They provide explicit solutions to least squares problems (without the solution of simultaneous equations) in terms of ratios of inner products, and moreover they provide numerically well-conditioned families of basis functions. They are much more powerful in fitting multivariate functions than many people realise, and indeed high degree approximations may be successfully computed to fit relatively complicated functions. It is a great advantage to have data which lie on a mesh, or alternatively on lines, planes, or other structures. Bennell and Mason [4] give a variety of powerful algorithms based on Chebyshev polynomials for bivariate approximation, which could certainly be extended to higher dimensions. Orthogonal polynomials have been adopted in a neural network context by [11]. Incidentally, multivariate homogenous polynomials can be generated from a basis of ridge functions, namely

$$(\underline{w}^T \underline{x})^k \tag{4.7}$$

Results of Light [26] and [12] establish that such basis functions span homogeneous polynomials and polynomials, respectively.

Most of the basis functions that we have discussed so far, namely (i), (ii), (iv) and (v), are defined globally. However, B-splines, which are the most popularly adopted basis functions for splines, have the potentially great advantage of being locally defined. Each B-spline has compact support and, for example, a cubic B-spline is only supported on 4 sub-intervals (connecting its knots) in each variable (see Figure 4.3). This gives two computational advantages. First, B-spline algorithms lead to simultaneous equations with sparse matrices (typically banded matrices), and, second, their evaluation is very economical since it only adopts local information. A detailed discussion of splines and their properties is given in such texts as [5, 49]. They have been adopted in a neural network context by [6, 7], who use tensor product B-splines and claim to exploit their local properties.

The early 'CMAC' device of [1] can now be classified as belonging to an analogous 'local' class [50, 7].

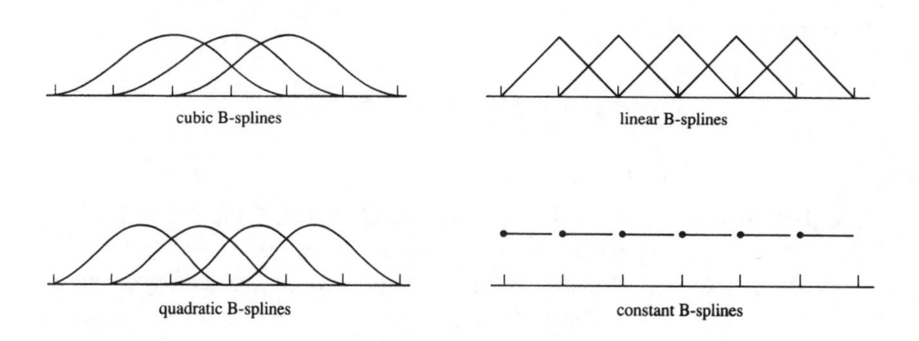

Figure 4.3 *Some local basis functions*

4.2.3 Existence of good approximations

Possibly the most important problem in approximation theory, especially in the context of fitting high dimensional nonlinear functions which occur in neural networks, is that of establishing that good approximations exist. More precisely we need to show that the approximation space, for sufficiently high orders n, is dense (in a given norm) in the function space. Unless otherwise stated the maximum absolute error (uniform norm) is adopted as norm.

A fundamental result, which is also of great importance in many areas of analysis, concerns algebraic polynomials:

Weierstrass's First Theorem states that algebraic polynomials are dense in the uniform norm in functions continuous on I_ℓ, the ℓ-dimensional cube $[0,1]^\ell$. A standard and constructive proof of this theorem is based on the representation of a continuous function in terms of Bernstein polynomials. An analogous result immediately follows for spline functions, when viewed as generalisations of polynomials.

In the context of neural networks, the main concern is centred on ridge functions and radial basis functions, and here significant results have been established recently.

Cybenko's Theorem [16]

Let σ be any continuous discriminatory sigmoidal function. Then ridge functions of the form of eqn. 4.4 are dense in $C(I_\ell)$.

Here σ is said to be discriminatory if, for a measure μ (which is a signed regular Borel measure),

$$\int_{I_\ell} \sigma(w^T x + \theta)\, d\mu(x) \quad = \quad 0$$

for all y in R^ℓ and θ in R implies that $\mu = 0$.

This theorem has been strengthened by Chui and Li [12], in the sense that, if the components of $\underline{w}$ and $\underline{\theta}$ are restricted to integer values, then the ridge functions are still dense in $C(I_\ell)$. Moreover, Chui and Li go beyond establishing existence, and indeed, for σ of bounded variation, they provide a construction of an arbitrarily accurate ridge function approximation (and hence a realisation of a neural network with one hidden layer of arbitrary accuracy). Their construction is based on the use of a ridge polynomial, using the basis eqn. 4.4, and the expression of a continuous function in terms of Bernstein polynomials and hence ridge polynomials.

Mhaskar and Micchelli [36] provide a more general result than that of Cybenko. They define *k*th-degree sigmoidal functions with the property that

$$\lim_{x \to -\infty} x^{-k}\sigma(x) = 0, \quad \lim_{x \to \infty} x^{-k}\sigma(x) = 1$$

and which are bounded by a polynomial of degree at most k on R. They then show that such kth degree sigmoidal functions are dense in continuous functions (on any compact set in R^{ℓ}).

Mhaskar [35] proceeds to discuss convergence for kth degree sigmoidal functions in multi-layer networks, establishing that as good an order of approximation as for classical free knot splines is possible. Indeed a geometric type of convergence, comparable to that for polynomials, can be obtained if the data can be viewed as coming from a function which may be extended to be analytic.

Radial basis functions have also been studied in some depth. Recent treatments are given by Light [26] and in [18, 19, 46]. Light's discussion is detailed on the question of denseness. He shows in particular that the basis functions (a), (d), (e) and (f) (i.e. all those listed above with the exception of $\phi(r) = r^3$ and thin plate splines), are each dense in continuous functions (on a compact set of R^{ℓ}). The latter two cases are discussed by Powell [46], who shows that it is necessary to augment the basis by adding linear polynomials in order to ensure denseness.

Thus all the forms that we have listed in Section 4.2.2 have the potential for approximating given multivariate functions (provided that in some cases some additional simple functions are added to the basis).

In concluding this section, it is probably appropriate to quote a famous result of Kolmogorov [24], that a continuous function of several variables may be represented by superpositions of continuous monotonic functions of one variable and the operation of addition. This result certainly supports the principle and potential of neural networks. However, it does not provide a realistically constructive realisation of a suitable network. The density results for specific forms, that we have quoted above, restrict the choice of network, adopt well understood forms of approximation and yet still guarantee the possibility of good approximations.

4.2.4 Approximation by interpolation

In the univariate case, the algebraic polynomials, trigonometric polynomials, and splines all have the property of providing unique interpolation on a set of distinct points equal in number to the number of approximation parameters. This is useful in itself in providing a potential algorithm, but it also implies that the basis forms a 'Chebyshev set' (or 'Haar system'), and hence that best approximations are unique in the ℓ_{∞} norm.

However, in the multivariate case it is not usually possible to guarantee a unique interpolant for the above basis functions, and hence a best approximation is not necessarily unique.

Radial basis functions are a glowing exception in this area. Indeed it has now been established that all six forms (a) to (f) (augmented by linear polynomials in cases (b) and (c)) uniquely interpolate any set of data on a distinct set of abscissae. All cases are covered in Light [26] and Powell [46]. The problem has a long history. Schoenberg [48] showed indirectly that there was a unique interpolant in case (a) ($\phi = r$), but the first author to solve the problem in some generality appears to have been Micchelli [37].

4.2.5 Existence and uniqueness of best approximation

Best approximations exist for all (linear) forms and norms that we have quoted above (the major requirement being that the search may be restricted to a compact set). Uniqueness also follows whenever the norm is 'strict', and this covers the ℓ_p norm eqn. 4.2 for $1 < p < \infty$ and in particular the ℓ_2 norm.

We are thus left with some uncertainty in both ℓ_1 and ℓ_∞ norms. It is clear that best ℓ_1 approximations are not in general unique, and indeed there are many situations in which a continuum of parameter sets provide an infinite number of best approximations. Best ℓ_∞ approximations are unique if the basis functions form a Chebyshev set, and this is the case for radial basis functions (as a consequence of their interpolation property). However, there is no guarantee of uniqueness in ℓ_∞ for other forms.

In practice, uniqueness is not necessarily a stumbling block in the implementation of algorithms and, for example, linear programming algorithms can be very effective for obtaining good ℓ_∞ approximations.

4.2.6 Parallel algorithms

Neural network approximation problems can involve very large numbers of parameters, possibly in the thousands or more. It is therefore highly desirable to consider adopting parallel processing procedures within the model and algorithm. Some useful progress has been made in this area, although, in our opinion, the subject is relatively poorly developed in the context of approximation algorithms. We shall therefore concentrate on drawing attention to relevant published work and potentially useful ideas, in the hope that this may lead to more mainstream neural networks approximation developments. Unless otherwise stated, all work is concerned with ℓ_2 norms.

A key parallel principle is that of domain decomposition, by which a domain is split up into many subdomains, on each of which a subproblem is solved in parallel. Such an approach, based at present on spline functions, has been studied in [39] (using L-splines and scattered data) and in [21] (using B-splines and meshes).

An important technique, which combines algorithm and architecture, is the use of a systolic array — a form of 'data flow architecture' in which data are processed without storage as they are fed through an array of processing elements. Such an architecture has been adopted successfully for least squares approximation with radial basis functions in [34], and should in principle be readily extendable to other forms of approximation.

More recently Harbour *et al.* [22] have developed a variety of parallel procedures, with an emphasis on radial basis functions, including a parallel algorithm for data whose abscissae lie on families of lines and a systolic array algorithm for smoothing noisy data by ordinary cross-validation.

Locally supported basis functions would appear to be particularly suitable for the exploitation of parallel processing algorithms (such as domain decomposition), and indeed Brown and Harris [6, 7] claim to be exploiting parallelism in their B-spline model.

We should conclude this section by noting that there is, of course, massive 'implicit parallelism' in a neural network. Great savings are immediately made if and when all operations at individual nodes in any layer are carried out in parallel.

4.2.7 Training/learning procedures and approximation algorithms

The fundamental design requirement in a neural network is to determine all the connection weights, such as w_{ij} and $c_j^{(k)}$ in Figures 4.2a and 4.2b, so that the generated output(s) match the required outputs(s) as closely as possible. This is an approximation problem, in that it in necessary to match $F(x_1,\ldots,x_\ell)$, over a set of data $\{(x_1,\ldots,x_\ell)\}$, by adopting an appropriate norm.

There are two distinct types of approximation which may be obtained, namely linear and nonlinear. If the weights w_{ij}, which link the first two layers, are regarded as fixed, and only the weights $c_j^{(k)}$ linking the second and third layers are allowed to vary, then the problem is a linear approximation problem. In that case, we may adopt a procedure such as the least squares method, and the problem then reduces to the least squares solution of an over-determined system of linear algebraic equations. (There is no shortage of good computer software packages for this problem, such as routine FO4 JGF in the NAG library, which obtains the ℓ_2 solution of minimal norm.) This is, for example, the type of approach proposed by Broomhead and Lowe [8] for radial basis functions. Here the fixing of the weights w_{ij} corresponds to the fixing of the centres $y^{(j)}$, and so these need to be placed appropriately in advance.

However, if w_{ij} are also allowed to vary, then the problem becomes a nonlinear approximation problem, and it is inevitably much harder to solve. In the case of the ℓ_2 norm, it is a nonlinear least squares problem, which may be tackled by an optimisation method designed for such problems (such as routine EO4 FDF in the NAG library). However, there is no guarantee that a global minimum of $\|f - F\|_2$

can be obtained by such a method in general, and many iterations may be needed even when convergence occurs.

The traditional approach in neural networks, which is in the end equivalent to solving the nonlinear approximation problem, is to select the weights w_{ij} and $c_j^{(k)}$ by an iterative 'training' procedure, starting from some initial choice, which may even be by random. The most popular technique is that of 'backpropagation', by which errors in the output are propagated back to modify the weights in a cyclic procedure — a detailed discussion is given, for example, in [30]. A closely related feedback approach is discussed, for example, in [10], and is based on a steepest descent or related procedure for solving the underlying optimisation problem. Again, many iterations may be needed and there is no absolute guarantee of success.

4.2.7.1 Linear or nonlinear approximation

Which of the two procedures should we adopt? Should we fix w_{ij} and solve a linear problem, or solve the full nonlinear problem by leaving w_{ij} free? This is a matter of opinion, depending to a great extent on the experience and/or knowledge that the user has in fixing the relevant parameters and on the ability of the resulting linear form to approximate the output well.

There is not as yet a great wealth of experience in the solution of the full nonlinear approximation problem in the cases of ridge functions (and other multi-layer perceptrons) or radial basis functions, and indeed much of the existing experience lies with neural networks researchers. It is open to question whether or not current learning algorithms can be used with any confidence. However, there is considerable experience amongst approximation specialists in the solution of the nonlinear problem of fitting a multivariate spline function, using a basis of tensor product B-splines (eqn. 4.6). Indeed, even in the one-dimensional case, the problem of optimising the approximation with respect to free choices of both knots ξ_j and coefficients c_j has been found to be an extremely difficult task from a numerical point of view. It has been found, in particular, that the determination of an optimal set of knots ξ_j is typically a highly ill-conditioned problem. A great deal of time has been wasted by researchers over the past 20 years or so in futile attempts to obtain good 'free knot' algorithms.

For this reason, the preferred approach to spline approximation is to fix knots by a selective or adaptive procedure, at each step of which a linear approximation problem is solved for the coefficients c_j. The problem of selecting good knots is not an easy problem either, but some progress has been made and, for example, knot insertion and deletion algorithms are offered by [14, 29].

Some success has recently been claimed by Loach [28] in solving the nonlinear approximation problem, in which splines are replaced by continuous piecewise polynomials (with less restrictive derivative continuity requirements). They adopt a dynamic programming procedure for solving the relevant nonlinear optimisation problem.

The lessons to be learned from splines are clear. Before embarking too enthusiastically on the nonlinear approximation problem, it is essential to study the conditioning of this problem. We are not aware that such studies have been undertaken as a routine matter in neural networks. In any case, there is much to be said in favour of the alternative use of selection procedures for determining good 'nonlinear parameters' w_i, such as radial basis centres, so that the resulting approximation problem may become linear. However, there are few such selection procedures at present for both radial basis functions and ridge functions, and we believe that this is an area that deserves far more attention. It is not at all obvious, for example, where to place radial basis centres in problems of high dimension. One attractive approach, when the data are plentiful, is to locate each radial centre in a distinct cluster of data abscissae, having previously clustered the data by a technique such as *k*-means, and successful implementations are discussed in [32, 52].

4.2.7.2 Economisation of parameters — parsimony

In addition to being highly nonlinear, the approximation problem (in many dimensions) is also a very large one, and it is not impossible to be faced with thousands of connection weights. Indeed, since massively parallel computation is envisaged, there is a temptation to adopt a significant number of nodes in each layer of the network. We are then faced with a very challenging task — the solution of an optimisation problem with many parameters — and our chances of success in a reasonable timescale are greatly reduced.

The procedure generally adopted by researchers, such as Chen *et al.* [10] has been to aim for 'parsimony', by progressively eliminating those parameters that appear to be making insignificant contributions to the approximation. Indeed Chen *et al.* have provided their own rigorous procedures for deciding when to eliminate certain parameters in the case of radial basis function expansions. This is certainly an intelligent approach, and probably an essential one, but there are also potential snags in the approach. For example, a parameter may make a small contribution at one stage of an (iterative) approximation procedure and then make a much larger contribution at a later stage. However, this is an area which also deserves considerable attention. Indeed, surprisingly little practical attention has been given by approximation specialists to the notion of parsimony, even for one-dimensional problems. There has, however, been some significant theoretical work on the use of 'incomplete polynomials', in which certain terms are missing, by workers such as Saff [47], and so there is some theoretical backing to such an approach.

4.2.8 Practicalities of approximation

In concluding this discussion of approximation, it is appropriate to point out various practical requirements which correspond to the various choices of forms of approximation, and which may influence the suitability of these choices. This leads us to attempt to compare the merits of different forms for use in neural networks.

Spline functions require a mesh of knots to be prescribed, and they effectively involve the subdivision of the approximation domain by some multivariate form of rectangulation. In a many-variable problem this can involve an enormous number of subdomains, and so the problem is potentially very cumbersome. On the other hand, B-splines have locally compact support, and so savings may be made by restricting computation to local areas of the network. Moreover, parallel algorithms may well be based on the inherent domain decomposition. (These points have essentially been noted in [6, 7].)

Ridge functions are interesting in that they effectively create a function of many variables from functions of one variable. They have a natural and historical role in neural networks and are relatively convenient to adopt. They are less well known in mainstream approximation theory than other forms and, from that viewpoint might be regarded as somewhat unorthodox. They have global, rather than local support.

Radial basis functions have some unique features that stand out in fitting multivariate functions. They can uniquely interpolate, they possess unique best ℓ_∞ approximations, and they do not require the division of the domain into a mesh of subdomains. What they do depend upon, however, is an appropriate choice of n centres $\underline{y}_j$ (where n is the number of parameters), and this is both their greatest strength and greatest weakness. If centres are well chosen, then a small number of parameters may be required. Indeed, Chen *et al.* [10] claim that their algorithm can reduce the number of centres to be selected by an 'orthogonal-forward-regression'.

Our instinct is that radial functions may just have the edge at present, since they are viewed with some favour by both neural network and approximation specialists. Ridge functions are popular in neural networks and may perhaps be the most natural choice, while spline functions are strongly supported in approximation theory for their local properties and versatility. These two latter forms are therefore strong contenders also.

Two other aspects which need to be taken into account are parallel processing and network training. The extent to which these aspects may be efficiently taken into account may well hold the balance in choosing finally between these three leading contenders.

Looking to the future, we should like to point out some possible advantages in using 'local' functions such as B-splines rather than 'global' functions such as sigmoids. 'Local' functions are known more technically as functions with 'compact support' which means that they are zero except in a limited finite interval in their argument (or arguments). Some one-dimensional examples are drawn in Figure 4.3. These examples include rectangular functions (i.e. degree zero B-splines), triangular functions (i.e. linear B-splines), and cubic B-splines. They may be used with more overlapping than in Figure 4.3 and may be generalised to two or more dimensions. Overlapping rectangular functions in ℓ dimensions are used in the 'cerebellar model articulation controller (CMAC)' developed by Albus [1]. (For a more recent exposition of his idea see Mischo *et al.* [38]).

Not only does this idea have a sound basis in the theory of spline approximation (see [5]) but in the learning process to adjust the coefficients of the linear combination of these functions, from which the estimate F of f is formed, only a finite number of weights is adjusted at each stage of this learning process. Moreover, a particularly simple projection algorithm, originally due to Kaczmarz [23], is used to adjust those weights, see [42]. The simplicity and speed of this algorithm lends itself to real-time on-line implementation, of the greatest importance for adaptive control using neural network techniques.

4.3 Convergence of weight training algorithms

The various schemes outlined in Section 4.2 can affect dramatically the convergence of the weight training algorithms. There is convincing evidence that modelling with 'local functions' as opposed to 'global functions' results in faster convergence of the weight training process. Figure 4.4, taken from [38], compares convergence of 'AMS' (associative memory systems) modelling using 32 overlapping local functions with a (2-5-1) neural network (i.e. 2 input nodes, 5 hidden nodes and 1 output node) using the well-known backpropagation weight training algorithm [33]. Here, a factor of 10^{-1} in convergence time is claimed in favour of 'local' functions. Moody and Darken [40] claim even higher factors e.g. 10^{-3}. This convergence speed could be a crucial deciding factor in on-line real-time closed loop use of ANNs in adaptive control loops, since we know that for more elementary linear control loops containing time delays, long delays usually lead to unstable closed loops.

4.4 Conclusions

What does approximation theory have to offer workers in neural networks?

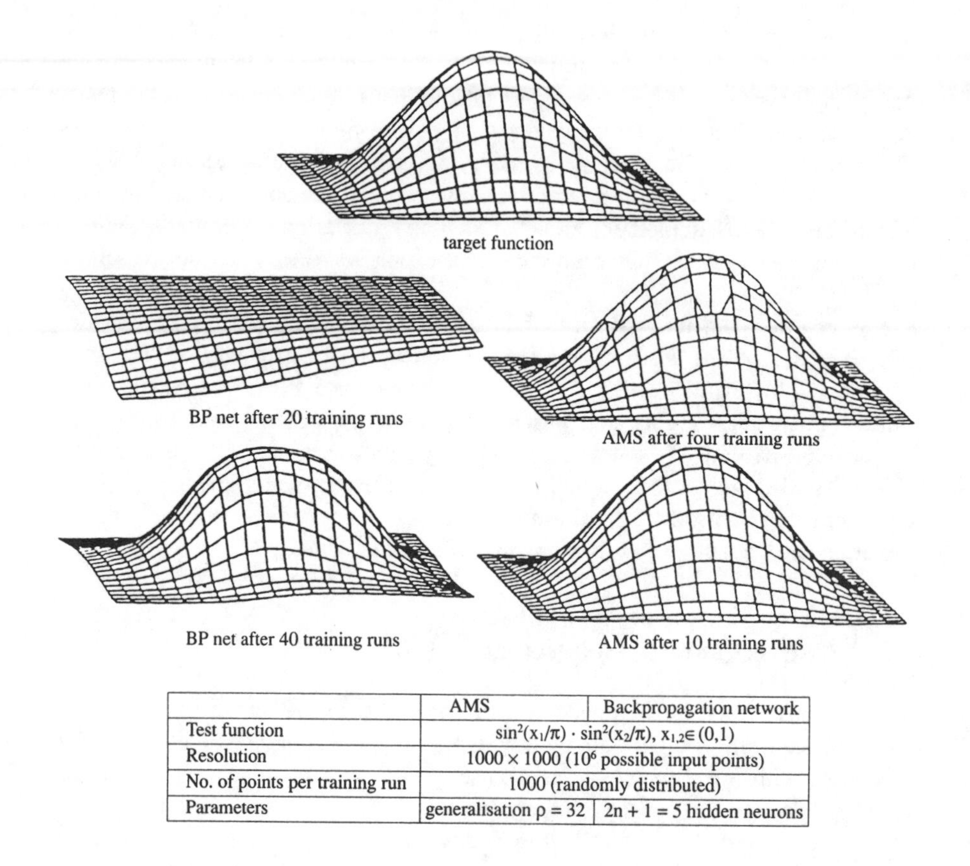

	AMS	Backpropagation network
Test function	$\sin^2(x_1/\pi) \cdot \sin^2(x_2/\pi)$, $x_{1,2} \in (0,1)$	
Resolution	1000×1000 (10^6 possible input points)	
No. of points per training run	1000 (randomly distributed)	
Parameters	generalisation $\rho = 32$	$2n + 1 = 5$ hidden neurons

Figure 4.4 *Comparison of convergence of AMS and backpropagation networks* (32 overlapping local functions against 5 sigmoids)

1) Approximation theory offers the 'alternative' approach of adopting a general purpose optimisation method to solve the relevant nonlinear approximation problem, in addition to the usual feedforward and backward propagation procedures.

2) Locally supported basis functions, such as B-splines, deserve to be given more attention. They have the potential for giving considerable savings, and they are also known to be versatile in fitting functions which vary in their behaviour locally over the domain.

3) For high-dimensional problems, radial basis functions are valuable, since they may be based on a limited number of centres, which do not have to be placed

on a grid throughout the domain (as splines generally do). However, this advantage is only a real one if a good selection procedure is available for centres.

4) Researchers in approximation theory and algorithms have significant contributions to make in neural networks. Existence of good/best approximations has been fundamental, but attention now needs to be directed towards (a) the development of algorithms, (b) the rate of convergence of the approximation (form) as the number of parameters increase, and (c) the integration of approximation methodology into feed-forward, backpropagation and related procedures.

4.5 References

[1] Albus, J.S., 1975, A new approach to manipulator control: the cerebellar model articulation controller (CMAC), *ASME Transactions Series G, J of Dynamic Systems, Measurement and Control*, Vol. 97, pp. 220-227.

[2] Barrodale, I. and Phillips, C., 1975, Solution of an overdetermined system of linear equations in the Chebyshev norm, *ACM TOMS 1*, pp. 264-270.

[3] Barrodale, I. and Roberts, F.D.K., 1974, Algorithm 478: solution of an overdetermined system of linear equations in the ℓ_1 norm, *Comm. ACM*, Vol. 17, pp. 319-320.

[4] Bennell, R.P. and Mason, J.C., 1991, Bivariate orthogonal polynomial approximation to curves of data, in: *Orthogonal Polynomials and their applications*, Brezinski, C. Gari, L. and Ronveaux. A., (Eds), J.C. Balzer Pub. Co., IMACS , pp. 177-183.

[5] de Boor, C., 1978, *A Practical Guide to Splines*, Springer Verlag, New York.

[6] Brown, M. and Harris, C.J., 1992, The B-spline neurocontroller, in: *Parallel Processing in Control*, Rogers, E. (Ed.), Prentice Hall.

[7] Brown, M. and Harris, C.J., 1994, *Neurofuzzy adaptive modelling and control*, Prentice-Hall.

[8] Broomhead, D.S. and Lowe, D., 1988, Multivariable functional interpolation and adaptive networks, *Complex Systems* Vol. 2, pp. 321-355.

[9] Buchanan, M.D. and Powell, M.J.D., 1990, Radial basis function interpolation on an infinite regular grid, in: *Algorithms for Approximation 2*, Mason, J.C. and Cox, M.G. (Eds), Chapman and Hall, London, pp. 146-169.

[10] Chen, S., Billings, S.A., Cowan, C.F.N. and Grant, P.M., 1990, Practical identification of NARMAX models using radial basis functions, *Int. J. Control* Vol. 52, pp. 1327-1350.

[11] Chen, S., Billings, C.A. and Luo, W., 1989, Orthogonal least squares methods and their application to nonlinear system identification, *Int. J. of Control* Vol. 50, pp. 1873-1896.

[12] Chui, C.K. and Li, X., 1991, Realization of neural networks with one hidden layer, *Report 244, Center for Approximation Theory*, Texas A and M University, March.

[13] Chui, C.K., Schumaker, L.L. and Ward, J.D. (Eds), *Approximation Theory 6*, Academic Press, New York.

[14] Cox, M.G., Harris, P.M. and Jones, H.M., 1990, A knot placement strategy for least squares spline fitting erased on the use of local polynomial approximations, in: *Algorithms for Approximation 2*, Mason, J.C. and Cox, M.G. (Eds), Chapman and Hall, pp. 37-45.

[15] Cox, M.G. and Mason, J.C. (Eds), 1993, Algorithms for Approximation 3, Special Volume, *Numerical Algorithms* Vol. 5, pp. 1-649.

[16] Cybenko, G., 1989, Approximation by superpositions of a sigmoidal function, *Math. Control Signals Systems* Vol. 2, pp. 303-314.

[17] Daman, A.E. and Mason, J.C., 1987, A generalised cross-validation method for meteorological data with gaps, in: *Algorithms for Approximation*, Mason, J.C. and Cox, M.G. (Eds), Clarendon Press, Oxford, pp. 595-610.

[18] Dyn, N., 1987, Interpolation of scattered data by radial functions in: *Topics in Multivariate Approximation*, Chui, C.K., Schumaker, L.L. and Utreras, F. (Eds), Academic Press, New York, pp. 47-61.

[19] Dyn, N., 1989, Interpolation and approximation by radial and related functions, in: *Approximation Theory 6*, Chui, C.K., Schumaker, L.L. and Ward, J.D. (Eds), Academic Press, New York, pp. 211-234.

[20] Franke, R., 1982, Scattered data interpolation: tests of some methods, *Math. and Computing* Vol. 38, pp. 181-200.

[21] Galligani, I., Ruggiero, V. and Zama, F., 1990, Solutions of the equality-constrained image restoration problems on a vector computer, in: *Parallel Processing 89*, Evans, D.J., Joubert, G.R. and Peters, F.J. (Eds), Elsevier Pub. Co..

[22] Harbour, S.K., Mason, J.C., Anderson, I.J. and Broomhead, D.S., 1994, Parallel Methods for Data Approximation with Radial Basis Functions, in: *Computer-Intensive Methods in Control and Signal Processing*, Kulhava, L., Karny, M. and Warwick, K. (Eds) IEEE Workshop, CMP 94, Prague, pp. 101-112 (preprint).

[23] Kaczmarz, S., 1937, Angenäherte Auflösung von Systemen Linearer Gleichungen, *Bull. Int. de l'Academie Polonaise des Sciences et des Lettres, Cl. d. Sc. Mathém A*, pp. 355-357, Cracovie.

[24] Kolmogorov, A.N., 1957, On the representation of continuous functions of many variables by superposition of continuous functions of one variable and addition, *Dokl. Akad. Nauk SSSR* Vol. 114, pp. 953-956.

[25] Kramer, M.A., 1991, Data analysis and system modelling with autoassociative and validity index networks, *Proc. Int. Symposium Neural Networks and Engineering Applications*, Newcastle, October.

[26] Light, W.A., 1992, Some aspects of radial basis function approximation, in: *Approximation Theory, Spline Functions and Applications*, Singh, S.P. (Ed.), Kluwer Math and Phys. Sci. Series, Vol. 356, pp. 163-190.

[27] Light, W.A., Xu, Y. and Cheney, E.W., 1993, Constructive methods of approximation by ridge functions and radial functions, *Numerical Algorithms* Vol. 4, pp. 205-223.

[28] Loach, P.D., 1990, Best least squares approximation using continuous piecewise polynomials with free knots, PhD Thesis, Bristol University.

[29] Lyche, T. and Morken, K., 1987, A discrete approach to knot removal and degree reduction algorithms for splines, in: *Algorithms for Approximation*, Mason, J.C. and Cox, M.G. (Eds), Clarendon Press, Oxford, pp. 67-82.

[30] Mansfield, A.J., 1990, An introduction to neural networks, in: *Scientific Software Systems*, Mason, J.C. and Cox, M.G. (Eds), Chapman and Hall, London, pp. 112-122.

[31] Mason, J.C., 1984, *BASIC Matrix Methods*, Butterworths.

[32] Mason, J.D., Craddock, R.J., Mason, J.C., Parks, P.C. and Warwick, K., 1994, Towards a stability and approximation theory for neuro-controllers, *Control 94, IEE Pub 389*, pp. 100-103.

[33] McClelland, J.L., *et al.*, 1986, *Parallel distributed processing*, 2 vols. MIT Press, Cambridge, Mass., USA.

[34] McWhirter, J.G., Broomhead, D.S. and Shepherd, T.J., 1993, A systolic array for nonlinear adaptive filtering and patters recognition, *J. of VLSI Signal Processing* Vol. 3, pp. 69-75.

[35] Mhaskar, H.N., 1993, Approximation properties of a multilayered feedforward artificial neural network, *Advances in Computational Maths* Vol. 1, pp. 61-80.

[36] Mhaskar, H.N. and Micchelli, C.A., 1992, Approximation by superposition of a sigmoidal function, *Adv. Appl. Math.* Vol. 13, pp. 350-373.

[37] Micchelli, C.A., 1986, Interpolation of scattered data: distance matrices and conditionally positive definite functions, *Constructive Approx.* Vol. 2, pp. 11-22.

[38] Mischo, W.S., Hormel, M. and Tolle, H., 1991, Neurally inspired associative memories for learning control: a comparison, in: *Artificial Neural Networks* (Kohonen, T., *et al.* (Eds)) Vol. 2, Elsevier Science Publishers, pp.1241-1244.

[39] Montefusco, L.B. and Guerrini, C., 1991, Domain decomposition methods for scattered data approximation on a distributed memory multiprocessor, *Parallel Computing* Vol. 17, pp. 253-263.

[40] Moody, J., and Darken, C., 1988, Learning with localised receptive fields, *Proc. 1988 Connectionist Models Summer School*, Morgan Kaufman Publishers, San Mafeo, California, pp. 1-11.

[41] Narendra, K.A. and Parthasarathy, K., 1990, Identification and control of dynamical systems using neural networks, *IEEE Trans, on Neural Networks*, Vol. 1, pp. 4-27.

[42] Parks, P.C., 1993, S. Kaczmarz (1895-1939), *Int. J. Control*, Vol. 57, pp. 1263-1267.

[43] Poggio, T. and Girosi, F., 1989, A theory of networks for approximation and learning, AI Memo No 1140, MIT AI Laboratory, July.

[44] Powell, M.J.D., 1981, *Approximation Theory and Methods*, Cambridge University Press.

[45] Powell, M.J.D., 1987, Radial basis functions for multivariable interpolation, in: *Algorithms for Approximation*, Mason, J.C. and Cox, M.G. (Eds), Clarendon Press, Oxford, pp. 143-168.

[46] Powell, M.J.D., 1991, Radial basis functions in 1990, in: *Advances in Numerical Analysis Vol. II*, Oxford University Press, pp. 105-210.

[47] Saff, E.B., 1983, Incomplete and orthogonal polynomials, in: *Approximation Theory 4*, Chui, C.K., Schumaker, L.L. and Ward, J.D. (Eds) Academic Press, pp. 219-256.

[48] Schoenberg, I.J., 1946, Contributions to the problem of approximation of equidistant data by analytic functions, *A B Quarterly Appl. Math.* Vol. 4, pp. 45-99 and pp. 112-141.

[49] Schumaker, L.L., 1981, *Spline Functions: Basic Theory*, Wiley.

[50] Tolle, H. and Ersü, 1992, Neurocontrol, *Lecture Notes in Control and Information Sciences*, No. 172, Springer-Verlag.

[51] Von Golitschek, M. and Schumaker, L.L., 1990, Data fitting by penalized least squares, in: *Algorithms for Approximation 2*, Mason, J.C. and Cox, M.G. (Eds), Chapman and Hall, London, pp. 210-227.

[52] Warwick, K., Mason, J.D. and Sutanto, E.L., 1995, Centre selection for radial basis function networks, *Proc. International Conference of Artificial Neural Networks and Genetic Algorithms,* Ales, France, Springer-Verlag, pp. 309-312.

Chapter 5

Electric power and chemical process applications

G.W. Irwin, P. O'Reilly, G. Lightbody, M. Brown and E. Swidenbank

5.1 Introduction

If current research efforts worldwide on neural networks are to gain recognition and continue at their current levels, it is essential that theoretical advances are accompanied by industrial applications where the advantages/disadvantages of this new technology can be properly assessed against more conventional techniques. The past few years have seen a marked shift towards practical, as opposed to purely simulation, studies in the field of control and systems engineering, which is a healthy sign both of a maturing technology and successful technology transfer. A good selection of such applications are included in this book and the present chapter contains the results of two studies, one concerned with nonlinear modelling of a 200 MW boiler in an electrical power station [1] the other with inferential estimation of viscosity, a key quality indication in a chemical polymerisation reactor [2, 3].

5.2 Modelling of a 200 MW boiler system

The simulation of the boiler system includes all of the important control loops together with typical subsystems of evaporation, heat exchangers, spray water attemperators, steam volume, turbines and water, and steam and gas properties. The model was designed and verified using power station tests and contains 14 nonlinear differential equations and 112 algebraic equations. It constitutes a direct representation of unit 5 at Ballylumford power station which consists of a 200 MW two-pole generator supplied by GEC, directly coupled to a three-stage turbine, driven by steam from a drum-type boiler. A more detailed description is contained in [3].

This electric power plant represents a highly nonlinear process and the use of a complex simulation is therefore a valuable precursor to actual tests on the real power plant.

5.2.1 Identification of ARMAX models

A seventh-order PRBS signal was superimposed on the main steam pressure setpoint and on the governor valve input, to generate plant data for identification purposes. A second-order, 4-input 4-output, ARMAX model of the form:

$$A(z^{-1})Y(k) = B(z^{-1})U(k) + C(z^{-1})E(k) \tag{5.1}$$

was then formed from the resulting data, where $A(z^{-1})$, $B(z^{-1})$ and $C(z^{-1})$ are polynomial matrices in the backward shift operator and:

$$Y(k) = \begin{bmatrix} y_1(k) \triangleq \textit{water level deviation} \\ y_2(k) \triangleq \textit{electrical power output deviation} \\ y_3(k) \triangleq \textit{main steam pressure deviation} \\ y_4(k) \triangleq \textit{main steam temperature deviation} \end{bmatrix} \tag{5.2}$$

$$U(k) = \begin{bmatrix} u_1(k) \triangleq \textit{feedwater flow deviation} \\ u_2(k) \triangleq \textit{governor valve input deviation} \\ u_3(k) \triangleq \textit{fuel flow deviation} \\ u_4(k) \triangleq \textit{attemperature spray deviation} \end{bmatrix} \tag{5.3}$$

$$E(k) = \left[4 \times 1 \textit{ vector of uncorrelated noise sources}\right] \tag{5.4}$$

Identified models were formed using training data at operating points of 100 MW and 200 MW. It was found that the linear models matched the plant quite closely around these operating points. However, Figure 5.1 shows the 180-step-ahead-prediction output of the 200 MW linear model when applied at the opposite end of the operating region. The poorer responses obtained, which is particularly marked on the graph of drum water level deviation, indicates that linear models are only valid around a small region of a particular operating point. When moving to a different operating point, the dynamics of the plant will change and, as expected, the original linear model is no longer valid.

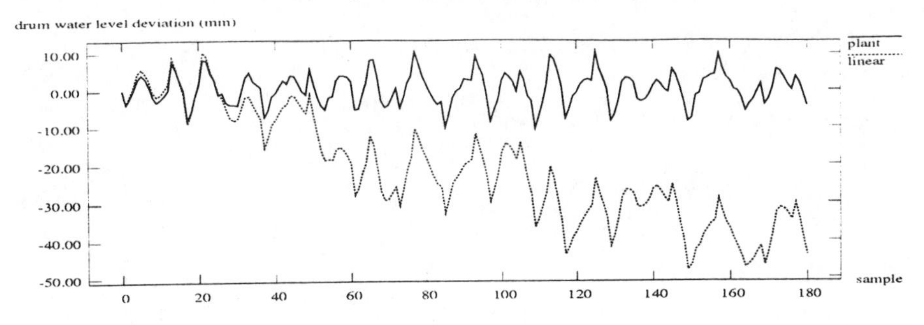

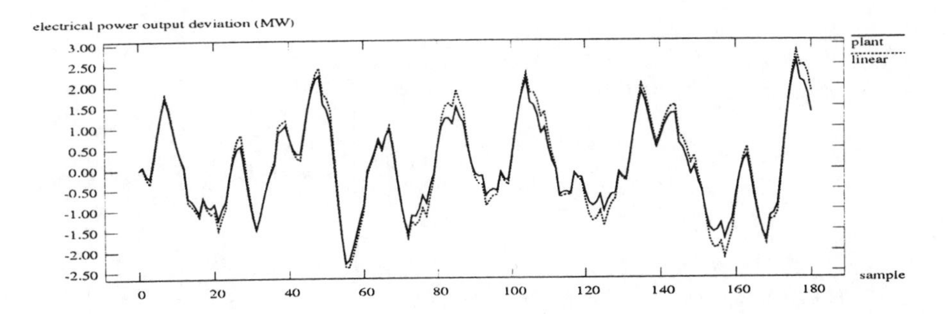

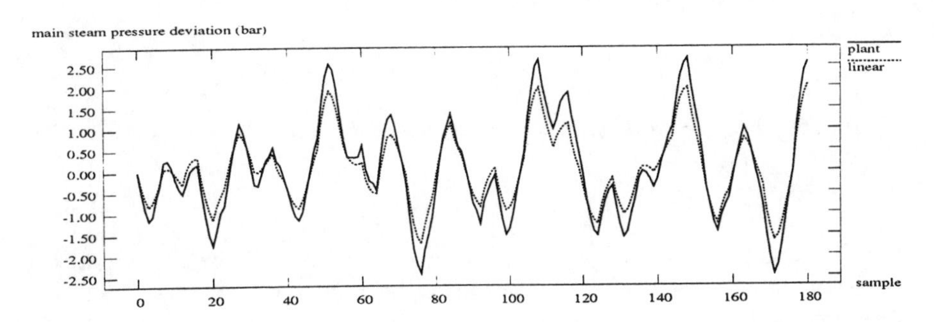

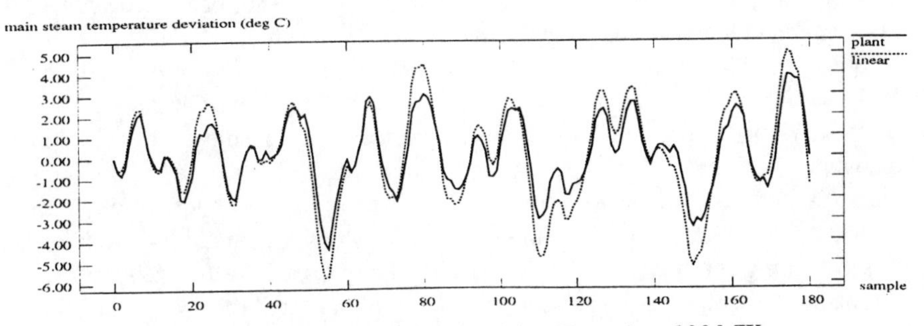

Figure 5.1 *200 MW linear model (180-sap) at 100 MW*

5.2.2 Neural boiler modelling: local models

A 4-input, 4-output, second-order, dynamic nonlinear model was simulated using a 16-24-4 multi-layer perceptron (MLP) as in Figure 5.2.

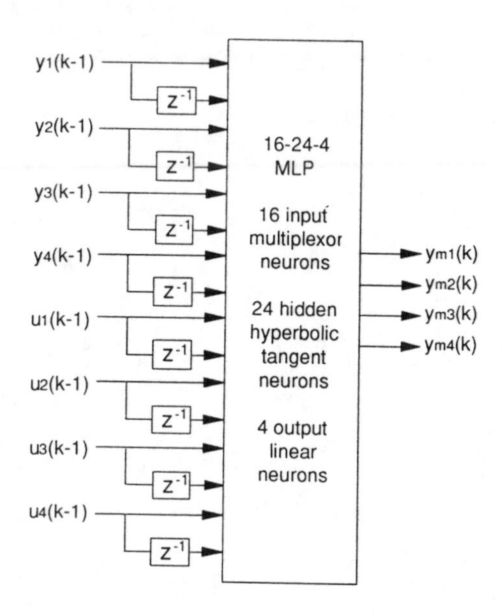

Figure 5.2 *Feedforward network dynamical model structure (variables defined in eqns. 5.2 and 5.3)*

The weights of the neural network were adjusted to minimise the sum of the one-step-ahead-prediction (sap) error squared over the training set, using a quasi-Newton technique called the Broyden-Fletcher-Goldfharb-Shanno (BFGS) algorithm [4]. This Hessian-based technique provides a significant acceleration of the training process compared with simple or batch backpropagation [5].

Since the plant data are contaminated with noise, extra care must be taken to ensure that only the dynamics contained in the data are learned and not the noise, as this lessens the predictive capability of the model and results in an over-trained network. With this in mind, the optimum number of presentations of the training data (or iterations) must be found. The effect of over-training is obvious by referring to Figure 5.3 which shows the evolution of the predictive error for the 1-sap training set, the 1-sap test set, the 10-sap test set, and the 180-sap test set. From these graphs, it can be seen that the optimum predictive capability of the network occurs around iteration 115, and that the predictive error (especially the 180-sap) increases significantly as the number of iterations increases. Notice also that the 1-sap over the training set always decreases as expected, since it was this error which was minimised during training.

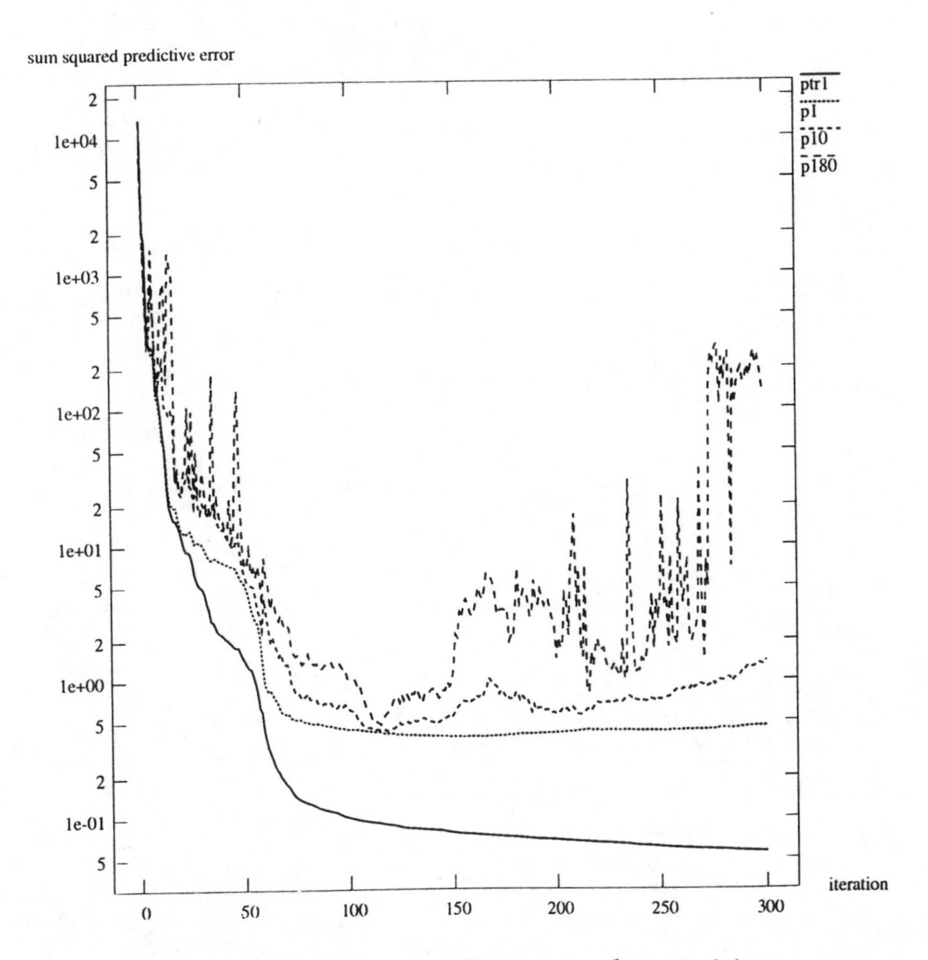

Figure 5.3 *Evolution of predictive errors during training*

Neural models of the Ballylumford boiler system were formed using the same training and test data as for the linear modelling. Figure 5.4 shows the 180-sap predictions for the 200 MW neural model in comparison with the plant responses. It can be seen that the predictive capability of the trained networks is excellent, despite the presence of noise on the data.

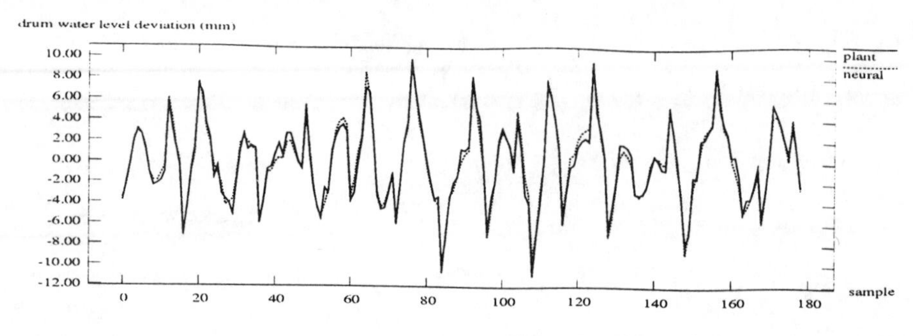

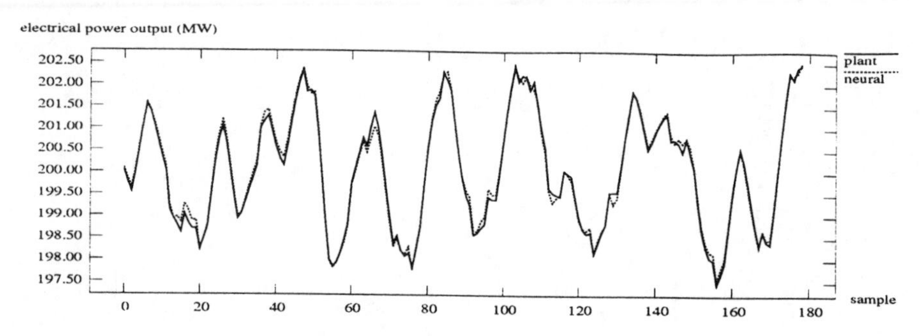

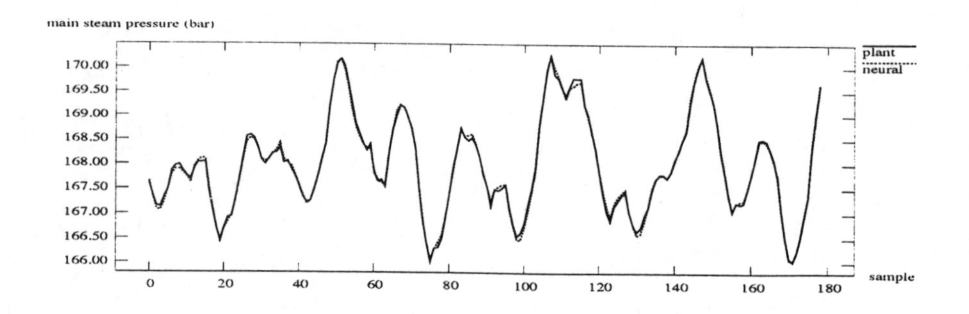

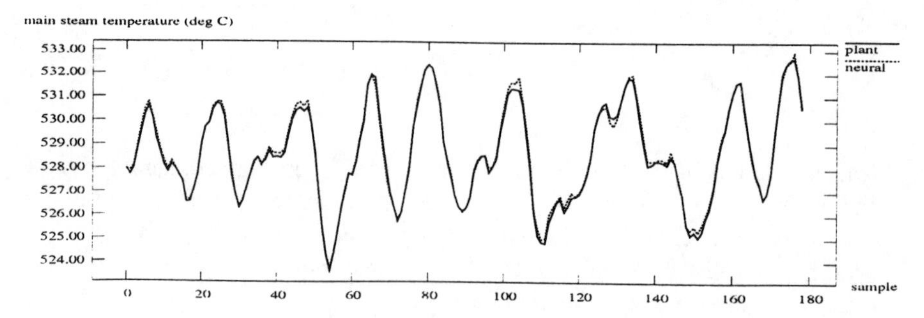

Figure 5.4 *200 MW neural model (180-sap) at 200 MW*

5.2.3 *Neural boiler modelling: global model*

The real power of the neural modelling technique comes from the network's ability to represent an arbitrary nonlinear function. With this in mind, an attempt was made to model the boiler system over its usual operating range (i.e. from 100-200 MW). To achieve this, data must be available that span the entire operating range of interest. These data were obtained by simulating a two-shifting operation where the boiler system was driven from half load to full load, whilst maintaining the PRBS signals as before. The simulated load profile is shown in Figure 5.5.

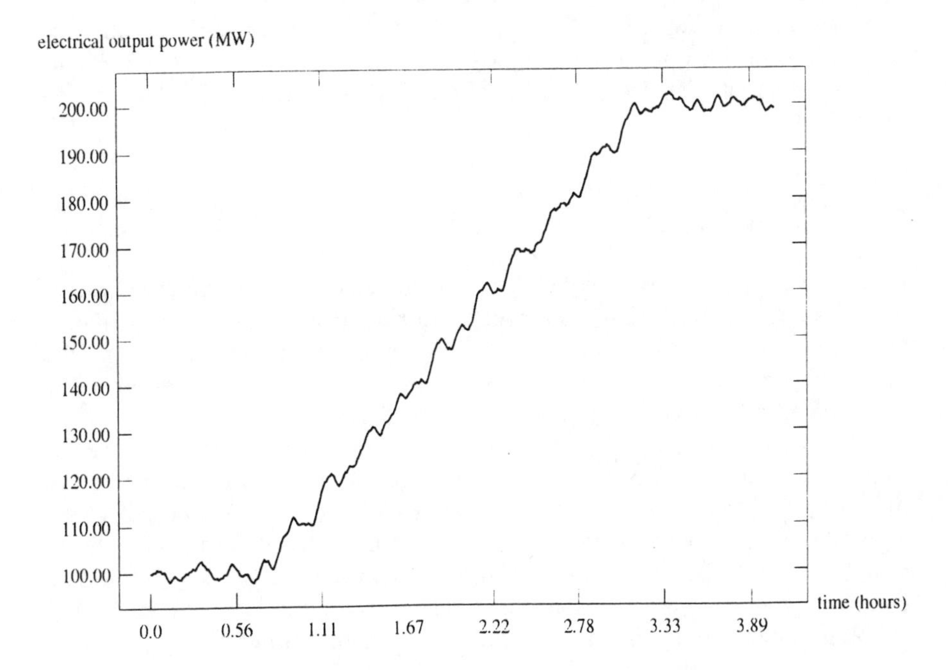

Figure 5.5 *Simulated load profile for training*

The neural network was trained and tested using data from the two-shift load profile, to produce a global neural boiler model. Further detailed examination of the prediction capability of the global model at 100 MW and 200 MW confirmed that the network had indeed captured the nonlinear system dynamics at the extremes of the operating region (Figure 5.6). Comparing Figures 5.6 and 5.1, particularly the graphs of drum water level deviation, illustrates clearly the improvements in predicted plant output which can accrue from nonlinear modelling.

The trained network was also used to predict the 180-sap responses of the boiler system to a +10 MW demand in power, followed by a –10 MW demand in power at a 100 MW operating point. Here again there was good agreement between the actual plant responses and the predicted outputs from the global neural model.

Further validation of the nonlinear global neural model of the boiler system was done by performing correlation tests on its inputs and outputs. These procedures, described in Chapter 11, are designed to detect the inadequacy of the fitted model which could arise from incorrect assignment of input nodes, insufficient hidden nodes or training that has not converged.

More recent work is concentrating on on-line, nonlinear control of turbogenerators using neural networks [6] and assessment of performance compared with fixed-parameter and adaptive control techniques. As in previous studies, testing on a laboratory system will be performed prior to implementation in the field.

5.3 Inferential estimation of viscosity in a chemical process

Inferential estimation involves the determination of difficult to measure process variables from easily accessible secondary information and is a topic of significant industrial interest. In addition to the capital costs of instrumentation, the downward pressure on manpower costs and overheads involved in maintenance make soft sensing attractive for industrial applications.

This section is concerned with viscosity control in a polymerisation reactor. Here the measurement from the viscometer is subject to a significant time delay but the torque from a variable speed drive provides an instantaneous, if noisy, indication of the reactor viscosity. The aim of the research was to investigate neural network based inferential estimation, where the network is trained to predict the polymer viscosity from past torque and viscosity data, thus removing the delay and providing instantaneous information to the operators.

5.3.1 The polymerisation process and viscosity control

The polymerisation plant consists of a continuously stirred tank reactor, into which a number of constituent ingredients are fed. The contents are stirred, using a variable speed drive, from which both measurements of speed and torque are available. On-line measurements are also provided for all the flow-rates and for the viscosity of the polymer. The function of the control system is viscosity regulation in the presence of disturbances, particularly due to feed-rate changes.

Two catalysts are added to the reactor: C_A, which promotes polymerisation, and C_B, which acts to inhibit the reaction. For this plant the flow-rate of catalyst C_A,

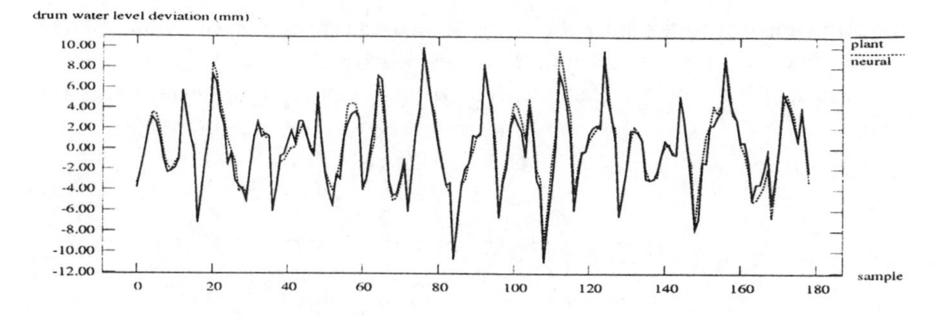

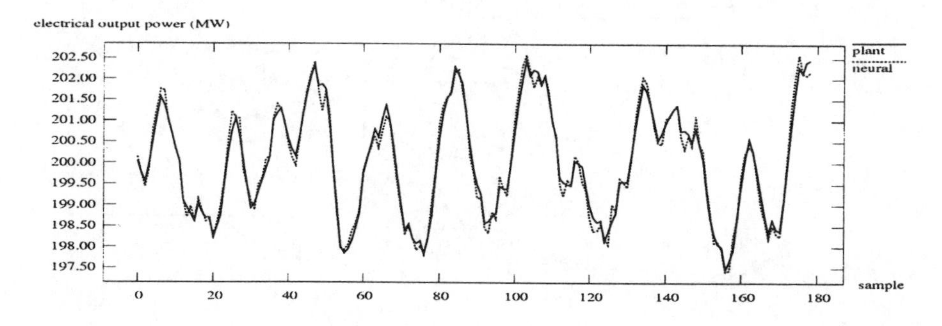

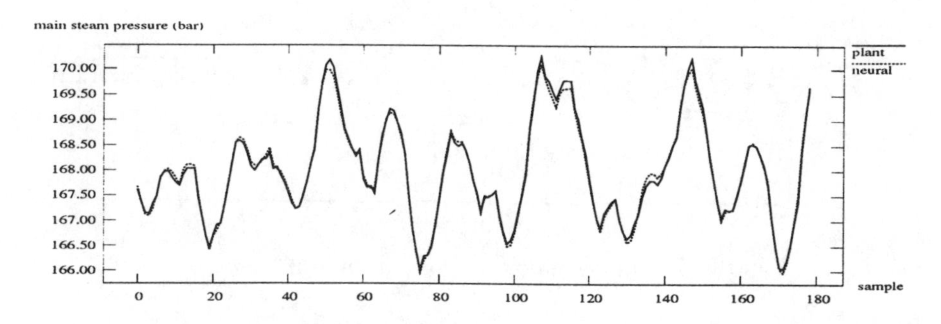

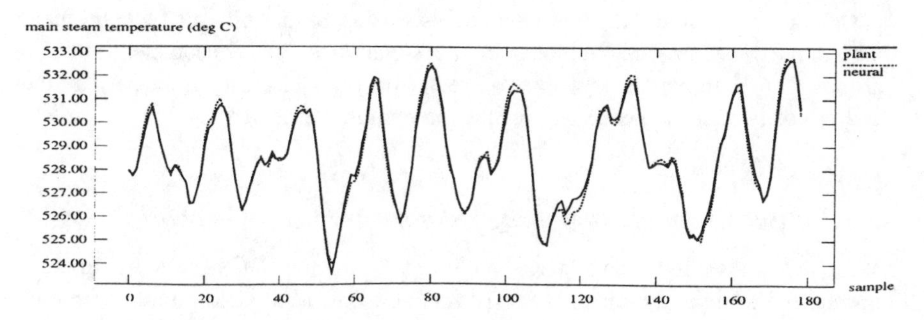

Figure 5.6 *Global neural model (180-sap) at 100 MW*

the flow-rates of all the other constituent compounds and the speed of the variable speed drive are all set for specific feed-rates, with the flow-rate of C_B manipulated to regulate the viscosity. The cascaded PID viscosity control system is shown in Figure 5.7.

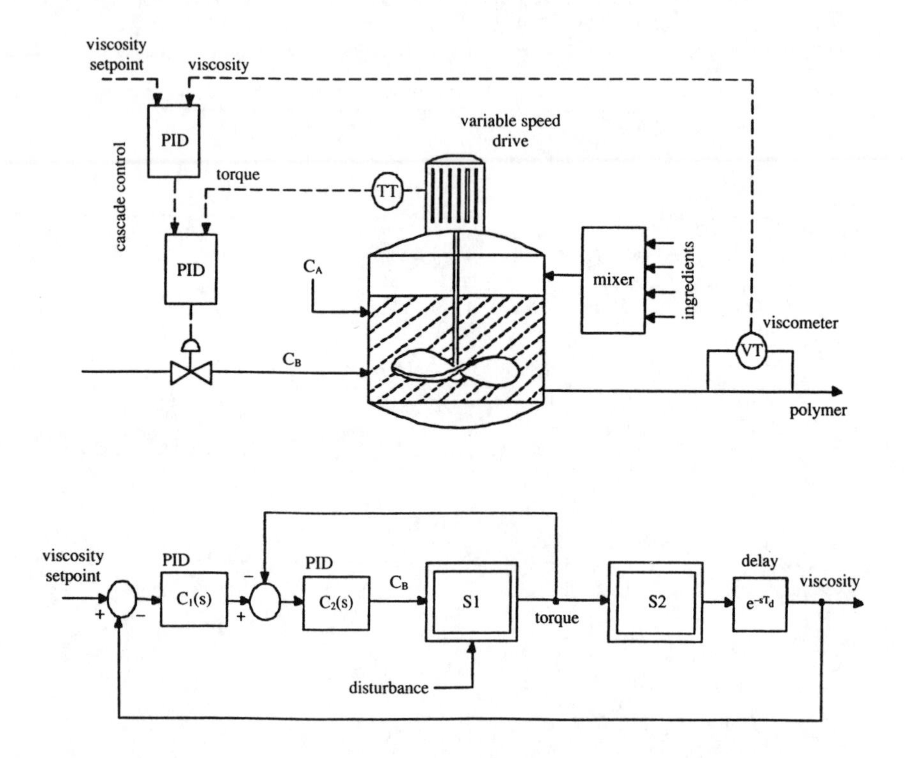

Figure 5.7 *Polymerisation reactor and viscosity control loops*

Detailed plant modelling and data analysis using Matlab revealed a significant pure time delay of three minutes in the signal from the viscometer. It was proposed to train neural networks to predict polymer viscosity from past torque and viscosity data and hence remove the three minute time delay.

5.3.2 Viscosity prediction using feedforward neural networks

Viscosity and torque data were collected from the DCS system, filtered, normalised to lie in the range [–1.0, 1.0] and decimated to yield a sample time of one minute. The predictive model structure of eqn. 5.5 was proposed, where $T(k)$

and $v(k)$ represent torque and viscosity, with a multi-layer perceptron being used to form the nonlinear function.

$$\hat{v}(k+3|k) = \hat{f}(\underline{\phi}(k),\ \underline{\varphi}(k))$$

$$\underline{\phi}(k) = [v(k), v(k-1), \ldots, v(k-n+1)]^T$$

$$\underline{\varphi}(k) = [T(k), T(k-1), \ldots, T(k-m+1)]^T \tag{5.5}$$

A 9-14-1 MLP feedback network with a linear output neuron was trained off-line using BFGS optimisation. The step size was chosen at each iteration using an efficient single-line search technique.

Figure 5.8 shows the response of the neural predictor over a test set of plant data and it is clear that, although the high and middle frequency dynamics are accurately reproduced, there is a low frequency or dc off-set.

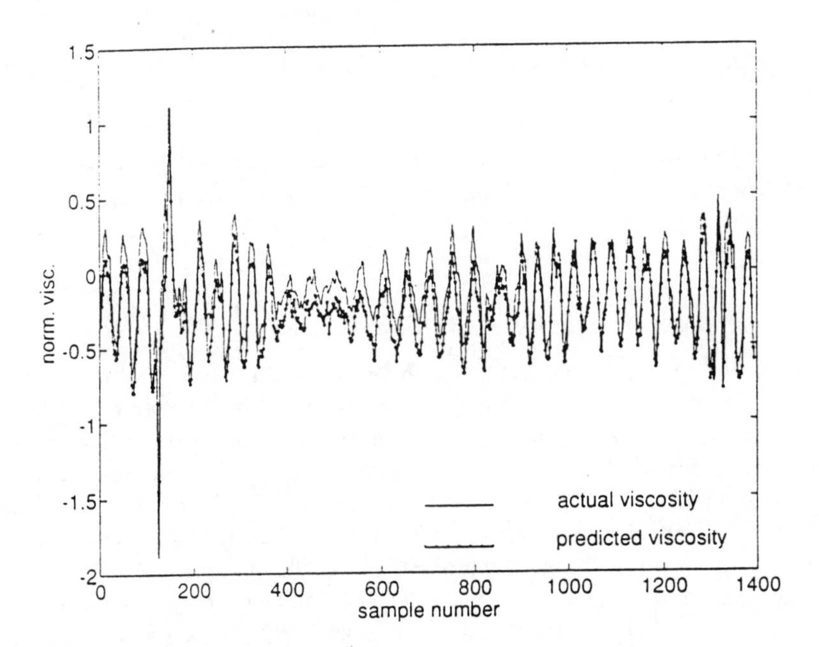

Figure 5.8 *Response of the neural viscosity predictor over the test set*

This was compensated for by utilising the present output of the plant, $v(k)$ and the predicted estimate of the viscosity $\hat{v}(k|k-3)$, to generate a correction term

$d(k)$ which was filtered to increase immunity to noise and to ensure that it reflected unmodelled low frequency errors. The complete neural predictive estimator is then given in Figure 5.9.

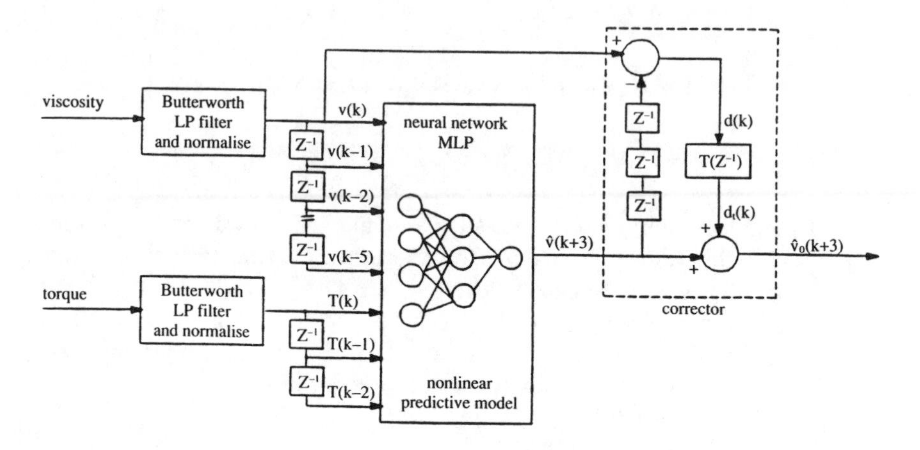

Figure 5.9 *Feedforward network for inferential estimation of viscosity*

When applied to the data of the test set this corrected predictor provided excellent results, predicting accurately over the measurement delay as shown in Figure 5.10.

5.3.3 Viscosity prediction using B-spline networks

The work described in the last section involves off-line training, with the network weights being fixed when applied to the plant. The limitation then is that parameter drift during plant operation cannot be tracked and over time, the network predictions may not match the plant behaviour. In order to solve this, the network needs to be implemented in such a way that it will adapt on-line to track plant variations. Neural networks with local support functions, and which are linear in the weights, are more appropriate, since they use computationally simple weight update rules. One such network is the B-spline neural network [7].

The major problem with B-spline networks is the 'curse of dimensionality', where the number of weights increases exponentially with the number of inputs to the network. Because of this, the structure of the predictor varies from that in Section 5.3.2. Instead of an explicit multi-step-ahead predictor, which calculates the prediction directly from the given inputs, the B-spline was used solely as a single step-ahead-predictor (sap), and a number of these were then cascaded to give implicit multi-step prediction (Figure 5.11).

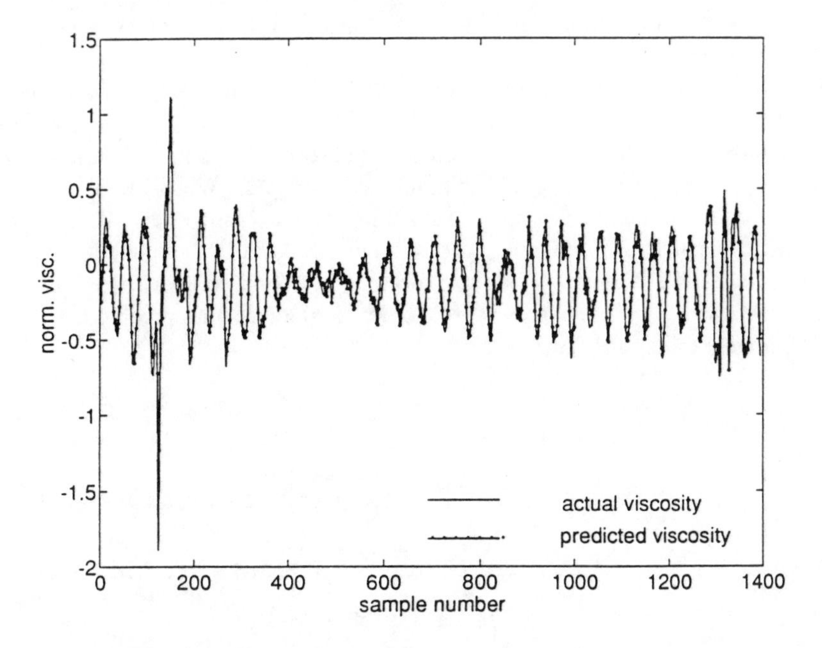

Figure 5.10 *Response of the corrected predictor over the test set*

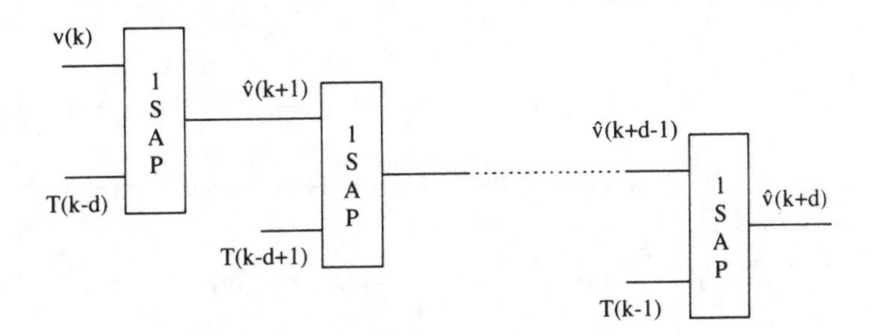

Figure 5.11 *B-spline predictor structure*

Linear system identification suggested that the plant had first-order dynamics and so the single step-ahead-predictor had only 2 inputs. The input space for each of these inputs was partitioned evenly, with 19 interior knots (at intervals of 0.1). The splines used were sixth-order and the training algorithm chosen was normalised stochastic LMS, [8]. The data were sampled at 30 second intervals, and thus the nominal time delay of three minutes was spanned by a six-step-ahead-prediction.

Figure 5.12 shows the performance of the predictor as applied to a typical normalised set of data over a 24 hour period. The network was slow to converge to an approximation of the plant dynamics, with the result that gross features, such as the disturbance just before 600 minutes, were missed. When the disturbance occurs, the input space changes. The resulting output is then zero, because the weights associated with this input space are identically zero, and the convergence to the actual value is slow. Over time this may not be important, since the input space will eventually be covered, but in order to give accurate predictions, some initial off-line training is desirable.

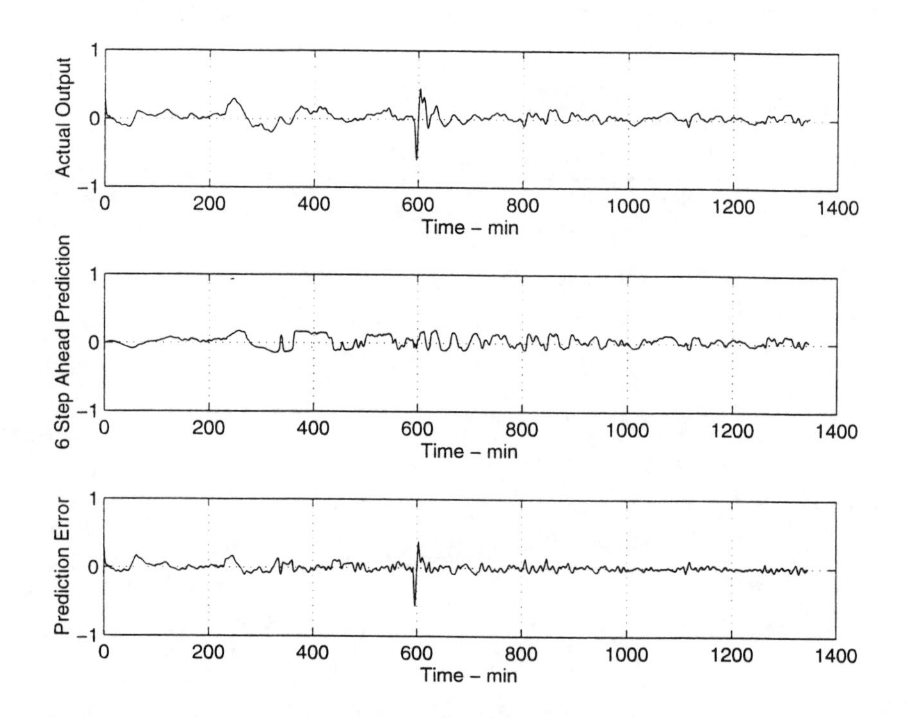

Figure 5.12 *Performance of the on-line predictor with no pre-training*

5.3.3.1 Pre-training the network

Working on the assumption that an inaccurate, linear prediction would be preferable to zero output in the case of new input spaces, it was decided to train the B-spline off-line to represent a simple, linear system model of the form,

$$v(k+1) = aT(k-d) + bv(k) \tag{5.6}$$

where d is the assumed time delay.

In order to train the network adequately across the whole input space, a random scatter of points across the whole space was generated, suitable values of a and b were found using ARX model identification, and the resulting model outputs for the random points generated. The B-spline predictor was then trained off-line to this set of points, and the weights saved for use by the on-line predictor.

Figure 5.13 shows the performance of the predictor on the same test data as Figure 5.8, with the weights initialised by pre-training. The pre-trained network now gives much better results. The errors are, in general, small and it can be observed that the predictions are, for the main part correct. However, the detail of the disturbance is still unacceptable, particularly the 'double peak' just after 600 minutes.

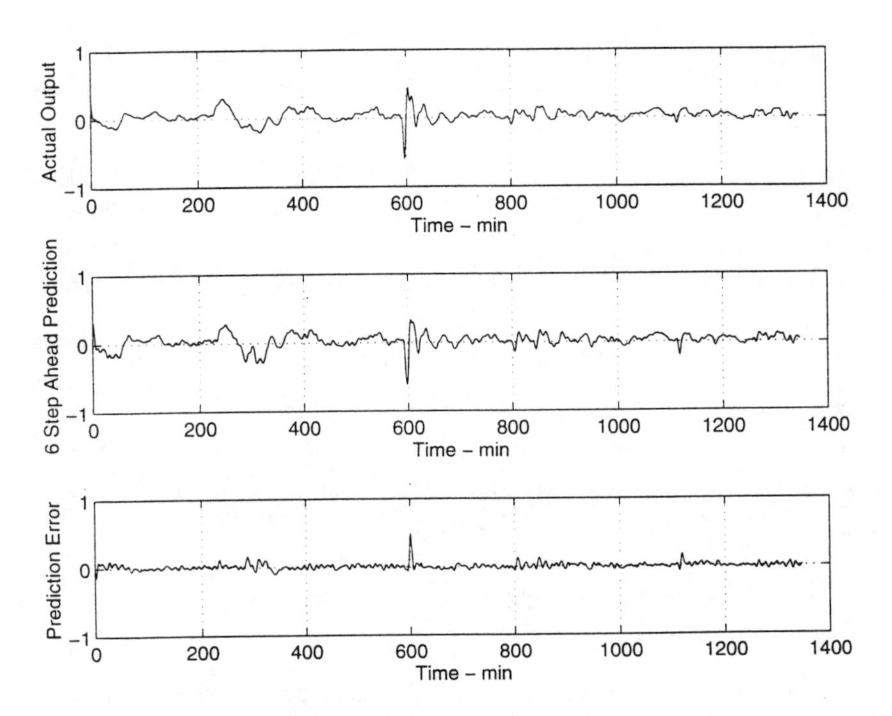

Figure 5.13 *Predictions from the pre-trained network*

5.3.3.2 Disturbance modelling

In order to improve the performance of the predictor with disturbances, a method suggested by Henson and Seborg [9] was used which involves the augmentation of the prediction at every step:

$$\hat{v}(k+m|k) = f(\hat{v}(k+m-1|k)), \quad T(k+m-d) + \lambda_d(v(k) - \hat{v}(k|k-1) \qquad (5.7)$$

This is based on the idea of step disturbances, and for these is proven to give unbiased predictions. Figure 5.14 shows that the predictions are in fact improved. The detail of the 'double peak' is hinted at, but is not strictly correct, because the disturbance is not strictly a step. However, the improvement suggests that some form of disturbance modelling is indeed useful.

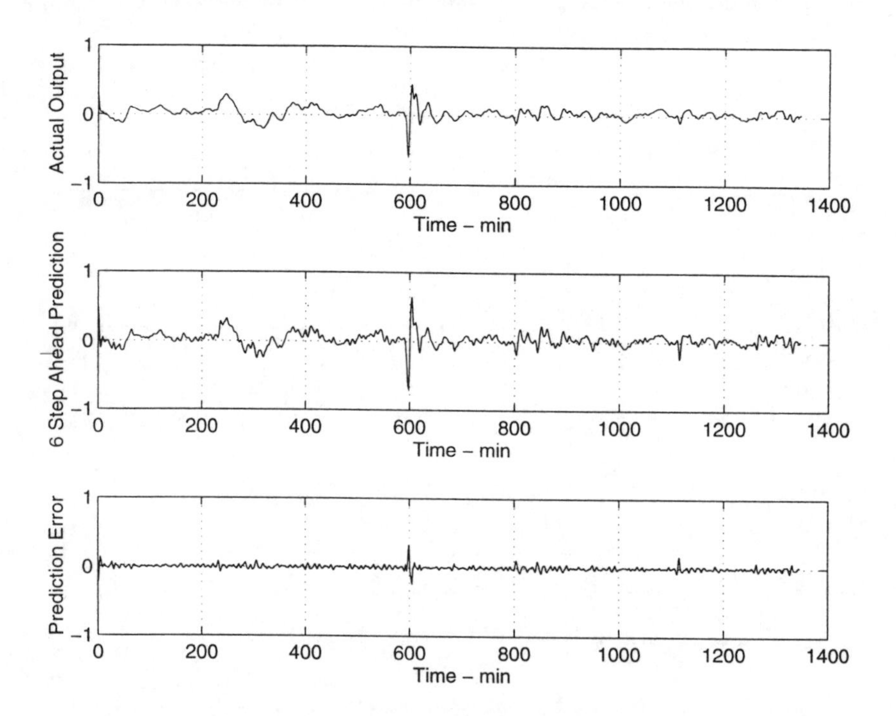

Figure 5.14 *Predictor with disturbance model*

The formulation of a more general disturbance model than that suggested by Henson and Seborg is under development. It is hoped that this will further improve the predictions, which at present are comparable with those obtained using feedforward networks trained off-line.

Long term testing of the on-line B-spline predictor on the plant is imminent, and it is planned to investigate the performance of the predictor for a wide range of operating conditions and set point changes. The use of the B-spline predictor within a nonlinear predictive control scheme is also under consideration.

5.4 Acknowledgment

The authors are pleased to acknowledge the financial support of DuPont (UK), the EPSRC (grant no. GR/H/82877) and the Industrial Research and Technology Unit, Department of Economic Development, Northern Ireland.

5.5 References

[1] Irwin, G.W., Brown, M., Hogg, B. and Swidenbank, E., Neural Network Modelling of a 200 MW Boiler System, to appear in *IEE Proceedings on Control Theory and Applications.*

[2] Lightbody, G.L. and Irwin, G.W., 1995, Neural Networks for the Modelling of Nonlinear Systems, To appear in *Journal of Fuzzy Sets and Systems*.

[3] Lightbody, G., Irwin, G.W., Taylor, A., Kelly, K. and McCormick, J., 1994, Neural network modelling of a polymerisation reactor, *Proc. IEE Int. Conf. Control '94*, Vol. 1, pp. 237-242.

[4] Battiti, R. and Masulli, F., 1990, BFGS Optimisation for Faster and Automated Supervised Learning, *Proc. Int. Neural Net. Conf.*, Vol. 2, pp. 757-760.

[5] Lightbody, G. and Irwin, G.W., 1992, A Parallel Algorithm for Training Neural Network Based Nonlinear Models, *Proc. 2nd IFAC Symp. on Algorithms and Architectures for Real-Time Control*, S. Korea.

[6] Wu, Q.H., Hogg, B.W. and Irwin, G.W., 1992, A Neural Network Regulator for Turbogenerators, *IEEE Trans. on Neural Networks*, Vol. 3, No. 1, pp. 95-100.

[7] Harris, C.J., Moore, C.G. and Brown, M., 1993, Intelligent Control: Some aspects of Fuzzy Logic and Neural Networks, World Scientific Press, London and Singapore.

[8] An, P.E., Brown, M. and Harris, C.H., 1994, Aspects of Instantaneous On-line Learning Rules, *Proc. IEE Int. Conf. Control '94*, Vol. 1, pp. 646-651.

[9] Henson, M.A. and Seborg, D.E., 1994, Time Delay Compensation for Nonlinear Processes, *Ind. Eng. Chem. Res.*, Vol. 33, pp. 1493-1500.

Chapter 6

Studies in artificial neural network based control

K. J. Hunt and D. Sbarbaro

6.1 Introduction

Artificial neural networks can be used as a representation framework for modelling nonlinear dynamical systems. It is also possible to incorporate these nonlinear models within nonlinear feedback control structures. Several possibilities for modelling and control of nonlinear dynamical systems are studied in this paper. We present case studies illustrating the application of these techniques. A more detailed coverage of the material in this paper may be found in the reviews [24, 56].

6.2 Representation and identification

In this Section we review the possibilities for using neural networks directly in nonlinear control strategies. In this context neural networks are viewed as a process modelling formalism, or even a knowledge representation framework; our knowledge about the plant dynamics and mapping characteristics is implicitly stored within the network.

The ability of networks to approximate nonlinear mappings is thus of central importance in this task. For this reason we first review a body of theoretical work which has characterised, from an approximation theory viewpoint, the possibilities of nonlinear functional approximation using neural networks.

Second, we discuss learning structures for training networks to represent forward and inverse dynamics of nonlinear systems. Finally, a number of control structures in which such models play a central role are reviewed. These established control structures provide a basis for nonlinear control using neural networks.

In the same way that transfer functions provide a generic representation for linear black-box models, ANNs potentially provide a generic representation for nonlinear black-box models.

6.2.1 Networks, approximation and modelling

The nonlinear functional mapping properties of neural networks are central to their use in control. Training a neural network using input-output data from a nonlinear plant can be considered as a nonlinear functional approximation problem. Approximation theory is a classical field of mathematics; from the famous Weierstrass Theorem [3, 46] it is known that polynomials, and many other approximation schemes, can approximate arbitrarily well a continuous function. Recently, considerable effort has gone into the application of a similar mathematical machinery in the investigation of the approximation capabilities of networks.

A number of results have been published showing that a feedforward network of the multi-layer perceptron type can approximate arbitrarily well a continuous function [7, 8, 13, 22, 4]. To be specific, these papers prove that a continuous function can be arbitrarily well approximated by a feedforward network with only a single internal hidden layer, where each unit in the hidden layer has a continuous sigmoidal nonlinearity. The use of nonlinearities other than sigmoids is also discussed.

These results provide no special motivation for the use of networks in preference to, say, polynomial methods for approximation; both approaches share the 'Weierstrass Property'. Such comparative judgements must be made on the basis of issues such as parsimony. Namely, do networks or polynomials require fewer parameters? For network approximators, key questions are how many layers of hidden units should be used, and how many units are required in each layer? Of course, the implementation characteristics of approximation schemes also provide a further basis for comparison (for example, the parallel, distributed, nature of networks is important). For the moment, however, we concentrate purely on approximation properties.

Although the results referred to above at first sight appear attractive they do not provide much insight into these practical questions. In fact, the degree of arbitrariness of the approximation achieved using a sigmoidal network with one hidden layer is reflected in a corresponding arbitrariness in the number of units required in the hidden layer; the results were achieved by placing no restriction on the number of units used. Cybenko himself says [8] 'we suspect quite strongly that the overwhelming majority of approximation problems will require astronomical numbers of terms'.

What is needed now is an indication of the numbers of layers/units required to achieve a specific degree of accuracy for the function being approximated. Some work along these lines is given in Chester [6]. That paper gives theoretical support to the empirical observation that networks with two hidden layers appear to provide higher accuracy and better generalisation than a single hidden layer network, and at a lower cost (i.e. fewer total processing units). Guidelines derived from a mix of theoretical and heuristic considerations are given in the introductory paper by Lippmann [31].

From the theoretical point of view the work of Kolmogorov [28] (see also Reference [34]) did appear to throw some light on the problem of exact approximation [43, 20, 31]. Kolmogorov's theorem (a negative resolution of Hilbert's thirteenth problem) states that any continuous function of N variables can be computed using only linear summations and nonlinear but continuously increasing functions of only one variable. In the network context the theorem can be interpreted as explicitly stating that to approximate any continuous function of N variables requires a network having $N(2N+1)$ units in a first hidden layer and $(2N+1)$ units in a second hidden layer. However, it has recently been pointed out [22, 16] that the practical value of this result is tenuous for a number of reasons:

1. Kolmogorov's theorem requires a *different* nonlinear processing functions for each unit in the network.
2. The functions in the first hidden layer are required to be highly non-smooth. In practice this would lead to problems with generalisation and noise robustness.
3. The functions in the second hidden layer depend upon the function being approximated.

It is clear that these conditions are violated in the practical situations of interest here.

From the foregoing discussion it is clear that the property of approximating functions arbitrarily well is not sufficient for characterising good approximation schemes (since many schemes have this property), nor does this property help in justifying the use of one particular approximation scheme in preference to another. This observation has been deeply considered by Girosi and Poggio [17] (see also the expository paper Reference [44]). These authors propose that the key property is not that of arbitrary approximation, but the property of *best approximation*. An approximation scheme is said to have this property if in the set of approximating functions there is one which has the minimum distance from the given function (a precise mathematical formulation is given in the paper). The first main result of their paper [17] is that multi-layer perceptron networks *do not* have the best approximation property. Secondly, they prove that radial basis function networks (for example, Gaussian networks) *do have* the best approximation property. Thus, although one must bear in mind the precise mathematical formulation of 'best', there is theoretical support for favouring RBF networks. Moreover, these networks may always be structured with only a single hidden layer and trained using linear optimisation techniques with a guaranteed global solution [2].

As with sigmoidal feedforward networks, however, there remain open questions regarding the network complexity required (i.e. the number of units). For RBF networks this question is directly related to the size of the training data. A related issue is that of choosing the centres of the basis functions (for results on automatically selecting the centres see Moody and Darken [38] and Chen *et al* [5]). Girosi and Poggio [17] and Broomhead and Lowe [2] discuss methods for achieving *almost best approximation* using restricted complexity RBF networks. This is discussed further in Sbarbaro and Gawthrop [47].

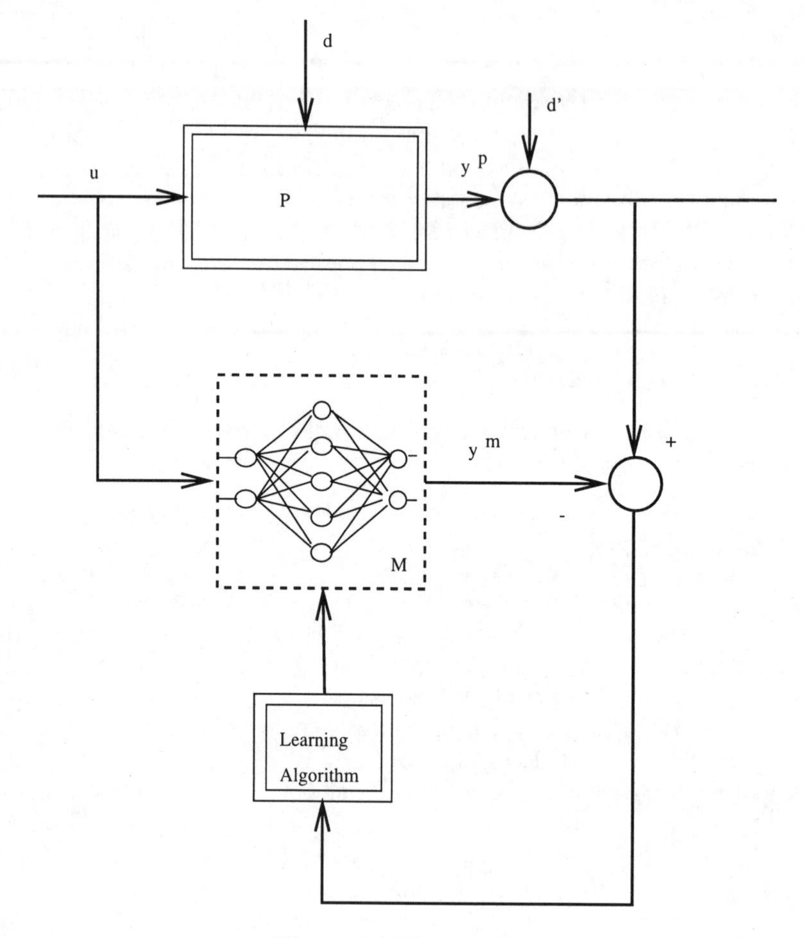

Figure 6.1 *Identification*

Thus, although RBF networks have the best approximation property, it is necessary to consider the cost of utilising this approach. For high dimensional input spaces the number of nodes needed grows with the dimension of the input space [19, 50]. This is in contrast to sigmoids which slice up the space into feature regions and hence are more economical with the number of nodes.

In summary, a large body of theoretical results relating to approximation using neural networks exists. These results provide theoretically important possibility theorems and deep insight into the ultimate performance of networks. They are not constructive results which define the type and structure of a suitable network for a given problem. Restraint must therefore be exercised when citing these results in support of the practical application of neural networks.

6.2.2 *Identification*

Although a number of key theoretical problems remain, the results discussed above do demonstrate that neural networks have great promise in the modelling of nonlinear systems. Without reference to any particular network structure we now discuss architectures for training networks to represent nonlinear dynamical systems and their inverses.

An important question in system identification is that of system identifiability (see References [33, 32, 48]), i.e. given a particular model structure, can the system under study be adequately represented within that structure? In the absence of such concrete theoretical results for neural networks we proceed under the assumption that all systems we are likely to study belong to the class of systems that the chosen network is able to represent.

6.2.2.1 *Forward modelling*

The procedure of training a neural network to represent the forward dynamics of a system will be referred to as *forward modelling*. A structure for achieving this is shown schematically in Figure 6.1. The neural network model is placed in parallel with the system and the error between the system and network outputs (the prediction error) is used as the network training signal. As pointed out by Jordan and Rumelhart [26] this learning structure is a classical supervised learning problem where the teacher (i.e. the system) provides target values (i.e. its outputs) directly in the output coordinate system of the learner (i.e. the network model). In the particular case of a multi-layer perceptron type network straightforward backpropagation of the prediction error through the network would provide a possible training algorithm.

An issue in the context of control is the *dynamic* nature of the systems under study. One possibility is to introduce dynamics into the network itself. This can be done either using recurrent networks [55] or by introducing dynamic behaviour into the neurons (see Reference [54]). A straightforward approach, and the one which for purposes of exposition will be followed here, is to augment the network input with signals corresponding to past inputs and outputs.

We assume that the system is governed by the following nonlinear discrete-time difference equation:

$$y^p(t+1) = f(y^p(t), \ldots, y^p(t-n+1); u(t), \ldots, u(t-m+1)) \tag{6.1}$$

Thus, the system output y^p at time $t+1$ depends (in the sense defined by the nonlinear map f) on the past n output values and on the past m values of the input u. We concentrate here on the dynamical part of the system response; the model does not explicitly represent plant disturbances (for a method of including disturbances see Chen *et al* [5], and Section 6.2.3).

The state space representation of the system described by eqn. 6.1, can be obtained using the following definition for the state variables

$$\begin{aligned} x_1^y(t) &= y(t) & x_1^u(t) &= u(t) \\ &\vdots & &\vdots \\ x_n^y(t) &= y(t-n+1) & x_m^u(t) &= u(t-m+1) \end{aligned} \tag{6.2}$$

Thus, the state space representation can be written as

$$\begin{bmatrix} x^y(t+1) \\ x^u(t+1) \end{bmatrix} = \begin{bmatrix} 0 & \dots & 0 & 0 & & & & \\ 1 & & & 0 & & & & \\ & \ddots & & \vdots & & & & \\ & & 1 & 0 & & & & \\ & & & & 0 & \dots & 0 & 0 \\ & & & & 1 & & & 0 \\ & & & & & \ddots & & \vdots \\ & & & & & & 1 & 0 \end{bmatrix} \begin{bmatrix} x^y(t) \\ x^u(t) \end{bmatrix} + \begin{bmatrix} f(x(t), u(t)) \\ 0 \\ \vdots \\ 0 \\ u(t) \\ 0 \\ \vdots \\ 0 \end{bmatrix} \tag{6.3}$$

Defining $x(t) = [x^y(t) \;\; x^u(t)]^T$, eqn. 6.3 has a general structure defined by

$$\begin{aligned} x(t+1) &= F[x(t), u(t)] \\ y(t) &= h(x(t)) \end{aligned} \tag{6.4}$$

Special cases of the model of eqn. 6.1 have been considered by Narendra and Parthasarathy [42] (see also Reference [40]). These authors consider particular cases where the system output is linear in either the past values of y^p or u.

An obvious approach for system modelling is to choose the input-output structure of the neural network to be the same as that of the system. Denoting the output of the network as y^m we then have

$$y^m(t+1) = \hat{f}(y^p(t), \dots, y^p(t-n+1); u(t), \dots, u(t-m+1)) \tag{6.5}$$

Here, $\hat{f}$ represents the nonlinear input-output map of the network (i.e. the approximation of f). Notice that the input to the network includes the past values of the *real* system output (the network has no feedback). This dependence on the system output is not included in the schematic of Figure 6.1 for simplicity. If we assume that after a suitable training period the network gives a good representation of the plant (i.e. $y^m \approx y^p$) then for subsequent post-training purposes the network output itself (and its delayed values) can be fed-back and used as part of the network input. In this way the network can be used independently of the plant. Such a network model is described by

$$y^m(t+1) = \hat{f}(y^m(t), \dots, y^m(t-n+1); u(t), \dots, u(t-m+1)) \tag{6.6}$$

The structure in eqn. 6.6 may also be used for training the network. This possibility has been discussed by Narendra [42, 40].

In the context of the identification of *linear* time invariant systems the two possibilities have been extensively considered by Narendra and Annaswamy [41]. The two structures have also been discussed in the signal processing literature (see Widrow and Stearns [53]). The structure of eqn. 6.5 (referred to as the series-parallel model by Narendra) is supported in the identification context by stability results. On the other hand, eqn. 6.6 (referred to by Narendra as the parallel model) may be preferred when dealing with noisy systems since it avoids problems of bias caused by noise on the real system output [53, 52].

6.2.2.2 *Inverse modelling*

Inverse models of dynamical systems play a crucial role in a range of control structures. This will become apparent in Section 6.5. However, obtaining inverse models raises several important issues which will be discussed.

Conceptually the simplest approach is *direct inverse modelling* as shown schematically in part (a) of Figure 6.2 (this structure has also been referred to as *generalised inverse learning* [45]). Here, a synthetic training signal is introduced to the system. The system output is then used as input to the network. The network output is compared with the training signal (the system input) and this error is used to train the network. This structure will clearly tend to force the network to represent the inverse of the plant. However, there are drawbacks to this approach:

- The learning procedure is not 'goal directed' [26]; the training signal must be chosen to sample over a wide range of system inputs, and the actual operational inputs may be hard to define *a priori*. The actual goal in the control context is to make the system *output* behave in a desired way, and thus the training signal in direct inverse modelling does not correspond to the explicit goal.

- Second, if the nonlinear system mapping is not one-one then an incorrect inverse can be obtained.

The first point above is strongly related with the general concept of *persistent excitation*; the importance of the inputs used to train learning systems is widely appreciated. In the adaptive control literature conditions for ensuring persistent excitation which will result in parameter convergence are well established (see, for example, Åström and Wittenmark [1] and the references therein). For neural networks, methods of characterising persistent excitation are highly desirable. A preliminary discussion on this question can be found in Narendra [40].

A second approach to inverse modelling which aims to overcome these problems is known as *specialised inverse learning* [45] (somewhat confusingly, Jordan and Rumelhart [26] refer to this structure as *forward modelling*). The specialised inverse learning structure is shown in part (b) of Figure 6.2. In this approach the network inverse model precedes the system and receives as input a training signal which spans the desired operational output space of the controlled system (i.e. it corresponds to the system reference or command signal). This learning structure also contains a

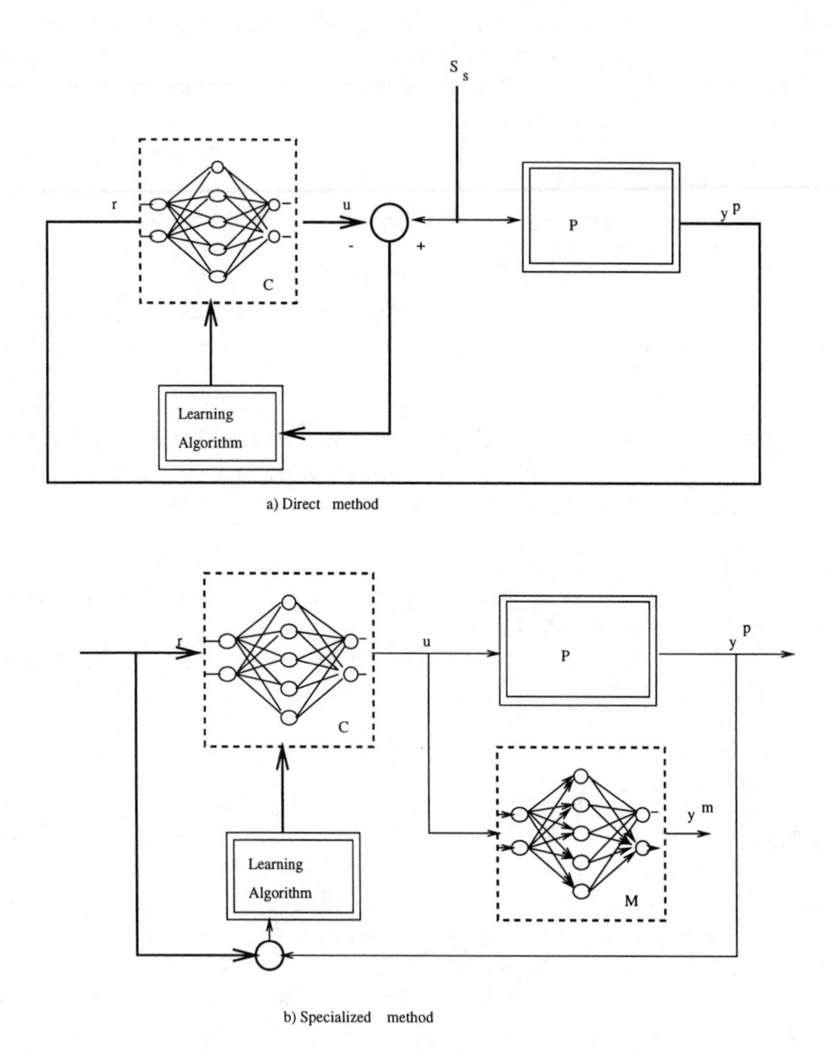

Figure 6.2 *Structures for inverse identification*

trained forward model of the system (for example, a network trained as described in Section 6.2.2.1) placed in parallel with the plant. The error signal for the training algorithm in this case is the difference between the training signal and the system output (it may also be the difference between the training signal and the forward model output in the case of noisy systems; this obviates the need for the real system in the training procedure which is important in situations where using the real system is not viable). Jordan and Rumelhart [26] show that using the real system output can produce an exact inverse even when the forward model is inexact; this is not the case when the forward model output is used. The error may then be propagated back through the forward model and then the inverse model; only the inverse network model weights are adjusted during this procedure. Thus, the procedure is effectively directed at learning an identity mapping across the inverse model and the forward model; the inverse model is learned as a side effect [26].

In comparison with direct inverse modelling the specialised inverse learning approach has the following features:

- The procedure is goal directed since it is based on the error between desired system outputs and actual outputs. In other words, the system receives inputs during training which correspond to the actual operational inputs it will subsequently receive.

- In cases in which the system forward mapping is not one-one a particular inverse will be found (Jordan and Rumelhart [26] discuss ways in which learning can be biased to find particular inverse models with desired properties).

We now consider the input-output structure of the network modelling the system inverse. From eqn. 6.1 the inverse function f^{-1} leading to the generation of $u(t)$ would require knowledge of the future value $y^p(t+1)$. To overcome this problem we replace this future value with the value $r(t+1)$ which we assume is available at time t. This is a reasonable assumption since r is typically related to the reference signal which is normally known one step ahead. Thus, the nonlinear input-output relation of the network modelling the plant inverse is

$$u(t) = \widehat{f^{-1}}(y^p(t), \ldots, y^p(t-n+1); r(t+1); u(t-1), \ldots, u(t-m+1)) \quad (6.7)$$

i.e. the inverse model network receives as inputs the current and past system outputs, the training (reference) signal, and the past values of the system input. In cases where it is desirable to train the inverse without the real system (as discussed above) the values of y^p in the above relation are simply replaced by the forward model outputs y^m.

The inverse model defines a recursive relation for the control signal $u(t)$, and the local stability of this model can be determined by the characteristic of its linear approximation.

6.2.3 *Modelling the disturbances*

It is important to include a reasonable representation of likely disturbances since the correct controller structure can then be deduced. As pointed out in Reference [15], it should not be necessary to force the controller to have a fixed strucure, i.e. integral action, but rather this structure should arise naturally from reasonable assumptions about the dynamics of the controlled system. Disturbances can be modelled by a set of basis functions representing the possible waveform components of the disturbance signal $w(t)$, i.e.

$$w(t) = c_1 f_1(t) + \ldots + c_m f_m(t) \tag{6.8}$$

where $f_i(t)$ are known functions, and c_i are unknown coefficients that may jump in value in a unknown manner [25]. In general a vector disturbance $w(t)$ can be expressed in a general form as

$$\begin{aligned} z(t+1) &= Dz(t) + \sigma(t) \\ w(t) &= Hz(t) \end{aligned} \tag{6.9}$$

H and D are known matrices, $z(t)$ is the state of the disturbance, and $\sigma(t)$ is a sequence of impulses with unknown arrival times and intensity. The basis functions f_i appear as principal models of the disturbance state model $z(t+1) = Dz(t)$.

6.2.3.1 *Output model*

This model considers disturbances only affecting the plant output

$$\begin{aligned} \begin{bmatrix} x(t+1) \\ z(t+1) \end{bmatrix} &= \begin{bmatrix} F(x(t), u(t)) \\ Dz(t) + \sigma(t) \end{bmatrix} \\ y(t) &= h(x(t)) + Hz(t) \end{aligned} \tag{6.10}$$

The output model can only represent disturbances which have linear dynamics and they are decoupled from the dynamic of the system.

6.2.3.2 *State model*

A more general representation of the effect of the disturbances with nonlinear dependencies is described by the following recursive equation

$$y(t+1) = f(y(t), \ldots, y(t-n+1); u(t), \ldots, u(t-m+1); w(t), \ldots, w(t-l+1)) \tag{6.11}$$

A simple approach to obtain a nonlinear input-output model, like eqn. 6.11, is to consider that the disturbances enter linearly at the first state variable in equation eqn. 6.4 [21], so that

$$\begin{aligned} \begin{bmatrix} x(t+1) \\ z(t+1) \end{bmatrix} &= \begin{bmatrix} F(x(t), u(t)) + G\,Hz(t) \\ Dz(t) + \sigma(t) \end{bmatrix} \\ y(t) &= h(x(t)) \end{aligned} \tag{6.12}$$

where $G = [1\,0\,, \ldots, 0]^T$.

6.3 Gaussian networks

The networks considered here are Gaussian feedforward networks with one hidden layer. We consider for the moment a network having many inputs and a single output (by duplication of this structure the approach can be generalised easily to multi-output systems). The number of input units, d, corresponds to the dimension of the network input vector, $u_{network} \in \mathbf{R}^d$.

The *linear* output unit is fully connected to the hidden units; the network output (the activation value of the output unit) is a weighted sum of the activation levels of the N hidden units:

$$y_{network} = \sum_{h=1}^{N} w_{yh} o_h \tag{6.13}$$

Here, w_{yh} is the connection weight between hidden unit h and the network output, $y_{network}$. o_h is the output (activation level) of hidden unit h.

In Gaussian units the activation level of a hidden unit depends only on the distance between the input vector $u_{network}$ and the centre of the Gaussian function of that unit. The centre of the function for hidden unit h is denoted by the vector $\mu_h \in \mathbf{R}^d$. For hidden unit h,

$$o_h = \exp\left(\frac{\|u_{network} - \mu_h\|^2}{-2\sigma_h^2}\right) \tag{6.14}$$

where σ_h is the width of the Gaussian function of unit h.

For nonlinear dynamic systems modelling purposes we assume that the plant is governed by the following nonlinear difference equation:

$$y^p(t+1) = f(y^p(t), \ldots, y^p(t-n+1); u(t), \ldots, u(t-m+1)) \tag{6.15}$$

where $y^p(.)$ is the plant output and $u(.)$ the input.

Following eqn. 6.5 we select the structure of the neural network to be trained to represent the plant as

$$y^m(t+1) = \hat{f}(y^p(t), \ldots, y^p(t-n+1); u(t), \ldots, u(t-m+1)) \tag{6.16}$$

Notice here that the input to the network includes the past values of the *real* plant output (the network has no feedback).

The nonlinear relation of the connectionist network modelling the plant inverse is given by

$$u(t) = \widehat{f^{-1}}(y^m(t), \ldots, y^m(t-n+1); r(t+1); u(t-1), \ldots, u(t-m+1)) \tag{6.17}$$

Here, we base the inverse model on the plant model (as indicated by the subscript m) rather than on the true plant (*cf* eqn. 6.7).

The inversion problem can be formulated as the problem of solving an operator equation; given the operator N represented by a neural network mapping $[y(t), \ldots, y(t-n), u^*(t), \ldots, u(t-m)]$ to $y(t+1) = y^*$;

$$Nu^* = y^*$$

Then, given $[y(t), \ldots, y(t-n), u(t-1), \ldots, u(t-m)]$ compute $u^*(t)$ to y^*. In Reference [11] two basic methods for solving the operator equation for u^* are described. One uses the contraction principle, the second involves the use of sensitivity functions and Newton's iteration method.

The advantages of the connectionist approach are, firstly, it provides differentiable models to be used as sensitivity functions and, secondly, association capabilities. In this way the information about u^* is stored in the network and it is not necessary to start the iterations from the same point every time. This network will finally represent the inverse of the model.

6.4 Learning algorithms

In Section 6.2.2 we outlined the two architectures used for plant modelling and plant inverse modelling. In this Section we present the learning laws used for training the networks.

6.4.1 *Modelling the plant: series-parallel model*

The plant is modelled using a network described by

$$y^m(t+1) = \sum_{i=1}^{N^m} c_i^m K_i^m \tag{6.18}$$

where

$$K_i^m = e^{-d_i^m(x^m(t), x_i^m, \Delta^m)} \tag{6.19}$$

Here, the m superscript indicates a variable related with the plant model. We choose the structure of the plant model to be the same as that of the plant, i.e. the model output is a nonlinear function of the present and past plant outputs, and the present and past plant inputs. The model input vector $x^m(t)$ is thus given as

$$x^m(t) = [y^p(t), \ldots, y^p(t-n+1), u(t), \ldots, u(t-m+1)]^T$$

We denote the centre of the Gaussian function of hidden unit i as

$$x_i^m = [y_{i,1}, \ldots, y_{i,n}, u_{i,1}, \ldots, u_{i,m}]^T$$

The parameters x_i^m and Δ^m are fixed to meet the interpolation conditions, i.e. the x_i^m are distributed uniformly over the input space and Δ^m is adjusted such that $\sum_{i=1}^{N^m} K_i = \text{const}$ over the input space. There are other possibilities for this [30].

The parameter vector c_i^m is adjusted to minimise the mean square error between the real plant and the model. That is,

$$c_i^m(t+1) = c_i^m(t) + \alpha K_i^m(y^p(t+1) - y^m(t+1)) \tag{6.20}$$

Here, α is a gain parameter. Using standard linear systems theory it can be shown that if the plant can be modelled as eqn. 6.18, the least mean square solution can be found by using eqn. 6.20 [47].

6.4.2 *Inverse model identification*

If the model of the plant is invertible then the inverse of the plant can be approximated in a similar way to the plant. This model is then used as the controller. For reasons described in Section 6.4.2.1 we choose to use the plant *model* inverse rather than the inverse of the real plant. We utilise a second network described by

$$u(t) = \sum_{i=1}^{N^C} c_i^C K_i^C \tag{6.21}$$

where,

$$K_i^C = e^{-d_i^C(x^C(t), x_i^C, \Delta^C)} \tag{6.22}$$

Here, the C superscript indicates a variable related with the controller. The inverse of the function f in eqn. 6.15 (required to obtain $u(t)$) depends upon the future plant output value $y^p(t+1)$. In order to obtain a realisable approximation we replace this value by the controller input value r. Finally, since we actually require to approximate the inverse of the plant model (as opposed to the plant itself), we define the controller network input vector $x^C(t)$ as

$$x^C(t) = [y^m(t), \ldots, y^m(t-n+1), r(t+1), u(t-1) \ldots, u(t-m+1)]^T$$

Here, the future value $r(t+1)$ is obtained at time t by suitable definition of the IMC filter F. The centre of the Gaussian function of hidden unit i is given by

$$x_i^C = [y_{i,1}, \ldots, y_{i,n}, r_i, u_{i,2} \ldots, u_{i,m}]^T$$

6.4.2.1 *Non-iterative methods*

The architecture used to adjust c_i^C is similar to the specialised learning architecture presented by Psaltis *et al.* [45] (the difference being that here we use the plant model, rather than the plant itself). The parameters in c_i^C are adjusted to minimise the mean square error between the output of the model and the input of the controller. This leads to the following learning algorithm:

$$c_i^C(t+1) = c_i^C(t) + \alpha K_i^C (r(t+1) - y^m(t+1)) \frac{\partial y^m(t+1)}{\partial u(t)} \tag{6.23}$$

Here, if the real plant were used in the learning procedure (as in Reference [45]) then $\frac{\partial y(t+1)}{\partial u(t)}$ would require to be estimated. This can be done using first order differences [45] changing each input to the plant slightly at the operating point and measuring the change at the output. By using the plant model, however, the derivatives can be calculated explicitly. From eqn. 6.18 we obtain

$$\frac{\partial y^m(t+1)}{\partial u(t)} = -2\Delta^m \sum_{i=1}^{N^m} c_i^m K_i^m (u(t) - u_{i,1}) \tag{6.24}$$

Proposition 1 *The learning algorithm defined by eqn. 6.23 converges to a global minimum of the index defined by*

$$J = \sum_j (r(j+1) - y^m(j+1))^2 \tag{6.25}$$

if the system is monotonically increasing with respect to $u(t)$.

Proof:

See Hunt and Sbarbaro [23].

Another approach involves the use of a synthetic signal [52]. This leads to the so called general learning architecture [45] as shown in Figure 6.2 (direct method). In this case the adaptation law for the weights does not depend on the derivatives of the plant:

$$c_i^C(t+1) = c_i^C(t) + \alpha K_i^C (S_s - u(t)) \tag{6.26}$$

Here, S_s is the synthetic signal.

Proposition 2 *If the system is invertible the algorithm defined by eqn. 6.26 converges to the best approximation of the inverse in the least square sense.*

Proof:

If the system is invertible then there exists an injective mapping which represents the inverse. Thus, from linear systems theory the algorithm defined by eqn. 6.26 converges to the least squares error [47].

As pointed out in Psaltis *et al.* [45] the specialised method allows the training of the inverse network in a region in the expected operational range of the plant. On the other hand, the generalised training procedure produces an inverse over the whole operating space. Psaltis *et al.* suggest a hybrid training procedure where the specialised and generalised architectures are combined.

6.4.2.2 *Iterative methods*

Iterative methods make use of a plant model to calculate the inverse. In this case a recursive method is used to find the inverse of the model in each operating point. This method is useful in singular systems which satisfy the invertibility conditions outlined earlier only locally and not for the whole operating space. This approach can also be used when it is necessary to have small networks due to memory or processing limitations. In this case the restricted accuracy of the trained network can be enhanced by using the network to provide stored initial values for the iterative method, establishing a compromise between speed of convergence and storing capacities.

At time t, the objective is to find an input u which will produce a model output $y^m(t+1)$ equal to $r(t+1)$. It is possible to use the method of successive substitution:

$$u^{n+1}(t) = u^n(t) + \gamma(r(t+1) - y^m(t+1))$$

where γ is a weight to be chosen.

According to the small gain theorem [10], the inverse operator is input-output stable if the product of the operator gains in the loop is less than 1:

$$\|I\| \|I - \gamma f\| \leq 1$$

The initial value $u^0(t)$ can be stored in a connectionist network.

6.5 Control structures

Models of dynamical systems and their inverses have immediate utility for control. In the control literature a number of well established and deeply analysed structures for the control of nonlinear systems exist; we focus on those structures having a direct reliance on system forward and inverse models. We assume that such models are available in the form of neural networks which have been trained using the techniques outlined above.

In the literature on neural network architectures for control a large number of control structures have been proposed and used; it is beyond the scope of this work to provide a full survey of all architectures used. In the sequel we give particular emphasis to those structures which, from the mainstream control theory viewpoint, are well established and whose properties have been deeply analysed. First, we briefly discuss two direct approaches to control: supervised control and direct inverse control.

6.5.1 *Supervised control*

There are many control situations where a human provides the feedback control actions for a particular task and where it has proven difficult to design an automatic controller using standard control techniques (e.g. it may be impossible to obtain an analytical model of the controlled system). In some situations it may be desirable to design an automatic controller which mimics the action of of the human (this has been called *supervised control* [51]).

An artificial neural network provides one possibility for this (as an alternative approach expert systems can be used to provide the knowledge representation and control formalisms). Training the network is similar in principle to learning a system forward model as described above. In this case, however, the network input corresponds to the sensory input information received by the human. The network target outputs used for training correspond to the human control input to the system. This approach has been used in the standard pole-cart control problem [18], among others.

6.5.2 *Direct inverse control*

Direct inverse control utilises an inverse system model. The inverse model is simply cascaded with the controlled system in order that the composed system results in an identity mapping between desired response (i.e. the network inputs) and the controlled system output. Thus, the network acts directly as the controller in such a configuration. Direct inverse control is common in robotics applications; the compilation book [37] provides a number of examples.

Clearly, this approach relies heavily on the fidelity of the inverse model used as the controller. For general purpose use serious questions arise regarding the robustness of direct inverse control. This lack of robustness can be attributed primarily to the absence of feedback. This problem can be overcome to some extent by using on-line learning: the parameters of the inverse model can be adjusted on-line.

6.5.3 *Model reference control*

Here, the desired performance of the closed-loop system is specified through a stable reference model M, which is defined by its input-output pair $\{r(t), y^r(t)\}$. The control system attempts to make the plant output $y^p(t)$ match the reference model output asymptotically, i.e.

$$\lim_{k \to \infty} \|y^r(t) - y^p(t)\| \leq \epsilon$$

for some specified constant $\epsilon \geq 0$. The model reference control structure for nonlinear systems utilising connectionist models is shown in Figure 6.3 [42]. In this structure the error defined above is used to train the network acting as the controller. Clearly, this approach is related to the training of inverse plant models as outlined above. In the case when the reference model is the identity mapping the two approaches coincide. In general, the training procedure will force the controller

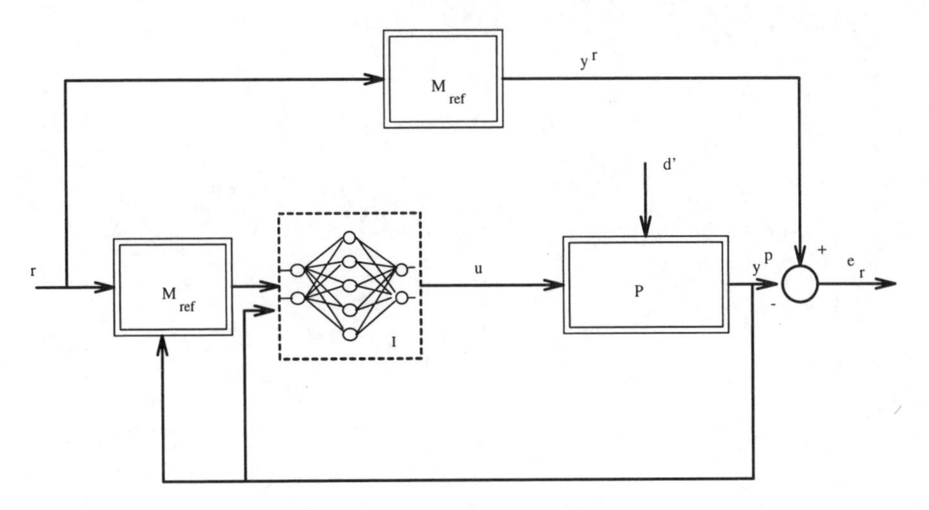

Figure 6.3 *Model reference structure*

to be a 'detuned' inverse, in a sense defined by the reference model. Previous and similar work in this area has been done considering linear in the control structures [29].

6.5.4 *Internal model control*

In internal model control (IMC) the role of system forward and inverse models is emphasised [14]. In this structure a system forward and inverse model are used directly as elements within the feedback loop. IMC has been thoroughly examined and shown to yield transparently to robustness and stability analysis [39]. Moreover, IMC extends readily to nonlinear systems control [11].

In internal model control a system model is placed in parallel with the real system. The difference between the system and model outputs is used for feedback purposes. This feedback signal is then processed by a controller subsystem in the forward path; the properties of IMC dictate that this part of the controller should be related to the system inverse (the nonlinear realisation of IMC is illustrated in Figure 6.4). The disturbances are modelled by an output model.

Given network models for the system forward and inverse dynamics the realisation of IMC using neural networks is straightforward [23]; the system model M and the controller C (the inverse model) are realised using the neural network models as shown in Figure 6.5. The subsystem F is usually a linear filter which can be designed to introduce desirable robustness and tracking response to the closed-loop system.

It should be noted that the *implementation* structure of IMC is limited to open-loop stable systems. However, the technique has been widely applied in process

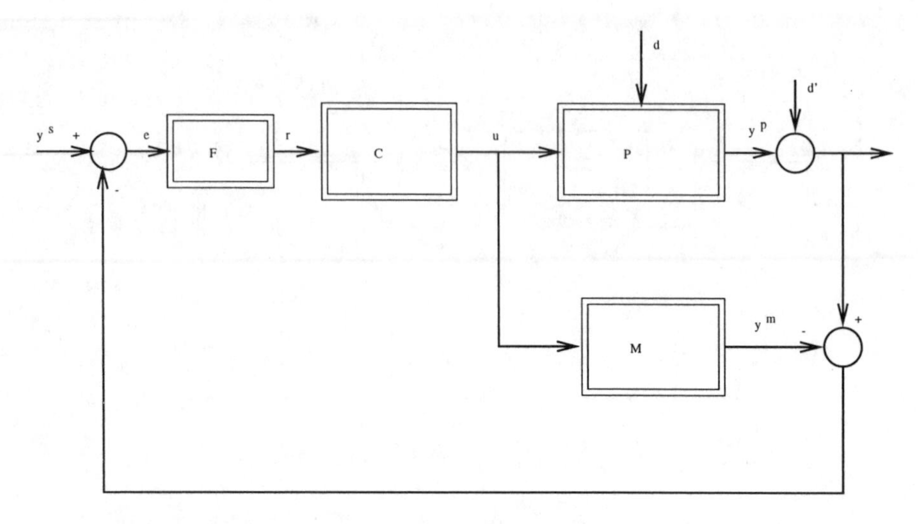

Figure 6.4 *Nonlinear IMC structure*

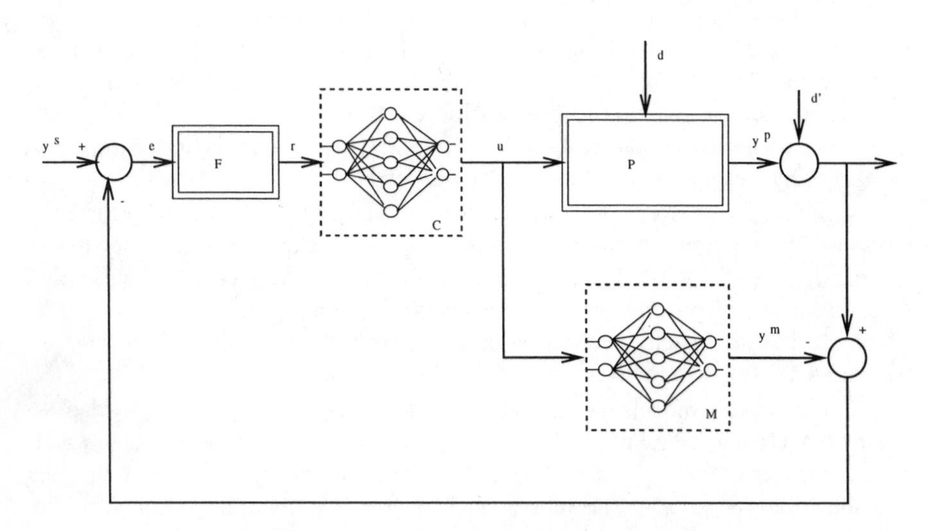

Figure 6.5 *Structure for internal model control*

control. From the control theoretic viewpoint there is strong support for the IMC approach. Examples of the use of neural networks for nonlinear IMC can be found in Hunt and Sbarbaro [23].

The IMC structure is now well known and has been shown to underlie a number of control design techniques of apparently different origin [14]. IMC has been shown to have a number of desirable properties; a deep analysis is given by Morari and Zafiriou [39]. Here, we briefly summarise these properties.

The nonlinear IMC structure is shown in Figure 6.4 (in this Section we follow Economou *et al.* [11]). Here, the nonlinear operators denoted by P, M and C represent the plant, the plant model, and the controller, respectively. The operator F denotes a filter, to be discussed in the sequel. The double lines used in the block diagram emphasise that the operators are nonlinear and that the usual block diagram manipulations do not hold.

The important characteristics of IMC are summarised with the following properties:

Assume that the plant and controller are input-output stable and that the model is a perfect representation of the plant. Then the closed-loop system is input-output stable. Assume that the inverse of the operator describing the plant model exists, that this inverse is used as the controller, and that the closed-loop system is input-output stable with this controller. Then the control will be perfect i.e. $y^p = y^s$. Assume that the inverse of the steady state model operator exists, that the steady state controller operator is equal to this, and that the closed-loop system is input-output stable with this controller. Then offset free control is attained for asymptotically constant inputs.

The IMC structure provides a direct method for the design of nonlinear feedback controllers. According to the above properties, if a good model of the plant is available, the closed-loop system will achieve exact set-point following despite unmeasured disturbances acting on the plant.

Thus far, we have not described the role of the filter F in the system. The discussion so far has considered only the idealised case of a perfect model, leading to perfect control. In practice, however, a perfect model can never be obtained. In addition, the infinite gain required by perfect control would lead to sensitivity problems under model uncertainty. The filter F is introduced to alleviate these problems. By suitable design, the filter can be selected to reduce the gain of the feedback system, thereby moving away from the perfect controller. This introduces robustness into the IMC structure. A full treatment of robustness and filter design for IMC is given in Morari and Zafiriou [39].

The significance of IMC, in the context of this work, is that the stability and robustness properties of the structure can be analysed and manipulated in a transparent manner, even for nonlinear systems. Thus, IMC provides a general framework for nonlinear systems control. Such generality is not apparent in alternative approaches to nonlinear control.

A second role of the filter is to project the signal e into the appropriate input space for the controller.

We propose a two step procedure for using neural networks directly within the IMC structure. The first step involves training a network to represent the plant response. This network is then used as the plant model operator M in the control structure of Figure 6.5. The architecture shown in Figure 6.1 provides the method for training a network to represent the plant. Here, the error signal used to adjust the network weights is the difference between the plant output and the network output. Thus, the network is forced towards copying the plant dynamics. Full details of the learning law used here are given in Section 6.4.1.

Following standard IMC practice (guided by Property P.2 above) we select the controller as the plant inverse model. The second step in our procedure is therefore to train a second network to represent the inverse of the plant. To do this we use one of the architectures shown in Figure 6.2. Here, for reasons explained in Section 6.4.2 (where full details of the learning law are given), we employ the plant model (obtained in the first learning step) in the inverse learning architecture rather than the plant itself. For inverse modelling the error signal used to adjust the network is defined as the difference between the (inverse modelling) network input and the plant model output. This tends to force the transfer function between these two signals to unity, i.e. the network being trained is forced to represent the inverse of the plant model. Having obtained the inverse model in this way this network is used as the controller block C in the control structure of Figure 6.5.

The final IMC architecture incorporating the trained networks is shown in Figure 6.5.

6.5.5 *Predictive control*

The receding horizon control approach can be summarised by the following steps:

1. predict the system output over the range of future times,
2. assume that the future desired outputs are known,
3. choose a set of future controls, $\hat{u}$, which minimise the future errors between the predicted future output and the future desired output,
4. use the first element of $\hat{u}$ as a current input and repeat the whole process at the next instant.

It has been shown that this technique can stabilise linear systems [9] and nonlinear systems as well [36].

The objective is to calculate the control such that the error over a future horizon is minimised, i.e. we consider a cost function of the following form:

$$J_{N_1,N_2,N_u} = \sum_{k=i+N_1}^{i+N_2} (y^r(k) - y^m(k))^2 + \sum_{k=i}^{i+N_u} \lambda_k(\Delta u(k))^2 \tag{6.27}$$

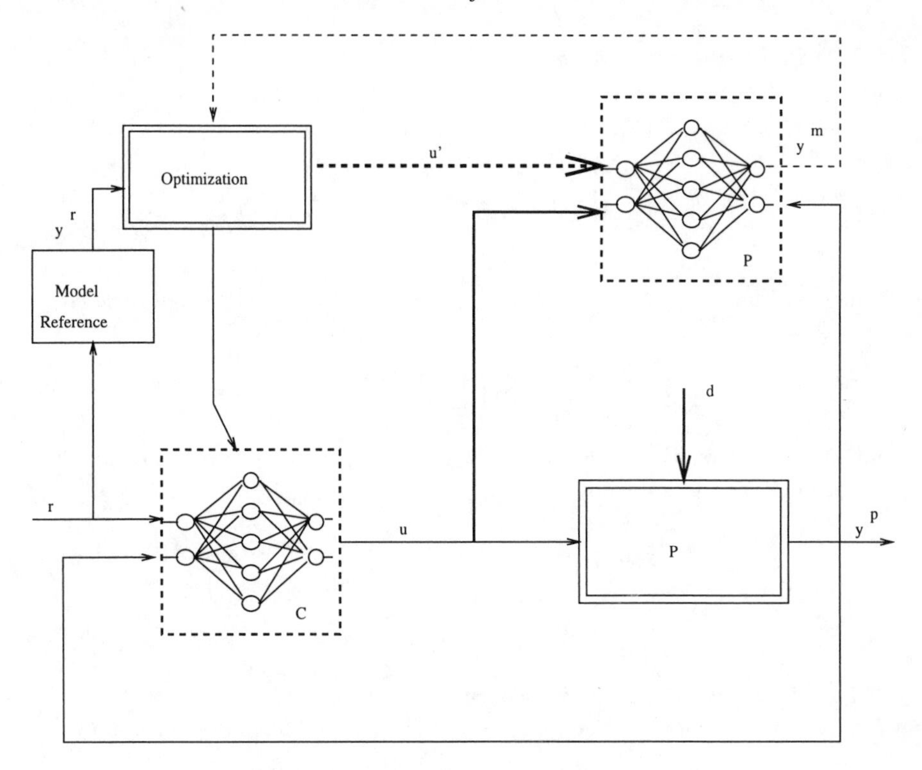

Figure 6.6 *Structure for predictive control*

where y^r represents the output of the reference model (i.e. the desired output) and y^m the output of the plant model. The first term of the cost function is a measure of the distance between the model prediction and the desired future trajectory. The second term penalises excessive movement of the manipulated variable.

The structure of the connectionist controller is shown in Figure 6.6. This is similar to a learning control structure previously proposed by Ersü and Tolle [12]. Those authors, however, consider only a one-step-ahead optimisation criterion. In our approach we propose the use of a trained connectionist model of the nonlinear plant M to predict the future plant outputs. This model is used to provide the prediction data to the optimisation algorithm. In principle, this optimised control value can then be applied to the plant. However, in a further step we also use this signal to train a controller network C, as shown in the figure. Training C in this way provides two important options: either to generate a useful starting value for the optimisation algorithm, or to act as a stand-alone feedback controller without recourse to the optimisation step. Of course, training C in the manner shown leads to the generation of an approximate plant inverse, depending on the cost function weights.

6.5.5.1 Statement of the problem

Consider a plant described as

$$\begin{aligned} y^p(t+1) &= f(y^p(t),\ldots,y^p(t-n+1); u(t),\ldots,u(t-m+1)) \quad t>i \\ [y^p(t),\ldots,y^p(t-n+1), u(t),\ldots,u(t-m+1)] &\in Z_t \subset R^{n+m} \quad t>i \\ y(i) &= a \end{aligned} \tag{6.28}$$

where f is a function defined as $f : R^{n+m} \to R$, such that the *moving horizon cost* at time i is

$$J_i = \sum_{k=i+N_1}^{i+T-1} \|e(k)\|^2 + \sum_{k=i}^{i+N_u} \|\Delta u(k)\|^2_{\lambda_i} \tag{6.29}$$

Here, $e(k)$ is defined as $e(k) = y^r(k) - y^m(k)$, $\Delta u(k) = u(k) - u(k-1)$, and the reference model is described by

$$\begin{aligned} y^r(t+1) &= g(y^r(t),\ldots,y^r(t-n+1); r(t),\ldots,r(t-m+1)) \\ [y^r(t),\ldots,y^r(t-n+1), r(t),\ldots,r(t-m+1)] &\in Z^r_t \subset R^{n+m} \\ t>i \quad y^r(i) &= b \end{aligned} \tag{6.30}$$

The problem is to determine a sequence $\{u(k)\}_{k=i}^{i+N_u}$ which minimises the cost given by eqn. 6.29, subject to eqn. 6.28, eqn. 6.30 and the alternative constraints

$$i \le k \le i+T-1 \quad , \quad y^p(i+T) = y^r(i+T) \tag{6.31}$$

This means that the reference model must have a settling time less than or equal to T, that is $y^r(i+T) = y^r_{ss}$. The demonstration of existence of the solution, stability of the closed loop system, and robustness are found in References [27, 36] and [35].

6.5.5.2 Minimising the functional

To minimise the functional (eqn. 6.29) a simple gradient algorithm was used, although more efficient, but at the same time more complex, algorithms can be applied. In this case the derivatives of the output against an input variable are estimated from the model and used to calculate the gradient in each iteration:

$$u|_{i+k}^{i+N_u} = u|_{i+k}^{i+N_u} + \eta(\nabla_{u|_{i+k}^{i+N_u}} I_y + \nabla_{u|_{i+k}^{i+N_u}} I_u) \tag{6.32}$$

where η fixes the step of the gradient, $I_y = \sum_{k=i+N_1}^{i+T-1}(y^r(k) - y^m(k))^2$, $I_u = \sum_{k=i}^{i+N_u} \lambda_k(\Delta u(k))^2$, and noting $N_2 = T-1$. The gradients are calculated as

$$\nabla_{u|_{i+k}^{i+N_u}} I_y = \begin{pmatrix} \frac{\partial y^m(i+N_1)}{\partial u(i)} & \cdots & \frac{\partial y^m(i+N_2)}{\partial u(i)} \\ \vdots & \ddots & \vdots \\ 0 & \cdots & \frac{\partial y^m(i+N_2)}{\partial u(i+N_u)} \end{pmatrix} \begin{pmatrix} y^r(i+N_1) - y^m(i+N_1) \\ \vdots \\ y^r(i+N_2) - y^m(i+N_2) \end{pmatrix} \tag{6.33}$$

$$\nabla_{u|_{i+k}^{i+N_u}} I_u = \begin{pmatrix} \lambda_1 & -\lambda_1 & \dots & 0 \\ 0 & \lambda_2 & \dots & 0 \\ \vdots & \vdots & \ddots & \vdots \\ 0 & 0 & \dots & \lambda_{N_u} \end{pmatrix} \begin{pmatrix} \Delta u(i) \\ \Delta u(i+1) \\ \vdots \\ \Delta u(i+N_u) \end{pmatrix} \quad (6.34)$$

Representing the model as a connectionist network $y^m(t+1)$ is given by

$$y^m(t+1) = \sum_{i=1}^{N^m} c_i^m K_i^m \quad (6.35)$$

Here, N^m represents the number of units in a connectionist network, c_i linear coefficients, and K_i^m is a Gaussian function defined as

$$e^{-d_i^m(x^m(t), x_i^m, \Delta^m)}$$

where

$$d_i^m(x^m(t), x_i^m, \Delta_i) = (x^m(t) - x_i^m)^T \Delta_i (x^m(t) - x_i^m) \quad (6.36)$$

$x^m(t) = [y^m(t), \dots, y^m(t-n+1), u(t), \dots, u(t-m+1)]^T$ is the input vector for the network, and x_i^m represents the centre of the unit i. Δ_i^m is a diagonal matrix with σ in its diagonal.

To calculate the partial derivatives the followings formulas were used for each control action k with $k = 1 \dots N_u$ and $j = N_1 \dots (T-1) - 1$:

$$\begin{aligned} D_{k,j} &= \tfrac{\partial y^m(i+j)}{\partial u(k)} \\ &= \begin{cases} \sum_{l=1}^{N} c_l^m K^m(x^m(i+j), x_l^m, \Delta^m)(u^m(k) - u_l^m)\Delta^m & k = j \\ D_{k,j-1} \sum_{l=1}^{N} c_l^m K^m(x^m(i+j), x_l^m, \Delta^m)(y^m(i+j) - y_l^m)\Delta^m & k < j \\ 0 & k > j \end{cases} \end{aligned} \quad (6.37)$$

At the end of the optimisation process it is necessary to verify that the condition $y^r(k+N_2) = y^m(k+N_2)$ is met.

For every starting point in the optimisation process, $x^m(t) = [y(t), \dots, y(t-n+1), u(t), \dots, u(t-m+1)]$, there is an associated optimal value of $u(t)$. Thus, $u(t)$ can be expressed as

$$u(t) = N_c(x^m(t), r(t)) \quad (6.38)$$

where N_c is an operator to be represented by a connectionist network. This network becomes the controller C in our overall control structure.

If independent random step disturbances affecting only the output are considered, then the augmented system is described by the following relationships

$$\begin{aligned} \begin{bmatrix} x(t+1) \\ z(t+1) \end{bmatrix} &= \begin{bmatrix} F(x(t), u(t)) \\ z(t) + \sigma(t) \end{bmatrix} \\ w(t) &= z(t) \\ y(t) &= h(x(t)) + w(t) \end{aligned} \quad (6.39)$$

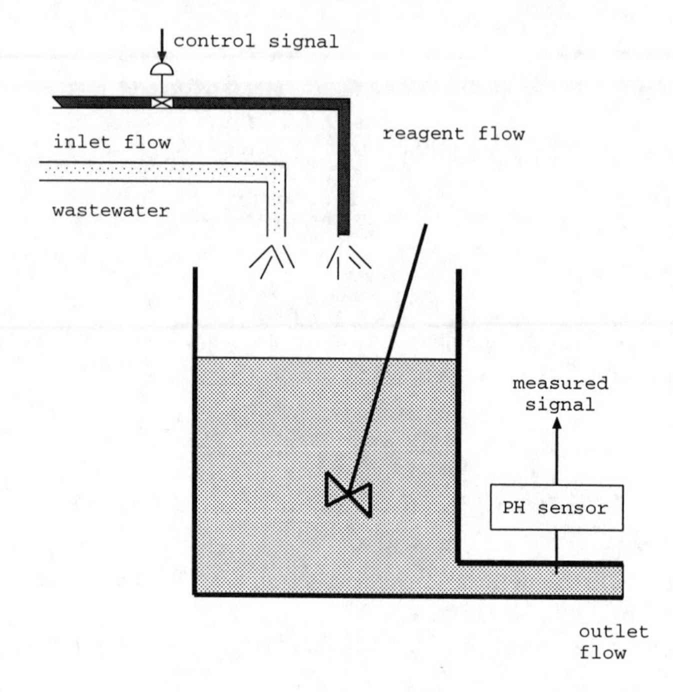

Figure 6.7 *Process diagram*

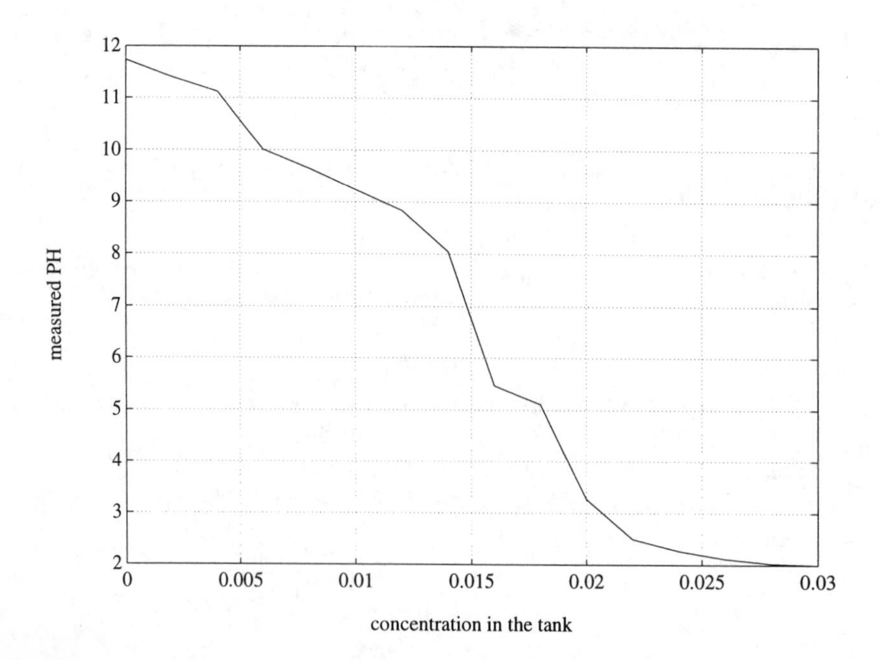

Figure 6.8 *Titration curve*

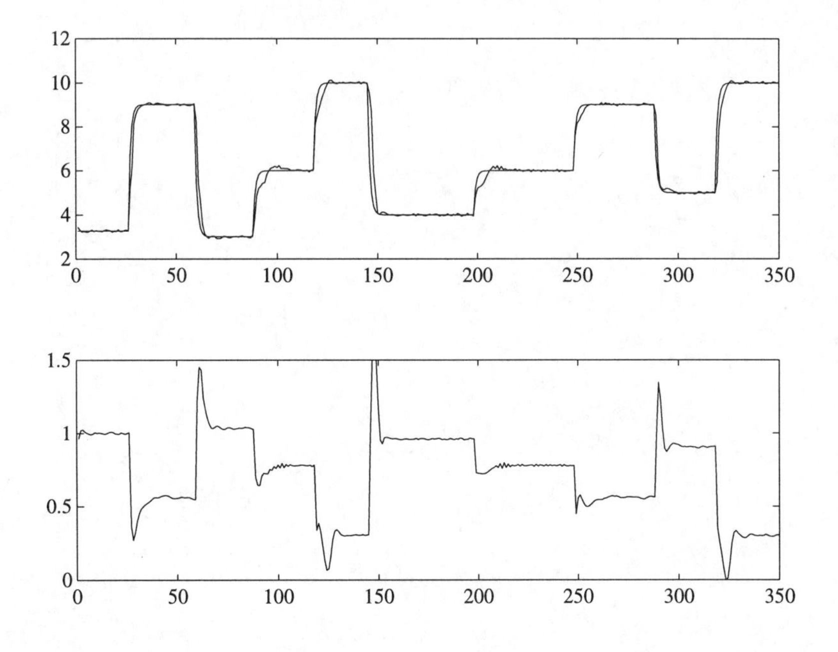

Figure 6.9 *IMC control*

A deadbeat observer can be designed to estimate the disturbance, such that

$$\hat{w}(t) = y^p(t) - y^m(t) \tag{6.40}$$

The general prediction model used to calculate the control signal can include an estimate of a step-like disturbance. Thus, the general predictive model is given by

$$y^m(t+1) = \sum_{i=1}^{N^m} c_i^m K_i^m + \hat{w}(t)$$

where

$$\begin{aligned} \hat{w}(t+1) &= \hat{w}(t) &&, t > i \\ &= (y^p(t) - y^m(t)) &&, t = i \end{aligned} \tag{6.41}$$

6.6 Example: pH control

pH control is an old problem involving significant nonlinearities and uncertainties. These features can be so severe that classical linear feedback control does not always achieve satisfactory performance. In its simplest form the process consists

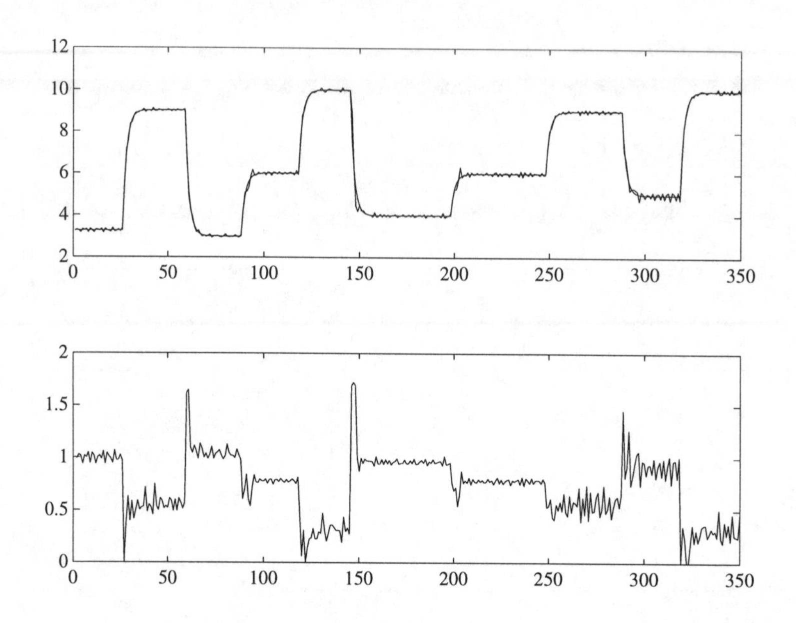

Figure 6.10 *Receding horizon control:* $\lambda = 0$

of a stirred tank in which waste water from a plant is neutralised using a reagent, usually a strong acid or a strong base (see Figure 6.7). The system is commonly modelled as the combination of a linear dynamic system followed by a nonlinear block representing the process titration curve. The nonlinear function representing the titration curve is shown in Figure 6.8.

In this case a neural network model was developed for the nonlinear system. The network was trained to learn the relation between the control variable and the process output. After presenting a set of training data consisting of 120 points (distributed over the operational range of the process variables) an absolute model mismatch error of 5% was obtained.

As discussed above, once the network has been trained to represent the nonlinear system it can be included in a nonlinear control loop. For this application we chose to utilise the internal model control and receding horizon control structures; these approaches directly incorporate a model of the process in the feedback loop. The neural network was therefore directly incorporated as part of the nonlinear controllers.

The set-point response of the system for the controlled pH and the corresponding control variable are shown in the figures: the results for IMC are shown in Figure 6.9, the results for receding horizon control with $\lambda = 0$ in Figure 6.10, and the results for receding horizon control with $\lambda = 1$ in Figure 6.11.

It can be seen that the nonlinear neural network controllers provide very close

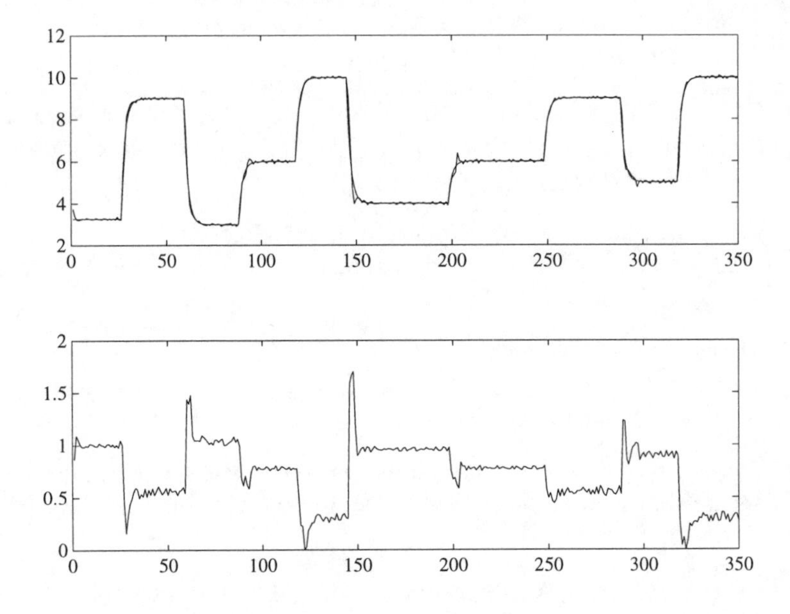

Figure 6.11 *Receding horizon control:* $\lambda = 1$

tracking performance over the range of pH variation; this is a considerable improvement over a linear controller on this loop.

6.7 Acknowledgments

The authors were formerly with the Control Group at the University of Glasgow, Scotland; the version of this chapter, which appeared in the first edition of this volume [49], was produced there. At that time Kenneth Hunt held a Royal Society of Edinburgh Personal Research Fellowship. Daniel Sbarbaro was supported by a Chilean government research scholarship.

6.8 References

[1] Åström, K. J. and Wittenmark, B., 1989, *Adaptive Control*. Addison-Wesley, Reading, Massachusetts

[2] Broomhead, D. S. and Lowe, D., 1988, Multivariable functional interpolation and adaptive networks. *Complex Systems* 2: 321–355

[3] Burkill, J. C. and Burkill, H., 1970, *A Second Course in Mathematical Analysis*. Cambridge University Press, Cambridge, England

[4] Carrol, S. M. and Dickinson, B. W., 1989, Construction of neural nets using the Radon transform. In *Proceedings of the International Joint Conference on Neural Networks IJCNN'89*

[5] Chen, S., Billings, S. A., Cowan, C. F. and Grant, P. M., 1990, Practical identification of NARMAX models using radial basis functions. *International Journal of Control* 52: 1327–1350

[6] Chester, D. L., 1990, Why two hidden layers are better than one? In *Proceedings of the International Joint Conference on Neural Networks IJCNN'90*, edited by Caudill, M., pp. A 265–A 268. Lawrence Erlbaum Associates

[7] Cybenko, G., 1988, Continuous valued neural networks with two hidden layers are sufficient. Technical report, Department of Computer Science, Tufts University

[8] Cybenko, G., 1989, Approximation by superposition of a sigmoidal function. *Mathematics of Control, Signals, and Systems* 2: 303–314

[9] Demircioglu, H. and Gawthrop, P. J., January 1991, Continuous-time generalised predictive control. *Automatica* 27, no. 1: 55–74

[10] Desoer, C. A. and Vidyasagar, M., 1975, *Feedback Systems : Input-Output Properties*. Academic Press, London

[11] Economou, C. G., Morari, M. and Palsson, B. O., 1986, Internal model control. 5. Extension to nonlinear systems. *Industrial & Engineering Chemistry Process Design and Development* 25: 403–411

[12] Ersü, E. and Tolle, H., 1984, A new concept for learning control inspired by brain theory. *Proceedings of the 9th World Congress of IFAC* 7: 245–250

[13] Funahashi, K., 1989, On the approximate realization of continuous mappings by neural networks. *Neural Networks* 2: 183–192

[14] Garcia, C. E. and Morari, M., 1982, Internal model control—1. A unifying review and some new results. *Industrial & Engineering Chemistry Process Design and Development* 21: 308–323

[15] Gawthrop, P. J., 1987, Hybrid self-tuning control. In *Encyclopaedia of Systems and Control*, edited by Singh, Pergamon

[16] Girosi, F. and Poggio, T., 1989, Representation properties of networks: Kolmogorov's theorem is irrelevant. *Neural Computation* 1: 465–469

[17] Girosi, F. and Poggio, T., 1990, Networks and the best approximation property. *Biological Cybernetics* 63: 169–176

[18] Grant, E. and Zhang, B., 1989, A neural net approach to supervised learning of pole balancing. In *IEEE International Symposium on Intelligent Control 1989*, pp. 123–129. IEEE Control Systems Society

[19] Hartman, E. and Keeler, J. D., 1991, Predicting the future: Advantages of semilocal units. *Neural Computation* 3, no. 4: 566–578

[20] Hecht-Nielsen, R., 1987, Kolmogorov's mapping neural network existence theorem. In *Proceedings of the International Joint Conference on Neural Networks*, vol. 3, pp. 11–14

[21] Hernandez, E. and Arkun, Y., 1993, Control of nonlinear systems using polynomial ARMA models. *AICHE Journal* 39: 446–460

[22] Hornik, K., Stinchcombe, M. and White, H., 1989, Multilayer feedforward networks are universal approximators. *Neural Networks* 2: 359–366

[23] Hunt, K. J. and Sbarbaro, D., 1991, Neural networks for non-linear Internal Model Control. *IEE Proceedings Part D* 138: 431–438

[24] Hunt, K. J., Sbarbaro, D., Żbikowski, R. and Gawthrop, P. J., 1992, Neural networks for control systems: A survey. *Automatica* 28, no. 6: 1083–1112

[25] Johnson, C., 1982, Discrete-time disturbance-accommodating control theory for digital control of dynamical systems. In *Control and Dynamic Systems. Advances in Theory and Applications*, edited by Leondes, C. T., vol. 18, pp. 224–315, Academic Press, New York

[26] Jordan, M. I. and Rumelhart, D. E., 1992, Forward models: Supervised learning with a distal teacher. *Cognitive Science* 16, no. 3: 307–354

[27] Keerthi, S. S. and Gilbert, E. G., 1986, Moving-horizon approximations for a general class of optimal nonlinear infinite-horizon discrete-time systems. In *Proceedings of the 20th Annual Conference Information Science and Systems, Princeton University*, pp. 301–306. Princeton University Press

[28] Kolmogorov, A. N., 1957, On the representation of continuous functions of many variables by superposition of continuous functions of one variable and addition. *Dokl. Akad. Nauk SSSR* 114: 953–956, (English translation: American Mathematical Society Translations, Vol. 28)

[29] Kosikov, V. S. and Kurdyukov, A. P., 1987, Design of a nonsearching self-adjusting system for nonlinear plant. *Avtomatika i Telemekhanika* 4: 58–65

[30] Lee, S. and Kil, R. M., 1991, A gaussian potential function network with hierarchically self-organizing learning. *Neural Networks* 4, no. 2: 207–224

[31] Lippmann, R. P., 1987, An introduction to computing with neural nets. *IEEE ASSP Magazine* pp. 4–22

[32] Ljung, L., 1987, *System Identification — Theory for the User*. Prentice-Hall, Englewood Cliffs, NJ

[33] Ljung, L. and Söderström, T., 1983, *Theory and Practice of Recursive Identification*. MIT Press, London

[34] Lorentz, G. G., 1976, *Mathematical Developments Arising from Hilbert's Problems*, vol. 2, pp. 419–430. American Mathematical Society

[35] Mayne, D. Q. and Michalska, H., 1990, An implementable receding horizon controller for stabilization of nonlinear systems. Report IC/EE/CON/90/1, Imperial College of Science, Technology and Medicine

[36] Mayne, D. Q. and Michalska, H., 1990, Receding horizon control of nonlinear systems. *IEEE Transactions on Automatic Control* 35: 814–824

[37] Miller, W. T., Sutton, R. S. and Werbos, P. J., 1990, *Neural Networks for Control*. MIT Press, Cambridge, Massachusetts

[38] Moody, J. and Darken, C., 1989, Fast learning in networks of locally-tuned processing units. *Neural Computation* 1: 281–294

[39] Morari, M. and Zafiriou, E., 1989, *Robust Process Control*. Prentice-Hall, Englewood Cliffs, NJ

[40] Narendra, K. S., 1990, *Neural Networks for Control*, chap. 5, pp. 115–142. MIT Press

[41] Narendra, K. S. and Annaswamy, A. M., 1989, *Stable Adaptive Systems*. Prentice-Hall, Englewood Cliffs, NJ

[42] Narendra, K. S. and Parthasarathy, K., 1990, Identification and control of dynamic systems using neural networks. *IEEE Transactions on Neural Networks* 1: 4–27

[43] Poggio, T., 1982, *Physical and Biological Processing of Images*, pp. 128–153. Springer-Verlag

[44] Poggio, T. and Girosi, F., 1990, Networks for approximation and learning. *Proceedings of the IEEE* 78: 1481–1497

[45] Psaltis, D., Sideris, A. and Yamamura, A. A., 1988, A multilayered neural network controller. *IEEE Control Systems Magazine* 8: 17–21

[46] Rudin, W., 1976, *Principles of Mathematical Analysis, 3rd Edition*. McGraw-Hill, Auckland

[47] Sbarbaro, D. G. and Gawthrop, P. J., 1991, Self-organization and adaptation in gaussian networks. In *9th IFAC/IFORS symposium on identification and system parameter estimation. Budapest, Hungary*, pp. 454–459

[48] Söderström, T. and Stoica, P., 1989, *System Identification*. Prentice-Hall, Hemel Hempstead

[49] Warwick, K., Irwin, G. R. and Hunt, K. J. (editors), 1992, *Neural Networks for Control and Systems: Principles and Applications*. Control Engineering Series, Peter Peregrinus

[50] Weigand, A. S., Huberman, B. A. and Rumelhart, D. E., 1990, Predicting the future: A connectionist approach. *International Journal of Neural Systems* 3: 193

[51] Werbos, P. J., 1990, Backpropagation through time: What it does and how to do it? *Proceedings of IEEE* 78: 1550–1560

[52] Widrow, B., 1986, Adaptive inverse control. In *Preprints of the 2nd IFAC workshop on Adaptive Systems in Control and Signal Processing, Lund, Sweden*, pp. 1–5, Oxford, Pergamon Press

[53] Widrow, B. and Stearns, S. D., 1985, *Adaptive Signal Processing*. Prentice-Hall, Englewood Cliffs

[54] Willis, M. J., Massimo, C. D., Montague, G. A., Tham, M. T. and Morris, A. J., 1991, On artificial neural networks in process engineering. *IEE Proceedings Part D* 138: 256–266

[55] Żbikowski, R., July 1994, *Recurrent Neural Networks: Some Control Problems*. Ph.D. Thesis, Department of Mechanical Engineering, Glasgow University, Glasgow, Scotland, (Available from FTP server `ftp.mech.gla.ac.uk` as `/rafal/zbikowski_phd.ps`)

[56] Żbikowski, R., Hunt, K. J., Dzieliński, A., Murray-Smith, R. and Gawthrop, P. J., 1994, A review of advances in neural adaptive control systems. Technical Report of the ESPRIT NACT Project TP-1, Glasgow University and Daimler-Benz Research, (Available from FTP server `ftp.mech.gla.ac.uk` as `/nact/nact_tp1.ps`)

Chapter 7

Applications of dynamic artificial neural networks in state estimation and nonlinear process control

P. Turner, J. Morris and G. Montague

7.1 Abstract

Since the mid-1980s interest in a major area of strategic research has emerged: that of artificial neural networks. The much wider availability and power of computing systems, together with new theoretical research studies, is resulting in expanding areas of application. It is particularly significant in these circumstances that the extremely important aspects involved in developing complex industrial process applications is emphasised, especially where safety-critical perspectives are prominent. Additionally, in complex processes it is important to understand that conventional feedforward networks imply that the manipulated process inputs directly affect the plant outputs. This is not true in complex processes where some manipulated inputs affect internal states that go on to affect the system outputs. A further complication in complex industrial processes is the display of direction-dependent dynamics.

The studies described here focus on the application of a dynamic network topology that is capable of representing the directional dynamics of a complex chemical process. An application of neural networks to the on-line estimation of polymer properties in an industrial continuous polymerisation reactor is presented. This approach leads to the implementation of an inferential control scheme that significantly improves process performance to market-driven grade changes. The generic properties of the approach are then demonstrated by transferring the technology to a totally different plant. The application is to the nonlinear predictive control of the pressure of a highly nonlinear, high purity distillation tower.

7.2 Introduction

System representation, modelling and identification are fundamental to process engineering where it is often required to approximate a real system with an appropriate model given a set of input-output data. The widespread interest in input-output mapping has been given impetus by the increasing attention being focused on pattern processing techniques and their potential application to a vast range of complex and demanding real-world problems. These include, for example, image and speech processing, inexact knowledge processing, natural language processing, event classification and feature detection, sensor data processing, control, forecasting and optimisation. The model structure needs to have sufficient representation ability to enable the underlying system characteristics to be approximated with an acceptable accuracy and in many cases the model also needs to retain simplicity. For linear, time-invariant systems these problems have been well studied and the literature abounds with many useful methods, algorithms and application studies. Widely used structures are the autoregressive moving average (ARMA), the autoregressive with exogeneous variables (ARX) and the autoregressive moving average with exogeneous variables (ARMAX) representations.

In practice, most systems encountered in industry are nonlinear to some extent and in many applications nonlinear models are required to provide acceptable representations. Nonlinear system identification is, however, much more complex and difficult, although the nonlinear autoregressive moving average with exogeneous variables (NARMAX) or nonlinear autoregressive with exogeneous variables (NARX) descriptions have been shown to provide useful unified representations for a wide class of nonlinear systems. Efficient parameter identification procedures are particularly important with nonlinear systems so that parsimonious model structures can be selected. The work of Billings and colleagues [1–5] provides seminal work in nonlinear system studies.

The problem of identifying, or estimating, a model structure and its associated parameters can be related to the problem of learning a mapping between a known input and output space [6,7]. A classical framework for this problem can be found in approximation theory. Almost all approximation, or identification, schemes can be mapped into, i.e. expressed as, a network. For example, the well known ARMAX model can be represented as a single layer network with inputs comprising of lagged system input-output data and prediction errors. In this context a network can be viewed as a function represented by the conjunction of a number of basic functions.

7.3 Dynamic neural networks

References [8,9,10,11] provide a good background to neural networks and their applications in system modelling and control. The basic feedforward network performs a nonlinear transformation of input data in order to approximate output data. This results in a static network structure. In some situations such a steady-state model may be appropriate. However, a significant number of systems require a description of the underlying dynamics. The most straightforward way to extend this essentially steady-state mapping to the dynamic domain is to adopt an approach similar to that taken in linear ARMA (auto-regressive moving average) modelling. Here, a time series of past process inputs (u) and outputs (y) are used to predict the present process outputs. Important process characteristics such as system delays can be accommodated for by utilising only those process inputs beyond the dead time. Additionally, any uncertainty in delay can be taken account of by using an extended time history of process inputs. Inevitably a significant number of network inputs result, especially for systems described by sets of 'stiff equations'.

The use of network models as predictors may be problematical if care is not taken in specifying the network topology correctly and then training the network properly. It is important to understand that conventional feedforward networks imply that the manipulated process inputs directly affect the plant outputs. This is not true in complex processes where some manipulated inputs affect internal states that go on to affect the system outputs. In addition, if the models are used to predict more than one-step-ahead in time, and have only been trained for such a task, then the ARMA time series approach alone is not appropriate. The one-step-ahead ARMA network model does not capture the process dynamics. Essentially, the autoregressive nature of this form of network results in the need to predict $y(t+n)$ from the estimates of $y(t+n-1)$. Errors in the estimate of y thus accumulate as the prediction horizon increases. The problem of the ARMA network approach can, however, be overcome by minimising the network prediction error over time. That is the network training minimises the estimate of $y(t)$ and all other output predictions up to a specified prediction horizon [12]. This approach is called 'backpropagation-in-time'. Recent studies have shown that the incorporation of dynamics into the network in this manner is highly beneficial in many real process application studies. Although perhaps the most concise network representation of a dynamic system is obtained by using network inputs comprised of past input and output data, the requirement to model processes over a wide dynamic range (models containing both very large and very small time constants) can result in large network structures leading to network training and convergence problems.

An alternative philosophy is to modify the neuron processing and interconnections in order to incorporate dynamics inherently within the network. Since a dynamic network model is not autoregressive the prediction problem does not arise. In addition to the sigmoidal processing of nodes, the neurons (or transmission between neurons) can be given dynamic characteristics, for example [13]:

$$\frac{N(s)}{D(s)} \exp(-st_d)$$

where $N(s)$ and $D(s)$ are polynomial functions in the Laplace operator s, and the time delay is represented by a Pade approximation. In this way the model of the function of the individual neuron is extended to account for dynamics and dead-time. Other workers have modelled the temporal properties using finite impulse response filters.

Finally, a relatively recent development has been the increasing attention being paid to recurrent networks. Although these can take different forms they are all capable of capturing temporal behaviour and provide multi-step-ahead predictions. Examples are the backpropagation-in-time networks [14], the Elman network [15] where the hidden neuron outputs at the previous time step are fed back to its inputs through time delay units; locally recurrent network representations [16,17] where each neuron has one or more delayed feedback loops around itself; and globally recurrent networks where the network output(s) are fed back to the inputs through time delay units. One difficulty with recurrent networks, however, is to be able to determine the best network architecture (i.e. number of hidden units). A new method has recently been developed which provides a systematic way of training recurrent networks and uses an orthogonal decomposition approach to sequentially minimise the prediction error [18,19]. Scott and Ray [20,21] show good performance of the Elman network in nonlinear systems modelling. A dynamic, filter based, network [22,23], which includes delayed filters in the neuron interconnections has been shown to provide very good multi-step-ahead predictions.

7.4 Assessment of model validity

A number of model validity tests for nonlinear model identification procedures have been developed. For example, the statistical chi-squared test [24], the final prediction error criterion [25], the information theoretic criterion [25] and the predicted squared error (PSE) criterion [26]. The PSE criteria, although originally developed for linear systems, can be applied to feedforward nets providing that they can be approximated by a linear model.

$$\text{final prediction error (FPE)} = (E/2N)\,(N + N_w)/(N - N_w)$$

$$\text{information theoretic criterion (AIC)} = \ln(E/2N) + 2\,N_w/N$$

$$\text{predicted squared error (PSE)} = E/2N + 2\times(s^2)N_w/N$$

where s^2 is the prior estimate of the true error variance and is independent of the model being considered.

It can be seen that these tests make use of functions that strike a balance between the accuracy of model fit (average squared error over N data points, $E\sqrt{2N}$) and the number of adjustable parameters or weights used (N_w). Because of this, minimisation of these test functions results in networks (models) that are neither under- nor over-dimensionalised. A procedure involving 'train-test-validate' experiments is used with different network dimensions to minimise a selected validation function.

NARMAX modelling approaches have used a number of model validation tests based on the autocorrelation of the residuals [1,2]. These correlation tests should also help with neural network model validation:

$$\Phi_{ee(t)} = E[e(t-\tau)e(t)] = \delta(t) \quad \forall t$$

$$\Phi_{ue(t)} = E[u(t-\tau)e(t)] = 0 \quad \forall t$$

$$\Phi_{u2'e(t)} = E[[u^2(t-\tau)-u^{-2}(t)]e(t)] = 0 \quad \forall t$$

$$\Phi_{u2'e2(t)} = E[[u^2(t-\tau)-u^{-2}(t)]e^2(t)] = 0 \quad \forall t$$

$$\Phi_{e(eu)(t)} = E[e(t)[e(t-1-\tau)u(t-1-\tau)]] = 0 \quad t>0$$

Normalisation, to provide all the tests with a range of ±1, and approximate 95% confidence bands at $1.96\sqrt{N}$ allows the tests to be independent of the excitation signal amplitudes and hence easier to interpret. Validation of the identified network model can also be checked by observing its predictive qualities for both one-step-ahead and multi-step-ahead predictions.

7.5 Data collection and pre-processing

Although there have been a number of successful modelling and control studies using neural networks, some approaches still appear to ignore good parameter identification practice and preclude any derivation of a model structure. Data collection involves a number of important tasks. First the modelling objectives need to be defined, together with the independent and dependent variables. The size of the excitation disturbances needs to be fixed by aiming for a signal-to-noise ratio of around six-to-one. Then, having obtained the plant settling time, the data length can be fixed to around ten times the settling time of the number of independent variables. Each independent variable will need to be changed around twenty times with an input disturbance sequence that at least satisfies all the well-known conditions associated with multi-input, multi-output parameter identification — that is, all the system modes must be excited with signals of appropriate power. Persistent excitation alone is not sufficient. Random excitation is required with a magnitude covering the whole dynamic range being considered and density that is sufficient to encapsulate the whole input domain of interest. The excitation required for good identification of an adequate model will be closely related to the properties of the system being modelled. This in turn is related to the distribution of the training data set(s). Care needs to be taken that operator intervention does not introduce correlations between independent variables and on-line controllers need to be working correctly. After data collection, periods of bad data, unusual events and outliers need to be removed. The data may need to be centrally filtered. All data should be pre-processed using standard statistical procedures: cross correlation tests on both independent and dependent variables to reveal time delays, independency of variables, plotting independent variables against each other to reveal strong underlying relationships, noise filtering, etc. Data interrogation using principal component analysis and projection to latent structures can also be very revealing. The importance of understanding the process being modelled, of understanding the information in the data, and screening of the collected data prior to modelling cannot be overstressed; it can be the key to success or failure. Data in the training set are also pre-processed to have zero mean and unit variance. This is necessary to prevent inputs with large average values in a certain dimension from overshadowing inputs in some other dimensions with small average values. Care must be taken, however, to ensure that unwanted artefacts are not introduced into the training data.

7.6 Neural networks in state estimation

A problem of present and increasing industrial interest where artificial neural networks may be of benefit is in providing a means by which to improve the quality of on-line information available to plant operators for plant control and optimisation. Major problems exist in the chemical and biochemical industries (paralleled with the food processing industries) concerned with the on-line estimation of parameters and variables that quantify process behaviour. The fundamental problem is that the key 'quality' variables cannot be measured at a rate which enables their effective regulation. This can be due to limited analyser cycle times or a reliance upon off-line laboratory assays. An obvious solution to such problems could be realised by the use of a model along with secondary process measurements, to infer product quality variables (at the rate at which the secondary variables are available) that are either very costly or impossible to measure on-line. Hence, if the relationship between quality measurements and on-line process variables can be captured, then the resulting model can be utilised within a control scheme to enhance process regulation. The concept is known as inferential estimation or soft-sensing [27]. Historically, with varying degrees of success, linear models, adaptive models and process specific mechanistic models have been used to perform this task. The industrial applications of neural network modelling in the general area of process engineering have demonstrated their significant potential in process modelling and control [22], especially in bioprocessing [28,29,30].

7.6.1 Estimation of polymer quality in an industrial continuous polymerisation reactor

In this application a continuous polymerisation reactor is required to meet operating schedules demanding significant polymer grade changes to meet differing market demands. The melt flow index (MFI) of the product was estimated using measurements of reactor feed rate, reactor coolant flow rate and hydrogen concentration above the reaction mass. The secondary variables were measured at 10 minute intervals, while measures of polymer quality (MFI) were obtained from laboratory analyses every 4 hours. Figure 7.1 shows the measured plant data (stepped response) used for training and the dynamic neural network representation (dotted response). Figure 7.2 compares the prediction capabilities of the dynamic model with laboratory assays (stepped response) using 'unseen' data indicating a very acceptable representational ability.

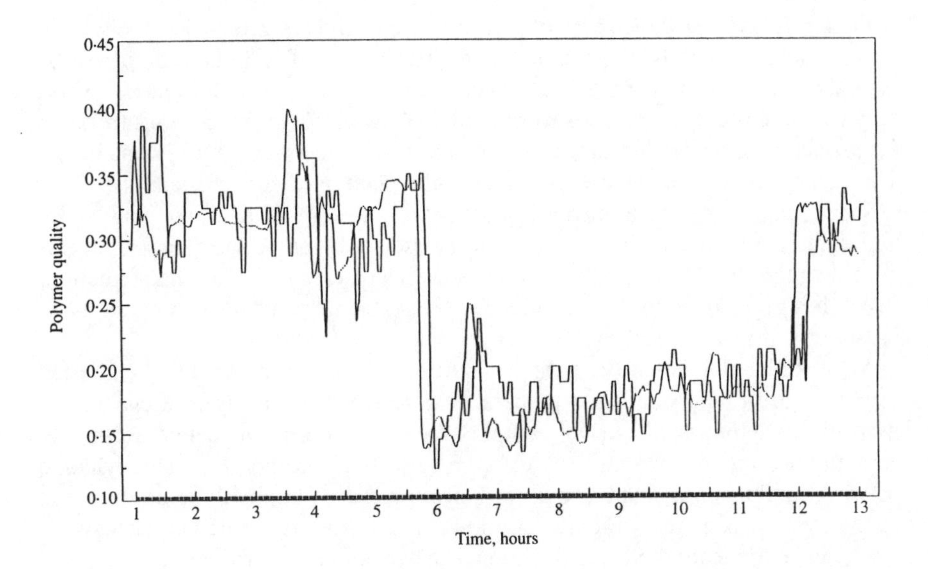

Figure 7.1 *Polymer quality estimation (training)*
········ model prediction ——— off-line data

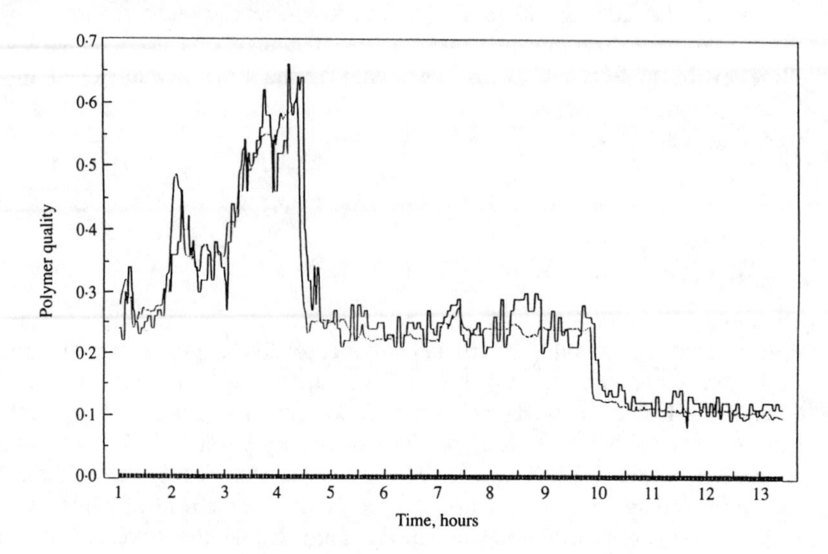

Figure 7.2 *Polymer quality estimation (testing)*
········ model prediction —— off-line data

Figure 7.3 shows the results of a more recent process run which included a major change in polymer grade at around 100 hours. The figure compares the laboratory assayed melt flow index (stepped response) and its estimates. Since only two laboratory samples per work-shift were available the predictions from the neural network model led the 'true' laboratory measurements by several hours. The tighter control that is now possible on the plant has led to the halving of the off-specification production during grade changes. In addition, the study highlights the fallibility of some laboratory assays. It can be seen that on several occasions the 'true' laboratory assay has shown a discrepancy with the prediction, which has given operators the confidence to use the network prediction and have laboratory assays re-checked.

Many industrial processes exhibit directionality in their dynamics and such responses are often found in polymerisation reactors. In many artificial neural network modelling studies such complex dynamics are not modelled accurately with the network 'averaging' the different directional responses. The dynamic network architectures used in our work are capable of implicitly taking account of this type of response. Figure 7.4 shows data from a dynamic neural network model development where the directional dynamics of the process can easily be observed.

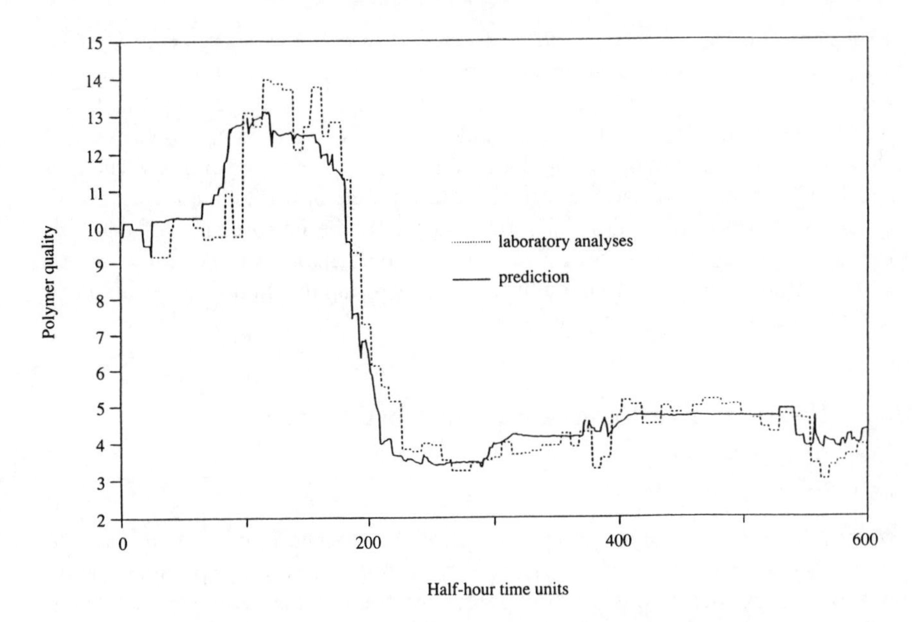

Figure 7.3 *On-line prediction of polymer quality*
——— model prediction ------- off-line data

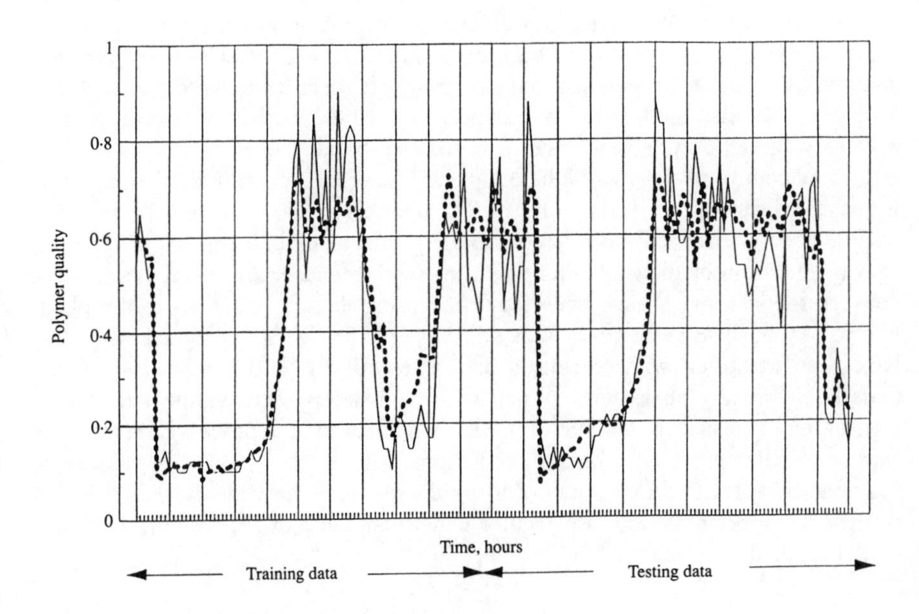

Figure 7.4 *Prediction of direction dependent dynamics*
.... model prediction ——— off-line data

Since continuous processes tend to operate at essentially steady state, it could be argued that the estimation or inferential measurement problem was roughly linear. In this case a fixed model, or adaptive, linear time series representation would be quite adequate. The same cannot be said of continuous processes that exhibit major changes in operating conditions, such as the polymer grade changes considered above, or in batch reactions. Here the nonlinearities associated with the different phases of reactor operation can cause the performance of adaptive mechanisms to significantly degrade. This is where neural network based process representations (models) provide a method of capturing the nonlinear dynamics in a relatively straightforward manner.

7.7 Neural networks in feedback process control

7.7.1 Inferential control

Whilst inferential estimation schemes, operating in 'open-loop', can be used to assist process operators with the availability of fast and accurate product quality estimates, the possibility of closed-loop inferential control becomes very appealing. Such a control strategy could be implemented manually, with conventional regulators, or adaptively. Here the inferred estimates of the controlled output are used directly for feedback control. The effective elimination of a time delay caused by the use of an on-line analyser or the need to perform off-line analysis affords the opportunity of tight product control, even through the use of standard industrial controllers. Consequently reductions in product variability caused by process disturbances, and corresponding reductions in off-specification product, can be achieved. Whilst the use of a secondary variable is, strictly speaking, not necessary, given a suitable choice of secondary variable the resulting adaptive inferential control strategy will inherently provide for feedforward regulation. This is because disturbance effects will manifest themselves first in secondary variable responses and hence will be reflected in the primary output estimate.

A potential problem with using neural network model-based estimators is that network models might have been identified using data collected from the plant which may have some control loops still closed. The resulting model will then have been identified with correlated data and will not be representative of the underlying process behaviour. When such a model is used within a feedback control loop (manual or automatic) it will be subject to new process data that are further correlated and the model predictions will degrade. In this case it is important to identify a new network using the loop data now available and when the new network predictions are deemed better than those of the previous network model it should replace the 'old' model.

7.7.2 Auto-tuning feedback control

Well over 90% of the controllers/regulators on existing process control loops are of the proportional-integral-derivative (PID) type. Since the number of PID regulators found in the process industries is so significant, the tuning of such controllers becomes a very important issue. Indeed the importance attached to PID controller tuning and the use of automatic techniques of achieving this is evidenced by the number of publications addressing the subject. It is not the place of this chapter to review the multitude of methods available, especially since such review articles are available. Recent studies, however, have addressed the use of neural networks to provide nonlinear models for PID controller tuning. One approach is to adopt a controller modelling and supervision philosophy where the use of a nonlinear model is shown to be able to provide for better controller tuning than that possible using other linear approaches [31].

7.8 Neural networks in nonlinear model based control

Given a dynamic neural network model of the process, it can be employed in a number of ways within a model based control strategy [9,13,22]. One very important dynamic modelling approach is that of inverse modelling. Inverse modelling plays a key role in process control. The simplest approach is direct inverse modelling or generalised inverse modelling [32]. Here, a synthetic training signal is introduced into the system and the system output used as an input to a neural network placed in the feedforward part of the control loop. The network is then trained on the error between the system input training signal and the system output. Such a structure forces the network to represent the plant inverse. There are, however, a number of drawbacks with this approach relating to persistency of excitation and the possibility that an incorrect, or even unstable, inverse might be modelled [9,32].

One approach which aims to overcome the problems of the direct inverse method is known as specialised inverse learning [33]. Here, the neural network inverse model in the control loop forward path is trained on the error between the plant output and a training signal replacing the control loop reference set-point. A trained feedforward model (trained using standard feedforward methods) is also included in the inverse network training structure and located in parallel with the plant. If the plant signals are noisy the network can be trained on the error between the training signal and the feedforward model output. The overall training procedure effectively learns the identity mapping across the inverse model and the feedforward model.

Another important approach is that of model reference control. This requires that the plant output follows a stable reference model. A nonlinear approach using neural networks has been proposed [34], which is related to the inverse model training approaches briefly reviewed above. The network training will force the controller to appear to be a de-tuned inverse model controller, the de-tuning resulting from the dynamics of the reference model.

Internal model control (IMC) [35] provides a control loop design philosophy that has wide-ranging implications in the area of process control. In the IMC approach the roles of the inverse model and the feedforward model have been analysed and discussed at length by Morari and colleagues [35]. Here, feedforward and inverse models are used directly within the feedback loop and can be linear or nonlinear [36]. In internal model control the plant (feedforward) model is placed in parallel with the plant, with the plant control input as the feedforward network input. The difference between the real plant output and the plant model output is used for feedback control. The network in the forward path of the control loop, and in series with the plant, is related to the inverse plant model. This form of nonlinear control, like its linear counterpart, is only applicable to open loop stable systems.

Although it is not the remit of this chapter to survey in detail all potentially applicable neural network based controllers, since other surveys are available [9], it is useful to look in a little more detail at one approach that is receiving increasing industrial attention: model based predictive control.

Since the early 1970s numerous model based predictive control algorithms have been proposed, with the prime development objective being to increase performance and robustness of process regulation. A strategically very useful control algorithm is one which minimises future output deviations from the desired set-point, whilst taking suitable account of the control sequence necessary to achieve this objective. This concept is not new and is common to most predictive control algorithms. However, the attraction of using the neural network instead of other model forms, such as autoregressive moving average and nonlinear autoregressive moving average representations within the control strategy, is the ability to represent complex nonlinear systems effectively.

Good examples of predictive control are model algorithmic control, dynamic matrix control and generalised predictive control [37]. Algorithms such as dynamic matrix control have been found to provide major cost benefits on many industrial systems with pay-backs achieved in relatively short time scales. In almost all such control schemes a linear description of the process is assumed. If the system dynamics are relatively linear around the operating region, then the use of a linear model based control algorithm may lead to acceptable performance. However, in situations where the process is highly nonlinear, the linearity assumption could well be detrimental to control system robustness. In this situation a common approach has been to adopt an adaptive control policy. Although the techniques for on-line adaptation are fairly standard, in a ‘real’ process environment the demands placed upon adaptive estimation schemes by everyday process occurrences can be extremely severe. Jacketing procedures are used to provide algorithm robustness. Whilst effective control system jacketing is essential, the consequences of failure in an adaptive scheme have resulted in their industrial application being far from common. In many situations a fixed linear model is used even if the system is known to be nonlinear. As a result performance may be sacrificed in order to maintain robustness in the face of model/process mismatch. The predictive control algorithm is centred around an iterative solution of the following cost function:

$$J = \sum_{i=1}^{N_L} \sum_{n=N_{1,i}}^{N_{2,i}} [w_i(t+n) - y_i(t+n)]^2 + \sum_{i=0}^{N_{u,i}} [\lambda_i \Delta u_i(t+i)]^2$$

where $y_i(t)$, $u_i(t)$ are the controlled output, manipulated input and set-point sequences of control loop 'i' respectively. $N_{1,i}$ and $N_{2,i}$ are the minimum and maximum output prediction horizons. $N_{u,i}$ is the control horizon and q_i is a weighting which penalises excessive changes in the manipulated input of loop 'i'. N_L is the number of individual control loops.

In the performance function, the terms $y_i(t+n)$, $n = N_1, ..., N_2$, represent a 'sequence' of future process output values for each respective loop, $i = 1, N_L$, which are unknown. Thus during minimisation, the sequence of process outputs is replaced by their respective n-step-ahead predictions. With the ability to predict the future outputs $y(t+n|t)$, together with known future set-points, the future controls which will minimise the performance function can be determined. In common with most predictive control strategies, beyond the control horizon, N_u, it is assumed that the control action remains constant. The optimisation algorithm therefore searches for N_u control values in order to minimise the performance function. Since a neural network model is nonlinear, an analytical solution of the cost function is not possible and a nonlinear programming problem needs to be solved at every time step.

There are a number of methods available to minimise the cost function, J. Most algorithms employ some form of search technique to scan the feasible space of the objective function until an extremum point is located. The search is generally guided by calculations on the objective function and/or the derivatives of this function. The various procedures available may be broadly classified as either 'gradient based' or 'gradient free'. It is important to note that care needs to be taken in the choice of an appropriate optimisation routine in that a quadratic minimisation of J will only provide a single 'global' solution. The requirement, however, is for a truly global solution in real situations where multiple solutions are possible. A gradient free method is adopted which enables the extremum of a function to be located using a sequential unidirectional search procedure. Starting from an initial point, the search proceeds according to a set of conjugate directions generated by the algorithm until the extremum is found.

Since processes are normally affected by physical or operational limitations, any advanced process control algorithm should have the ability to account for such constraints. Failure to do so can result in poor closed-loop performances, and possibly even system instability. Although it is usual for the control signals, say, to be rate and magnitude limited, such constraints have been commonly implemented by limiting or clipping the changes and the magnitude of the control signal: a technique which has worked satisfactorily within the framework of conventional control policies. With long range predictive control algorithms however, a sequence of control moves is usually calculated, i.e. $N_U > 1$. Although the control may be implemented in a receding horizon manner, and the implemented control signal may have been 'clipped', other moves in the sequence may still lie outside the acceptable bounds. Indeed, control signal clipping could, in some cases, lead to an unstable closed loop. In the case of multivariable systems, simple clipping of calculated control moves can also cause degeneracy in the controller decoupling properties. To maintain integrity of control and to fulfil overall operational requirements, it is therefore necessary to consider all process constraints explicitly in the formulation of the control cost function. In fact limits on process outputs can only be implemented when they are explicitly taken into account in the cost function minimisation. In a similar manner to output constraints, constraints can be imposed on auxiliary outputs (i.e. those outputs that are important but not controlled).

7.8.1 Application to an industrial high purity distillation tower

In this study, the performance of four controllers were compared by application to the control of column pressure which is a critical variable for high performance process operation. The important variables, identified by multivariate statistical analysis of column operating data, made up from 27 measured variables, were column feed, cooling water temperature, column vent, bottoms level and reboiler energy. A dynamic, filter-based, artificial neural network topology was used with a single hidden layer. The network was trained using the Levenberg-Marquardt nonlinear algorithm [38]. Comparison of a linearised model of column pressure with the neural network model (Figures 7.5 and 7.6) demonstrates the inadequacy of adopting a linear approach. The neural network model provides an excellent representation with an average error of ± 1.5% on the test data.

The four controllers compared were a de-tuned PI controller, a high-gain PI controller, a linear model-based predictive controller, and a neural network based predictive controller. The two PI controllers were tuned using a robust controller design package. Figure 7.7 compares the control performance of the four controllers.

The performance of the high-gain controller is seen to require excessive levels of reboiler activity which are unacceptable for long-term operation. The neural network model predictive controller is seen to outperform all the other controllers, reducing the standard deviation about set-point of the linear controller by 50% and the high-gain controller by 25%. The results demonstrate that any form of model predictive control is only suitable if a reliable and representative dynamic model can be developed from the process data. The linear model had an error of ± 10% on the test data set but was outperformed by the robust PI regulator. The neural network model had an error of ± 0.5% and outperformed the conventional PI regulator by 25%. It is an appreciation of this balance that needs to be addressed by those involved in model predictive control and the acceptance that neural networks can successfully be used to model, with similar effort to that required in developing linear models, nonlinear process dynamics.

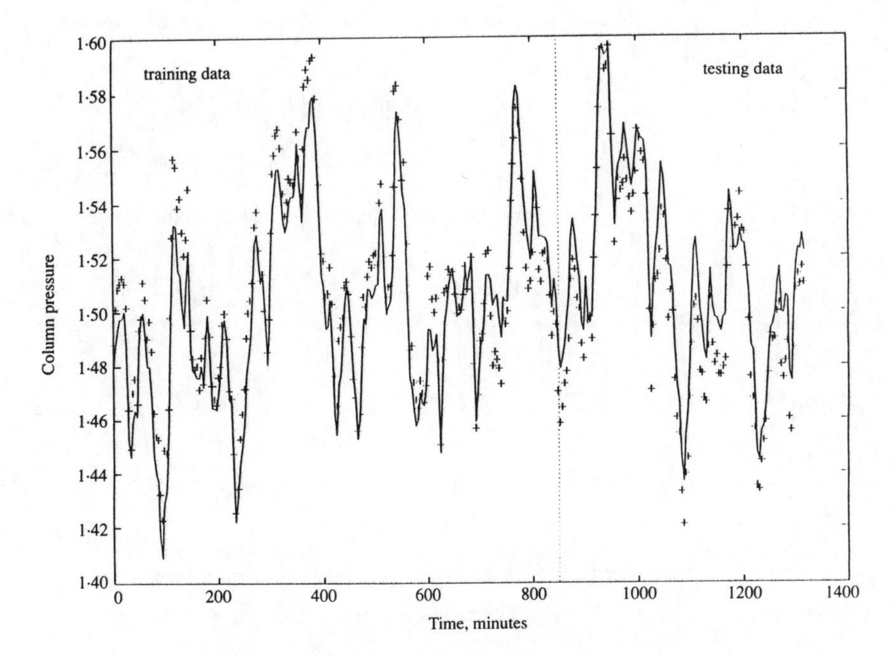

Figure 7.5 *Linear model of column pressure*
——— model prediction +++++ plant data

7.9 Concluding remarks

Artificial neural networks provide an exciting opportunity for the process engineer. It is possible rapidly to develop models of complex operations that would normally take many man-months to model using conventional structured techniques. It is essential to understand, however, that neural network modelling is no replacement for a good understanding of process behaviour.

The dynamic modelling capabilities of artificial neural networks have been reviewed showing that care needs to be exercised in selection of network structure and network training. This was highlighted by considering the consequences of using autoregressive structures for dynamic modelling. A low error in single step-ahead predictions can be deceptive as an indicator of 'quality' of the process model. Given that a representative dynamic model can be obtained, then it is relatively straightforward to incorporate the artificial neural network within an industrially acceptable multivariable control framework. As the control solution is iterative in nature the approach may only be suitable for systems which are not time critical. It is believed that the artificial neural network approach to generic system modelling, if necessary coupled with inferential and predictive control approaches, will result in improved process supervision and control performance without needing to undertake major model development programmes.

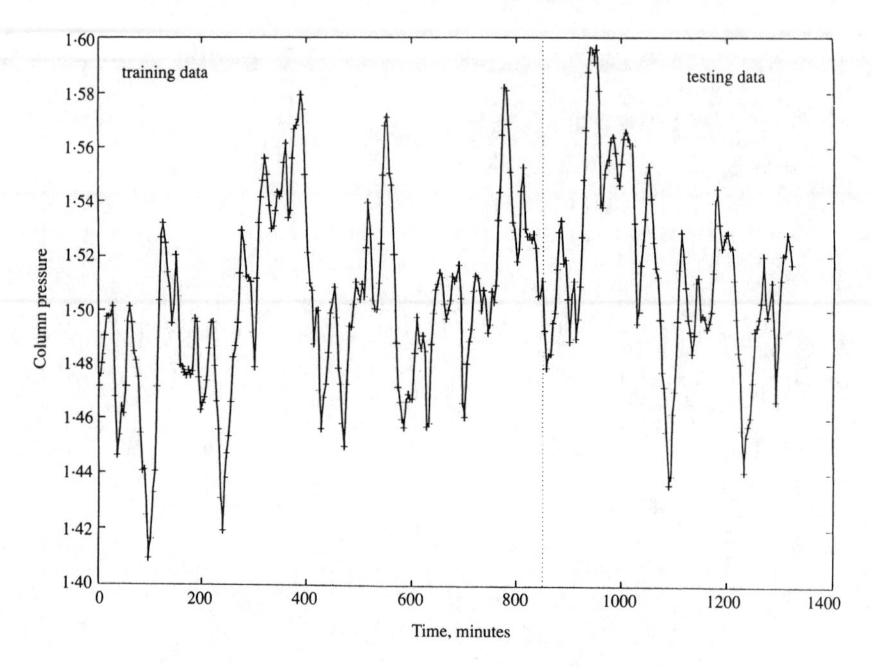

Figure 7.6 *Neural network model of column pressure*
——— model prediction +++++ plant data

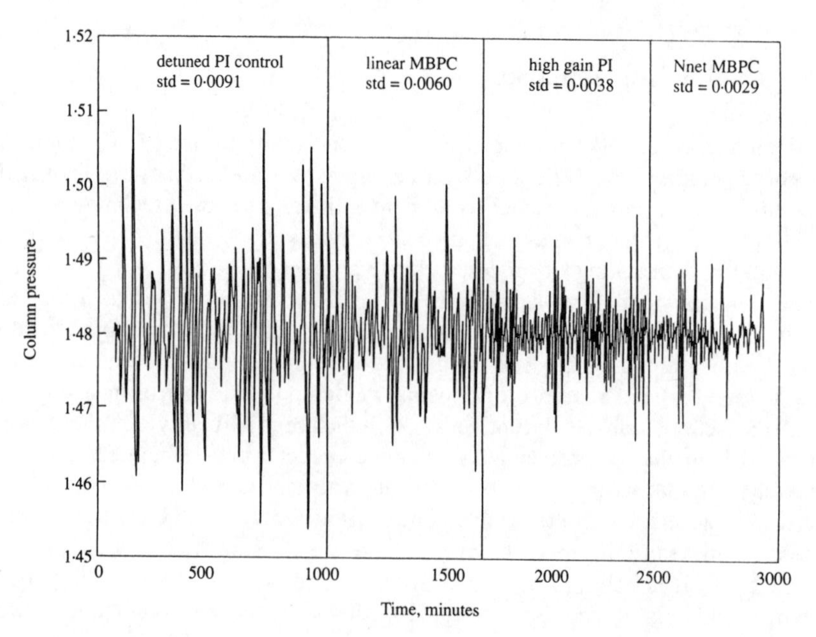

Figure 7.7 *Comparison of four controllers on column pressure loop*

It is stressed, though, that the field is still very much in its infancy and many questions still have to be answered. Determining the 'best' network topology is one example. What is best for one problem may well not be best for another. Currently, rather *ad hoc* procedures tend to be used. This is not good and this arbitrary facet of an otherwise promising philosophy is a potential area of active research. A formalised technique for choosing the appropriate network topology is desirable. More analysis of network structure is needed to explore the relationship between the number of neurons required, to characterise totally the important process information, and the dimensionally finite topological space. There is also no established methodology for determining the robustness and stability of the networks. This is perhaps one of the most important issues that has to be addressed before their full potential might be realised.

7.10 Acknowledgments

The support of the Department of Chemical and Process Engineering, University of Newcastle, and the Teaching Company Directorate are gratefully acknowledged. Paul Turner additionally acknowledges the support he received from the Fellowship of Engineering that enabled him to carry out neural network studies in Australia. The many enlightening discussions with colleagues in the research group are also gratefully acknowledged.

7.11 References

[1] Billings, S.A. and Voon, W.S.F., 1983, Structure detection and model validity tests in the identification of non-linear systems, *Proc. IEE* Pt. D, 130, pp. 193-199.

[2] Billings, S.A. and Voon, W.S.F., 1986, Correlation based model validity tests for non-linear models, *Int. J. Control*, Vol. 44, No. 1, pp. 235-244.

[3] Billings, S.A., Jamaluddin, H.B. and Chen, S., 1991, Properties of neural networks with applications to modelling non-linear dynamical systems, *Int. J. Control*.

[4] Chen, S., Billings, S.A. and Grant, P.M., 1990, Nonlinear system identification using neural networks, *Int. J. Control*, Vol. 51, No. 6, pp. 1191-1214.

[5] Chen, S., Billings, S.A., Cowan, C.F.N. and Grant, P.M., 1990, Practical identification of NARMAX models using radial basis functions, *Int. J. Systems Science*, Vol. 21, pp. 2513-2539.

[6] Wang, Z., Tham, M.T. and Morris, A.J., 1992, Multilayer neural networks: approximated canonical decomposition of nonlinearity, *Int. J. Control*, Vol. 56, pp. 665-672.

[7] Wang, Z., Di Massimo, C., Montague, G.A. and Morris, A.J., 1993, A procedure for determining the topology of feedforward neural networks, *Neural Networks*, Vol. 7, No. 2, pp. 291-300.

[8] IEEE (1988, 1989, 1990). Special Issues on Neural Networks, *Control Systems Magazines* - Nos. 8, 9, 10.

[9] Hunt, K.J., Sbarbaro, D., Zbikowski, R. and Gawthrop, P.J., 1992, Neural networks for control systems - a survey, *Automatica*, Vol. 28, No. 6, pp. 1083-1112.

[10] Thibault, J. and Grandjean, B.P.A., 1991, Neural networks in process control - a survey, *IFAC Conference on Advanced Control of Chemical Processes*, Toulouse, France, pp. 251-260.

[11] Morris, A.J., Montague, G.A. and Willis, M.J., 1994, Artificial neural networks: studies in process modelling and control, *Trans. IChemE*, Vol. 72, Part A, pp. 3-19.

[12] Bhat N. and McAvoy, T.J., 1990, Use of neural nets for dynamic modelling and control of process systems, *Computers Chem. Eng.*, Vol. 14, No. 4/5, pp. 573-583.

[13] Montague G.A., Willis M.J., Morris A.J. and Tham M.T., 1991, Artificial neural network based multivariable predictive control, *ANN'91*, Bournemouth, England.

[14] Su, H.T., McAvoy, T.J. and Werbos, P., 1992, Long term prediction of chemical processes using recurrent neural networks: a parallel training approach, *Ind. Eng. Chem. Res.*, Vol. 31, pp. 1338-1352.

[15] Elman, J.L., 1990, Finding structures in time, *Cognitive Science*, Vol. 14, pp. 179-211.

[16] Frasconi, P., Gori, M. and Soda, G., 1992, Local feedback multilayered networks, *Neural Computing*, Vol. 4, pp. 2460-2465.

[17] Tsoi, A.C. and A.D. Back, 1994, Locally recurrent globally feedforward networks: a critical review of architectures, *IEEE Trans. on Neural Networks*, Vol. 5, No. 2, pp. 229-239.

[18] Zhang, J., Morris, A.J. and Montague, G.A., 1994, Dynamic system modelling using mixed node neural networks, *Proc. IFAC Symposium ADCHEM'94*, Kyoto, Japan, pp. 114-119.

[19] Zhang, J. and Morris, A.J., 1995, Sequential orthogonal training of locally recurrent neural networks, submitted to *Neural Networks*.

[20] Scott, G.M. and Ray, W.H., 1993, Creating efficient nonlinear network process models that allow model interpretation, *Journal of Process Control*, Vol. 3, No. 3, pp. 163-178.

[21] Scott, G.M. and Ray, W.H., 1993, Experiences with model-based controllers based upon neural network process models, *Journal of Process Control*, Vol. 3, No. 3, pp. 179-196.

[22] Willis, M.J., Di Massimo, C., Montague, G.A., Tham, M.T. and Morris, A.J., 1991, Artificial neural networks in process engineering, *Proc. IEE*, Pt D., Vol. 138, No. 3, pp. 256-266.

[23] Montague G.A., Tham, M.T., Willis, M.J. and Morris, A.J., 1992, Predictive control of distillation columns using dynamic neural networks, *3rd IFAC Symposium DYCORD+ '92*, Maryland, USA, April, pp. 231-236.

[24] Leontaritis, I.J. and Billings, S.A., 1987, Model selection and validation methods for non-linear systems, *Int. Journal of Control*, Vol. 45, pp. 311-341.

[25] Akaike, H., 1974, A new look at the statistical model identification, *IEEE Trans. Auto. Cont.*, Vol. AC-19, No. 6, pp. 716-723.

[26] Barron, A.R., 1984, Predicted squared error: a criterion for automatic model selection, *Self Organising Methods*, Farlow, S.J. (Ed.), pp. 87-103.

[27] Tham. M.T., Morris, A.J., Montague, G.A. and Lant, P.A., 1991, Soft sensors for process estimation and inferential control, *J. Proc. Control*, Vol. 1, pp. 3-14.

[28] Lant, P., Willis, M.J., Montague, G.A., Tham, M.T. and Morris, A.J., 1990, A comparison of adaptive estimation with neural network based techniques for bioprocess application, *Proceedings of the American Control Conference*, San Diego, USA, pp. 2173-2178.

[29] Di Massimo, C., Willis, M.J., Montague, G.A., Tham, M.T. and Morris, A.J., 1991, Bioprocess model building using artificial neural networks, *Bioprocess Engineering*, Vol. 7, pp. 77-82.

[30] Di Massimo, C., Montague, G.A., Willis, M.J., Tham, M.T. and Morris, A.J., 1992, Towards improved penicillin fermentation via artificial neural networks, *Computers and Chemical Engineering*, Vol. 16, No. 4, pp. 283-291.

[31] Willis, M.J. and Montague, G.A., 1993, Auto-tuning PI(D) controllers with artificial neural networks, *Proc. IFAC World Congress*, Sydney, Australia, Vol. 4, pp. 61-64.

[32] Hunt, K.J. and Sbarbaro, D., 1991, Neural networks for internal model control, *IEE Proceedings-D*, Vol. 138, No. 5, pp. 431-438.

[33] Psaltis, D., Sideris, A. and Yamamura, A.A., 1988, A multilayered neural network controller, *IEEE Control Systems Magazine*, April, pp. 17-21.

[34] Narendra, K.S. and Parthasarathy, K., 1990, Identification and control of dynamical systems using neural networks, *IEEE Trans Neural Networks*, Vol. 1, No. 1, pp. 4-27.

[35] Garcia, C.E. and Morari, M., 1982, Internal model control - 1. A unifying review and some new results, *Ind. Eng. Chem. Process Des. Dev.*, Vol. 21, pp. 308-323.

[36] Economou, C.G., Morari, M. and Palsson, B.O., 1986, Internal model control - 5. Extension to nonlinear systems, *Ind. Eng. Chem. Process Des.*, Vol. 25, pp. 403-411.

[37] Ricker, N.L., 1991, Model predictive control: state of the art, Preprints CPC IV, South Padre Island, Texas, USA.

[38] Marquardt, D., 1963, An algorithm for least squares estimation of nonlinear parameters, *SIAM J. Appl. Math.*, Vol. 11, p. 431.

Chapter 8

Speech, vision and colour applications

R.J.Mitchell and J.M.Bishop

8.1 Introduction

In this chapter some examples are presented of systems in which neural networks are used. This shows the variety of possible applications of neural networks and the applicability of different neural network techniques. The examples given also show some of the varied work done in the Cybernetics Department at the University of Reading.

8.2 Neural network based vision system

On-line visual inspection is an important aspect of modern-day high speed quality control for production and manufacturing. If human operatives are to be replaced on production line monitoring, the inspection system employed must exhibit a number of human-like qualities, such as flexibility and generalisation, as well as retaining many computational advantages, such as high speed, reproducibility and accuracy. An important point in considering human replacement though, is total system cost, which should include not only the hardware, software and sensor system, but also commissioning time and problem solving.

A neural network learning system based on an n-tuple or weightless network (see Chapter 2) is appropriate, as these are easy to use and can be implemented in hardware and so can operate very quickly. In fact the system produced, as described here, operates at video frame rate, i.e. 50 frames per second. This speed of operation is more than adequate for most production line inspection, in that up to 50 items per second can be checked as they pass a fixed point. The overall system is hardware based so that a very simple set-up and start procedure is needed in order to make the system operative. This merely involves a small number of front panel switch selections with no software requirements. The overall set-up is therefore low cost and simple to operate.

8.2.1 N-tuple vision system

A description of *n*-tuple networks is given in Chapter 2, and also by Aleksander *et al.* [1], but a brief summary follows. In *n*-tuple networks a neuron is modelled by a standard memory device as shown in Figure 8.1a, and the pattern to be analysed is presented to the address lines of the memory. If that pattern is to be remembered, a '1' is written at that address; but to see if that pattern has been learnt, the value at that address is read to see if it is a '1'. These memories can be used in a practical vision system by configuring them in the form shown in the block diagram in Figure 8.1b. The system analyses data patterns comprising *r* bits of information, there are *m* neurons (or memories) and each neuron has *n* inputs. Initially, '0' is stored in each location in each neuron.

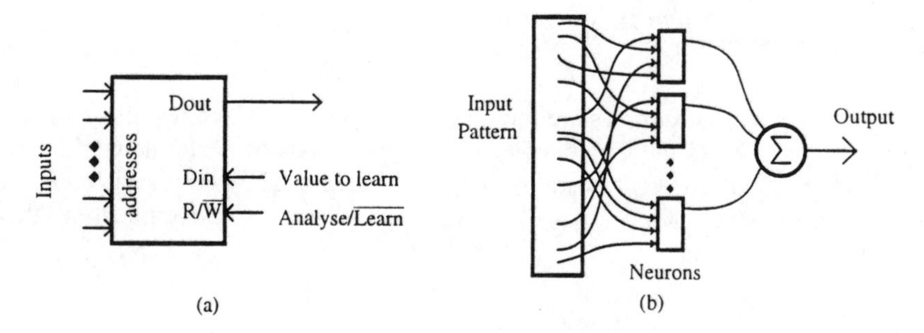

Figure 8.1 *RAM neuron and system*

In the learn phase, *n* bits are sampled from the input pattern (these *n* bits are called a tuple). If the input contains analogue information, then the data at any point are normally considered to be '1' if the analogue value at that point exceeds a given threshold value. This tuple is used to address the first neuron, and a '1' is stored at the specified address: effectively the neuron has 'learnt' the first tuple. Another *n* bits are sampled and these are 'learnt' by the second neuron. This process is repeated until the whole input pattern has been sampled. Note that it is usual to sample bits from the memory in some random manner, as this enables the system to utilise all of the spatial frequency spectrum of the image.

In the analyse phase, the same process is used to sample the input pattern so as to generate the tuples. However, instead of writing '1's into the neurons, a count is made to see how many of the neurons had 'learnt' the tuples presented to them: that is, how many neurons output a '1'. If the input pattern had already been taught to the system, the count would be 100%. If the pattern was similar, then the count would be slightly smaller.

In practice, many similar data patterns should be presented to the system, so each neuron would be able to 'learn' different tuples. Therefore the count when a pattern similar to but not identical with any member of the training set is shown to the system might still be 100% because, for example, the first tuple from the image might be the same as one from the third data pattern taught, and the second tuple might be the same as one from the seventh data pattern, etc. In general, the system is said to recognise an object if the count is greater than p%, where p is chosen suitably and is likely to depend on the application. This allows the system to deal with the inevitable noise which will affect the image.

Therefore such a system is capable of being taught and subsequently recognising data patterns. With a slight extension the system is able to be taught, recognise and discriminate between different classes of data. One set of neurons, as described above, is called a discriminator. If two sets of data are stored in one discriminator, the system will be able to recognise both sets, but not be able to discriminate between them. However, if the system has two discriminators and each data set is taught into its own discriminator, then when a data pattern is presented, a count of the number of recognised tuples is made for each discriminator, and the input pattern is most like the data class stored in the discriminator with the highest count.

An important property of the technique is that it processes all images presented to it in the same manner. Therefore it can be used in a great variety of applications.

8.2.2 A practical vision system

The basic aim in the design of the vision system is that it should be flexible and easy to operate. The *n*-tuple network method provides the flexibility, the careful system design provides the ease of operation. There are many vision systems on the market, and these often require much setting-up or programming, details of which are often specific to the application. For this vision system, the intention was to allow the user to set-up the system using a series of switches and light indicators. These are available on a simple front panel whose layout is shown in Figure 8.2.

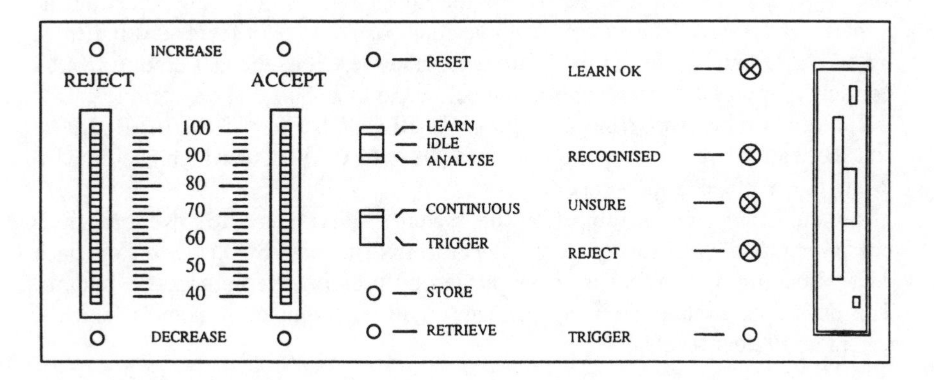

Figure 8.2 *Front panel of vision system*

Another aim of the system is low cost. Therefore in the configuration described here, there is only one discriminator: this should be enough for many applications. However, the system was designed in such a way that extra discriminators could be added easily, with a minimum of extra hardware.

The system basically operates in one of two modes: learn or analyse. In learn mode the image presented to the system is learnt, whereas in analyse mode the system indicates if the image being presented is sufficiently similar to the image or images it has been taught. Both actions are achieved by simply arranging for the object under test to be shown to the camera, the first stage of the vision system.

Each of these modes can be continuous, whereby the system processes an image at each frame time (50 frames per second), or in trigger mode, when the system only processes an image when it receives a suitable signal. These modes are selected using the switches shown in Figure 8.2. An important point when learning is to teach the system a number of slightly different images, and this is easily accomplished by allowing the object being taught slowly to pass the camera along a conveyor belt; for this continuous mode is useful. However, if an object can come down the production line in various orientations, it is important to train the system by showing it the object in many orientations. This can be achieved in trigger mode, by placing the object in front of the camera, learning it by pressing the trigger button, turning the object slightly, learning the object, etc.

It is important not to saturate the memory, as then the system will recognise almost anything. Therefore the system will automatically stop learning when the discriminator is about 10% full. A light on the front panel indicates when to stop learning.

As regards analysing the object, a decision has to be made as to what is an acceptable image, and what is unacceptable, that is what percentage of recognised tuples indicates a good object. A novel feature of this vision system is that it provides an extra degree of control: it allows the user to select one level above which the system accepts the image, and a second level below which the system rejects the object. Those objects in between are marginal, and perhaps need a human to examine them to determine the appropriate action. By selecting the same level for each, the system just responds pass/fail. The levels are indicated by two LED bars shown on the front panel, and these are adjusted up or down by pressing appropriate buttons above and below the LED bars. The system response is indicated on the front panel with lights for RECOGNISED/UNSURE/REJECT, and there are corresponding signals on the back of the system which can be used to direct the object appropriately.

Although teaching an object to the system is very easy, to save time it is possible to load a previously taught object from disk and, obviously, to store such data. Loading and saving is accomplished by pressing the appropriate buttons. The data files contain both the contents of the discriminator memory and the accept and reject levels.

8.2.3 Implementation

As the system requires hardware for acquiring and processing the input video data, as well as being able to process the front panel switches and access disk drives, it was felt necessary to include in the system a microprocessor to coordinate the action. However, most of the processing is accomplished by the special hardware, so the microprocessor need not be very powerful, and its choice was largely due to the available development facilities. The *n*-tuple network requires various hardware modules, and the system needs to access the disk and the front panel, and so it was considered sensible to make the device in modules. A convenient method of bringing these modules together is to use a bus system, and a suitable bus is the STE bus, about which the Department has considerable expertise [2]. The main processor board used is DSP's ECPC which provides an IBM-PC compatible computer on one board with all the necessary signals for controlling the disk, etc.

A block diagram of the system is shown in Figure 8.3. The object under study is placed in front of a standard video camera, and a monitor is used to indicate to the user that the camera is aligned correctly: this is only needed in learning mode. The output from the camera is then digitised using a module from the departmental video bus system VIDIBUS [3], and this is then fed to the *n*-tuple hardware.

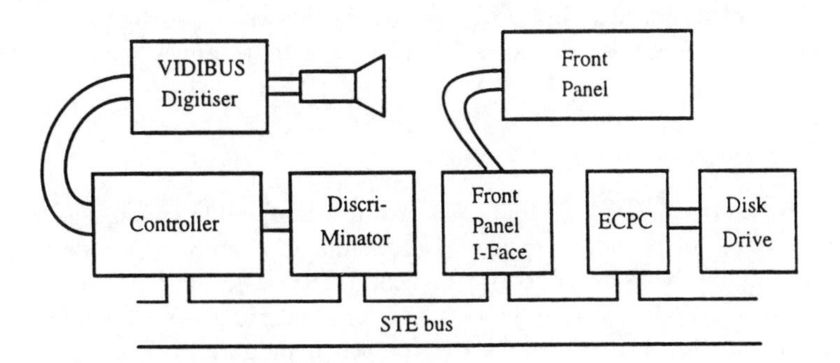

Figure 8.3 *STE based implementation*

There are two parts to the *n*-tuple network, the controller and the discriminator. The former samples the image, forms the tuples, etc., whereas the discriminator contains the memory in which the tuples are taught and counters which indicate whether the memory is saturated (in learn mode) and the number of tuples which have been recognised (in analyse mode). The discriminator is in a separate module so as to allow a multi-discriminator system to be produced relatively easily.

The ECPC is used to configure the controller, for example to tell it to process the next frame or to specify whether to learn or analyse, and to process the saturation and recognition counters. Such actions are determined by the controls on the front panel which the ECPC processes continually via a suitable interface. The last block is the disk drive, the controlling hardware/software for which is on the ECPC.

The software for the system is quite simple, as most of the processing is performed by special purpose hardware. Essentially the software monitors the front panel, configures the controller, checks the discriminator counters and processes the disk. This is all achieved by a simple program written in Turbo-C.

When developing the system, the ECPC was used as a standard PC running a compiler, etc. When the system was complete, however, EPROMs were blown containing the control program so that the system boots from the EPROM and immediately runs the program. Therefore the system does not require a disk, though it is useful for saving and re-loading taught information.

8.2.4 Performance

The system has been tested on a variety of household products, including coffee jars, packets of cleaning material and hair lotion, and has been shown to operate in the expected manner. One slight problem with the system is that some care is required in setting up the lighting conditions because of the crude thresholding technique used to sample the input video signal. The Minchinton cell [4] can solve this problem by inserting an extra module between the digitiser and the *n*-tuple network. More details on the Minchinton cell are given in Section 8.3.

8.2.5 Conclusion

The system described in this section is an easy-to-use, flexible vision system ideal for on-line visual inspection. Its ease of use is due to careful system design, yet it remains flexible because of its use of neural networks. The use of standard modules, like the ECPC STE system, makes for a simple low cost solution.

More recently the Department has developed a modular system for *n*-tuple networks: one module provides the digitiser, another stores the image, a third provides samples from the image, a fourth contains the discriminator RAMs, etc. These modules can be put together to form various configurations of *n*-tuple networks, not just those for basic recognition. More details on the modular system and on other configurations of network can be found in [5] and [6].

8.3 A hybrid neural network system

The system described above is a basic neural network usually used for analysing simple two-dimensional images. Although this is satisfactory for many applications, it has limitations. *N*-tuple networks in common with most other neural architectures, are good at recognising patterns of a given size and position, but cannot themselves directly handle patterns of varying size and position.

Published work on this problem has concentrated on rotation and scale invariance using complex log mappings [7] and is restricted to simple 2-D transformations. Other connectionist techniques, such as knowledge replication and Hinton mapping [8] have also been suggested. Recently, a novel technique has been developed, called stochastic diffusion networks (SDNs), which have been successfully applied to this and other problems [9]. The advantages of SDNs over the other methods are that: they are fast and have already been successfully applied to solving practical pattern recognition problems; they can be implemented in hardware; and they have the potential for extension to ever more complex domains.

Stochastic diffusion networks describe a family of neural architectures capable of efficiently searching a large space for the best fit to a given data model or archetype. For a search space of size S and model of size M, SDNs can be shown to converge on a solution in a time that has been empirically shown to be probabilistically bounded by log S for a given ratio of M/S [10].

Further research for the BT Connectionist project [11] demonstrated that an SDN, as part of a hybrid system, could locate the position of eyes on snapshots of human faces. This was achieved by using the SDN in conjunction with an n-tuple network to examine the input image, testing whether a selection of image pixels at each search point were part of an eye. The SDN was used to direct the search, as a simple sequential search for an eye in an image of 256×256 pixels will nominally require 65536 tests.

The simple n-tuple vision system described in section 8.2.2 sampled the incoming analogue video signal and produced a digital value of '0' or '1' dependent on whether the analogue signal exceeded a given threshold value. This rejects much information, and can cause problems when the lighting of the input image changes. The images of faces are grey level, and may occur over different lighting conditions, and so an appropriate technique is needed to handle such information. The usual method used to overcome this problem is thermometer coding [12] but this requires extra memory as more neurons are needed. An alternative method is to use the Minchinton cell [4]: this is also described below.

8.3.1 Stochastic diffusion networks

The stochastic diffusion networks are used to obtain the best fit of a given data model within a specified search space. For this application the problem was to find an eye in an image of a face, but the technique could be used, for example, to find the best fit of a character string in a larger string. The model and search space are defined in terms of features between which comparisons can be made; for the eye problem these features are the groups of pixels which form the tuples, but for the string problem the features could be one or more characters from the string.

Stochastic diffusion uses a network of variable mapping cells to test possible locations of model features within the search space. At each iteration of the search, every cell randomly selects an associated model feature for which it defines a mapping into the search space, that is, it specifies where to look for that particular feature in the search space. When a model feature is detected, its mapping cell becomes active and the mapping defined by that cell is maintained. If the model feature is not detected then the cell chooses a new mapping. This is accomplished by selecting another cell at random from the network, and copying its mapping if this cell is active; however, if this cell is inactive then the new mapping must be chosen randomly.

If the search conditions are favourable, a statistical equilibrium condition is soon reached whereby the distribution of mappings maintained by the cells remains constant. The most popular mapping then defines the position of the object within the search space.

Stochastic diffusion networks have been shown to be very successful. For example, on average a search of 60 000 images of size 256×256 found the best match of a specified image in 24.5 iterations.

8.3.2 Using analogue information

The Minchinton cell is a general purpose pre-processor connected between the analogue input image and the address lines of a memory neuron. In its simplest form it functions as a comparator between two analogue signals (Va and Vb), outputting a digital high signal if Va is greater than Vb. In this form the cell has the useful characteristic that its output is dependent on the relative amplitude of the point to the rest of the input, not its absolute amplitude. Thus, if the background lighting level increased such that the grey level values of all the points of the image were increased uniformly, then, providing there is no saturation, the output of the Minchinton cells would be unchanged from that at the lower lighting levels. In practice all points of an image would not change by the same amount, but if this cell is coupled to an n-tuple network, the system is made much more tolerant of changes in the background lighting levels than an n-tuple system using simple thresholding.

8.3.3 The hybrid system

The system for determining the position of eyes in faces thus consists of a stochastic diffusion network used to suggest possible mappings into the search space, and an n-tuple network using Minchinton cell pre-processing, which had been taught a number of images of eyes.

The prototype system worked well: it was one of many systems employed as part of the BT Connectionist project, and within 150 iterations correctly found the position of 60% of the eyes, and performed as well as, or better than the other prototypes. The hybrid system demonstrates the potential of mixing neural and other techniques in the solution of a particular problem.

8.4 The application of neural networks to computer recipe prediction

An important aspect of the quality control of manufacturing processes is the maintenance of colour of the product. The use of colour measurement for production and quality control is widespread in the paint, plastic, dyed textile and food industries. An industrial colour control system will typically perform two functions relating to the problems encountered by the manufacturer of a coloured product. First the manufacturer needs to find a means of producing a particular colour. This involves selecting a recipe of appropriate dyes or pigments which, when applied at a specific concentration to the product in a particular way, will render the required colour. This process, called recipe prediction, is traditionally carried out by trained colourists who achieve a colour match via a combination of experience and trial and error. Instrumental recipe prediction was introduced commercially in the 1960s and has become one of the most important industrial applications of colorimetry. The second function of a colour control system is the evaluation of colour difference between a batch of the coloured product and the standard on a pass/fail basis.

Conventional computer mechanisms for colorant formulation (recipe prediction) commonly employ the Kubelka-Munk theory to relate reflectance values to colorant concentrations. However, there are situations where this approach is not applicable and hence an alternative is needed. One such method is to utilise artificial intelligence techniques to mimic the behaviour of the professional colourist. The following section will describe research carried out at Reading University, originally sponsored by Courtaulds Research, to integrate neural techniques into a computer recipe prediction system.

8.4.1 Introduction

Since the early development of a computer colorant formulation method, computer recipe prediction has become one of the most important industrial applications of colorimetry. The first commercial computer for recipe prediction [13] was an analogue device known as the COMIC (colorant mixture computer) and this was superseded by the first digital computer system, a Ferranti Pegasus computer, in 1961 [14].

Several companies now market computer recipe prediction systems which all use digital computers. All computer recipe prediction systems developed to date are based on an optical model that performs two specific functions:-

i) The model relates the concentrations of the individual colorants to some measurable property of the colorants in use.

ii) The model describes how the colorants behave in mixture.

The model that is commonly employed is the Kubelka-Munk theory [15] which relates measured reflectance values to colorant concentration via two terms K and

S, which are Kubelka-Munk versions of the absorption and scattering coefficients respectively of the colorant. In order for the Kubelka-Munk equations to be used as a model for recipe prediction it is necessary to establish the optical behaviour of the individual colorants as they are applied at a range of concentrations. It is then assumed that the Kubelka-Munk coefficients are additive when mixtures of the colorants are used. Thus it is usual for a database to be prepared which includes all of the colorants which are to be used by the system and allows the calculation of K and S for the individual colorants.

The Kubelka-Munk theory is in fact an approximation of an exact theory of radiative transfer. Exact theories are well documented in the literature [15], but have rarely been used in the coloration industry. The Kubelka-Munk approximation is a two-flux version of the many-flux [16,17] treatment for solving radiative problems.

The Kubelka-Munk approximation is valid if the following restrictions are assumed:

i) The scattering medium is bounded by parallel planes and extends over a region very large compared to its thickness.
ii) The boundary conditions which include the illumination, do not depend upon time or the position of the boundary planes.
iii) The medium is homogenous for the purposes of calculation.
iv) The radiation is confined to a narrow wavelength band so that the absorption and scattering coefficients are constant.
v) The medium does not emit radiation (e.g. fluoresce).
vi) The medium is isotropic.

There are many applications of the Kubelka-Munk approximation in the coloration industry where these assumptions are known to be false. In particular, the applications to thin layers of colorants (e.g. printing inks [18]) and fluorescent dyestuffs [19,20] have generally yielded poor results.

The use of an approximation of the exact model has attracted criticism. For example, Van de Hulst [21] when discussing its application to paint layers comments: 'it is a pity that all this work has been based on such a crude approximation...'.

The popularity of the Kubelka-Munk equations is undoubtedly due to their simplicity and ease of use. The equations give insight and can be used to predict recipes with reasonable accuracy in many cases. In addition, the simple principles involved in the theory are easily understood by the non-specialist and rightly form the basis for study for those involved in the coloration industry [22].

The use of the exact theory of radiation transfer is not of practical interest to the coloration industry. The calculations generally require databases and spectrophotometers of a greater complexity than those suitable for Kubelka-Munk calculations.

8.4.2 Computer recipe prediction

The operation of a typical recipe prediction system is as follows (see Figure 8.4):

i) The computer makes an initial guess at a suitable recipe for the specified colour, based on historical data and the choice of dyes available.
ii) The reflectance values for the given recipe are computed using Kubelka-Munk theory.
iii) The colour difference (ΔE) between the generated recipe and the desired colour is computed using the CMC 2:1 colour difference equation.
iv) If the ΔE value is within specification the recipe is acceptable and is presented to the user. Otherwise, the recipe is modified (typically using a gradient descent optimisation technique) and the procedure repeated from step (ii).

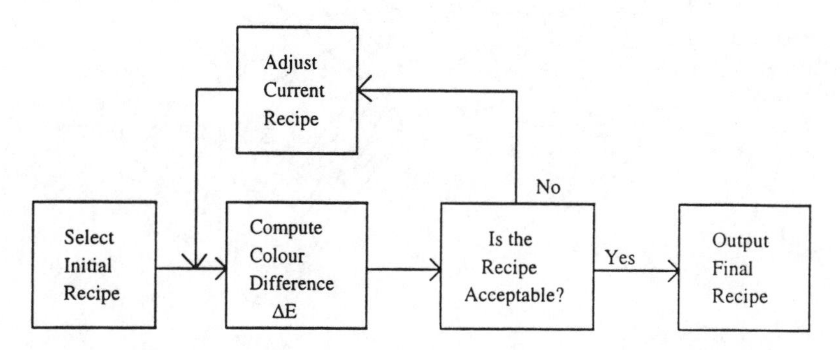

Figure 8.4 *Block diagram of a typical computer colour recipe prediction system*

8.4.3 Neural networks and recipe prediction

As discussed above, one of the problems with conventional recipe prediction is that the application of exact colour theory is not computationally practical and an approximation to it has to be employed. It was expected that a neural network approach to reflectance prediction would offer a novel new approach to this problem, since many problems in artificial intelligence (AI) involve systems where conventional rule based knowledge is not complete or the application of the pure theory is too computer intensive to be used in practical systems.

To investigate whether a neural network can mimic Kubelka-Munk theory and successfully predict reflectance values from specified recipes, a data set was obtained that contained a list of measured reflectance values (at 5 nm intervals between 400 and 700 nm i.e. 31 wavelengths) with their corresponding recipes (concentrations of colorants used to generate the samples). The total number of colorants in the set was 5, and recipes could be combinations of any number of these 5 dyes. Two of the five colorants were fluorescent, hence some reflectance values are greater than 1.0, and thus the data would present a considerable challenge to Kubelka-Munk theory.

A simple multi-layer perceptron architecture was chosen, as shown in Figure 8.5. The network has 5 input units, three hidden units in a single layer, and 31 output units. The network is therefore designed to map recipe to reflectance. It is known that this mapping is a many-to-one mapping since there may be more than one way of generating a set of reflectance values from a large set of colorants.

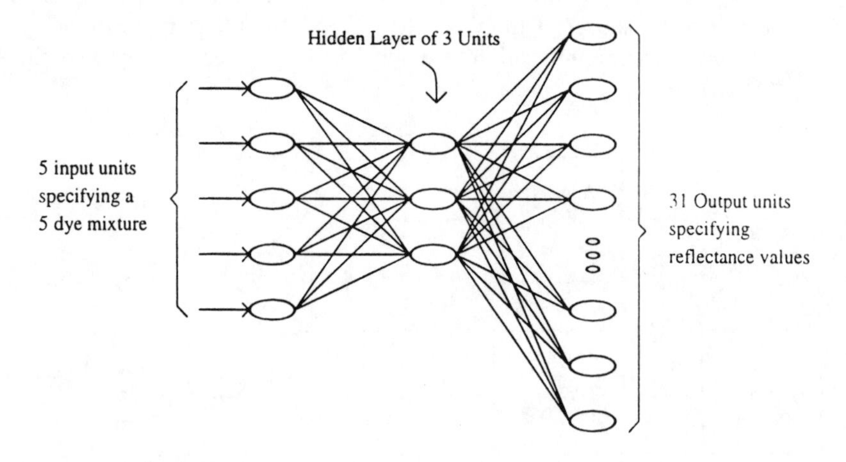

Figure 8.5 *Network structure for recipe prediction*

Input and output data were scaled before presentation to the network and 200 samples were used for training the network. The network was trained using the standard backpropagation algorithm. The learning rate (initially 0.3) and momentum (initially 0.5) terms were reduced during training at epochs (200 000; 400 000; 700 000) with the complete training lasting approximately 3 000 000 epochs (an epoch being defined as the presentation of a single sample chosen at random from the 200 sample data set).

Performance

After learning, a measure of the effectiveness of the network was made using the test data. The output data ranges from 0 to 100 and the average output error (averaged over all 31 output units and all 200 training samples) is 3.41. This corresponds favourably with the performance of the best Kubelka-Munk model of the data even though it is known that the training data were not evenly spread though colour space, making successful modelling of the data harder. Analysis of the weights in the trained network showed a trichromatic representation of reflectance data. One of the hidden units was responsible for the blue part of the spectrum, one for the red part, and one for the green part. A paper by Usui [23] showed that a network could be used to compress reflectance data into three hidden units; two units were insufficient and when four units were used a correlation was found between two of them.

8.4.4 Discussion

The performance of the neural network on the specified reflectance prediction task is very promising - training on a second confidential commercial data set being significantly better than the results quoted above. Earlier attempts at using neural networks in colour recipe prediction concentrated on the inverse problem of predicting dye concentrations from a colour specification [24,25]. Not only is this a more difficult task (it is a one-to-many mapping), but such a system, even if perfect, would have limited commercial appeal since the user needs to retain control of the colorants to be used. The method outlined above enables, for the first time, a neural solution to be embodied as a black box replacement for the Kubelka-Munk stage in any recipe prediction system, and hence promises to have widespread use within the colour industry.

The significance of the network's hidden nodes arriving at a trichromatic representation of reflectance data is currently being investigated.

8.5 Kohonen networks for Chinese speech recognition

Speech recognition is an area of research in which neural networks have been employed. In particular, Teuvo Kohonen, one of Europe's foremost contributors to neural computing, has successfully applied his Kohonen networks to this problem [26]. This work has been concerned largely with the understanding of Finnish speech, but this section describes briefly the work done in the Department by Ping Wu concerning Chinese speech. This is given in more detail in [27,28,29].

8.5.1 Review of Kohonen networks

A brief description of Kohonen networks is given here, but more details can be found in [30]. These networks belong to the class of unsupervised learning networks, that is they learn to identify classes of data without being explicitly instructed, whereas the *n*-tuple and multi-layer perceptron types require an external teacher. The basic structure and operation of these networks is as follows.

The network consists of a number of neurons, each neuron has a number of associated values (weights) and the input to the system is connected to each neuron. These neurons are effectively arranged in a rectangular grid, the outputs of which form a 'feature map'. These are shown in Figure 8.6. Initially the weights of the neurons are given random values.

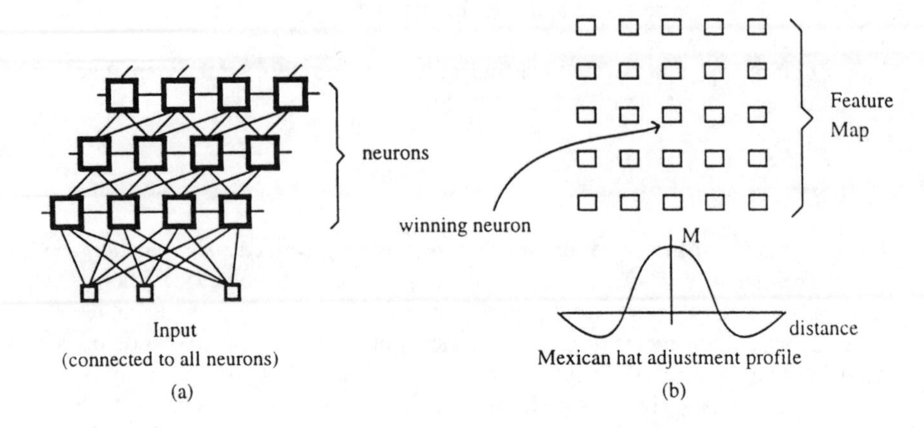

Figure 8.6 *Kohonen networks*

In use, the values associated with each neuron are compared with the input and the neuron which most closely resembles the input is the one which 'fires'. The weights of this neuron are then adjusted so that this neuron even more closely resembles the input. This assists the neuron to learn that pattern. However, the system must also be able to recognise similar patterns, so those neurons near the winning neuron also have their weights adjusted, but the amount by which any neuron is adjusted is proportional to the distance measured between it and the winning neuron. If the main neuron is adjusted by D, the other neurons are adjusted by $M \times D$, where M is defined by the 'Mexican hat' profile shown in Figure 8.6.

As a result of the learning process, certain adjacent neurons will respond to one type of input, other neurons to other inputs, etc. In such a 'feature map' there are distinct areas of 'classes'. Thus when a pattern is input to the network, one neuron will be most active, and the system will have classified that input according to the area of the feature map in which that neuron occurs.

8.5.2 Use in speech recognition

Many forms of speech analysis categorise speech by a series of phonemes. These are the basic speech sounds, and they are concatenated to form words and ultimately sentences. Kohonen networks are used therefore to classify various inputs as to their phonemes.

If a word or sentence is then input, various neurons will fire in a particular sequence. This appears as a trajectory in the feature map. Thus if the same word is spoken on different occasions, the system will follow approximately the same trajectory, and it is relatively easy to recognise such trajectories, and hence recognise words.

8.5.3 Chinese phoneme characteristics

Western languages concentrate on the pronunciation of complex words, so each word is formed by a number of phonemes. The Chinese language concentrates on the shape of Chinese square characters. Different shapes have different meanings, but utterance of each character takes a very simple form. There are 36 phonemes, and these can represent all of the Chinese character pronunciations. This makes Chinese a good choice for study, and has enabled an improvement to be made to the Kohonen technique.

In addition, there are only three rules determining the pronunciation of Chinese characters. The phonemes are grouped as follows:

Group 1 : b, p, m, f, d, t, n, l, g, k, h, j, q, x
Group 2 : zh, ch, sh, r, z, c, s
Group 3 : i, u, u"
Group 4 : a, o, e, ai, ei, ao, ou, an, en, ang, eng, ong

The first two are the initial consonant groups, the others are vowel sounds. The rules for pronouncing Chinese are:

A single expression is one phoneme from groups 2, 3 or 4.
A diphthong is a phoneme from group 3 followed by one from group 4.
Or, the character is one phoneme from group 1 or 2, followed by one from group 3 or group 4, or one from each group 3 and then one from group 4.

All Chinese characters obey these simple rules, so that no more than three basic phonemes are used for any character.

8.5.4 Neural network system

A neural network to recognise Chinese speech has been simulated on a SUN computer system. The input to the system is digitised natural speech stored as phonemes. The network consists of 20×20 neurons. The Kohonen learning strategy can then be used to classify the phonemes.

As described in [29], the normal feedback method used to reinforce learning (that is to adjust the winning neuron and others near it, according to the 'Mexican hat' profile), can be improved upon for Chinese speech.

The Mexican hat profile is symmetrical, and weights are adjusted purely according to the distance between a neuron and the winning neuron. However, as there are few rules specifying how the phonemes are grouped together, coupling of phonemes can be achieved. Thus if one phoneme fires, another particular phoneme is more likely next. Thus the feature map should be 'encouraged' to position this second phoneme near to the current one, and a 'bubble' shape is used in the adjustment.

Experiments have shown that the technique is successful in the recognition of Chinese speech. More details on the method can be found in the cited references, in particular in [27].

8.6 Conclusion

The examples described in this chapter show some of the many types of neural network used in a variety of different applications. Although networks cannot be used as a solution to all projects, and there are particular problems associated with networks to improve their performance, there is no doubt that neural networks can be a useful tool to aid the system designer when solving a particular problem.

8.7 References

[1] Aleksander, I., Thomas, W. and Bowden, P., 1984, WISARD, a radical new step forward in image recognition, *Sensor Review*, 120-4.

[2] Mitchell, R.J., 1989, *Microcomputer systems using the STE bus*, Macmillan.

[3] Fletcher, M.J., Mitchell, R.J. and Minchinton, P.R., 1990, VIDIBUS- A low cost modular bus system for real time video processing, *Electronics and Communications Journal*, Vol. 2, No. 5, pp. 195-201.

[4] Bishop, J.M., Minchinton, P.R. and Mitchell, R.J., 1991, Real Time Invariant Grey Level Image Processing Using Digital Neural Networks, *Proc IMechE Eurodirect '91*, London, pp. 187-188.

[5] Mitchell, R.J., Minchinton, P.R., Brooker, J.P. and Bishop, J.M., 1992, A Modular Weightless Neural Network System, *Proc. ICANN 92*, Brighton, pp. 635-638.

[6] Bishop, J.M. and Mitchell, R.J., 1995, Auto-associative memory using n-tuple techniques, *Intelligent Systems Engineering*, Winter 1994.

[7] Messner. R.A. and Szu. ?., 1985, An image processing architecture for real-time generation of scale and rotation invariant patterns, *CVGIP*, Vol. 31, pp. 50-66.

[8] Rumelhart, D.E. and McClelland, J.L., 1986, *Parallel Distributed Processing*, Vol. 1, Chapter 4, MIT Press, pp. 114-115.

[9] Bishop, J.M., 1989a, Stochastic Searching Networks, *Proc. 1st IEE Conf. Artificial Neural Networks*.

[10] Bishop, J.M., 1989b, Anarchic Techniques for Pattern Classification, PhD Thesis, University of Reading.

[11] Bishop, J.M. and Torr, P., 1992, The Stochastic Search, in *Neural Networks for Images, Speech and Natural Language*, Lingard, R. and Nightingale, C. (Eds.), Chapman Hall.

[12] Aleksander, I. and Stonham, T.J., 1979, A Guide to Pattern Recognition using Random Access Memories, *IEE Jrn.Cmp.Dig.Tch.*, Vol. 2, p. 1.

[13] Davidson, H.R., Hemmendinger, H. and Landry, J.L.R., 1963, A System of Instrumental Colour Control for the Textile Industry, *Journal of the Society of Dyers and Colourists*, Vol. 79, p. 577.

[14] Alderson, J.V., Atherton, E. and Derbyshire, A.N., 1961, Modern Physical Techniques in Colour Formulation, *Journal of the Society of Dyers and Colourists*, Vol. 77, p. 657.

[15] Judd, D.B. and Wyszecki, G., 1975, *Color in Business, Science and Industry*, 3rd ed., Wiley, New York, pp. 438-461.

[16] Chandrasekhar, S., 1950, *Radiative Transfer*, Clarendon Press, Oxford.

[17] Mudgett, P.S. and Richards, L.W., 1971, Multiple Scattering Calculations for Technology, *Applied Optics*, ?0?, pp. 1485-1502.

[18] Westland, S., 1988, The Optical Properties of Printing Inks, PhD Thesis, University of Leeds, U.K.

[19] Ganz, E., 1977, Problems of Fluorescence in Colorant Formulation, *Colour Research and Application*, Vol. 2, p. 81.

[20] McKay, D.B., 1976, Practical Recipe Prediction Procedures including the use of Fluorescent Dyes, PhD Thesis, University of Bradford, U.K.

[21] Van de Hulst, H.C., 1980, *Multiple Light Scattering; Tables, Formulas and Applications*, Academic Press, New York.

[22] Nobbs, J.H., 1986, *Review of Progress in Coloration*, The Society of Dyers and Colourists, Bradford.

[23] Usui, S., Nakauhci, S. and Nakano, M., 1992, Reconstruction of Munsell Color Space by a Five-Layer Neural Network, *Journal of the Optical Society of America*, Series A, Vol. 9, No. 4, pp. 516-520.

[24] Bishop, J.M., Bushnell, M.J. and Westland, S., 1990, Computer Recipe Prediction Using Neural Networks, *Proc. Expert Systems '90*, London.

[25] Bishop, J.M., Bushnell, M.J. and Westland, S., 1991, The Application of Neural Networks to Computer Recipe Prediction, *Color*, Vol. 16, No. 1, pp. 3-9.

[26] Kohonen, T., 1988, The Neural Phonetic Typewriter, *IEEE Computing Magazine*, Vol. 21, No. 3, pp. 11-22.

[27] Wu, P., 1994, Kohonen Self-Organising Neural Networks in Speech Signal Processing, PhD Thesis, University of Reading, U.K.

[28] Wu, P., Warwick, K. and Koska, M., 1990, Neural network feature map for Chinese phonemes, *Proc. Int. Conf. Neuronet 90*, Prague, pp. 357-360.

[29] Wu, P. and Warwick, K., 1990, A new neural coupling algorithm for speech processing, *Research and Development in Expert Systems VII*, Proc. of Expert Systems 90, London, pp. 65-69.

[30] Kohonen, T., 1984, *Self Organisation and Associative Memory*, Springer Verlag.

Chapter 9

Real-time drive control with neural networks

D. Neumerkel, J. Franz and L. Krüger

9.1 Introduction

Drive control is an important task of the Daimler-Benz subsidiary company *AEG Daimler-Benz Industrie*. In order to increase the performance of modern drive systems it is advantageous to exploit nonlinear control techniques, e.g. neurocontrol. Within the nonlinear control framework bounds on control, state and output variables can be taken into account. Different control objectives may also be pursued in different operating regions. This is especially important if safety requirements must be met. Furthermore, nonlinear dependencies such as friction, hysteresis or saturation can be included in the mathematical-physical modelling process and control scheme, if known. On the other hand these effects are very often neglected in drive control in order to design linear controllers, but this often results in poor control. If the mathematical structure (equations) representing a process is not known, very difficult to obtain or just too time-consuming to evaluate, learning systems may be engaged to improve the modelling process.

It is, however, necessary that the effects to be captured by learning systems are representable in input-output data and that these data are available. Figure 9.1 displays the process-knowledge hierarchy one is usually confronted with in practical applications assuming a correct sampling procedure. It will rarely be the case that the whole process information is represented in measured input-output data; it is most likely to generate false process information through measurement noise. This can only be suppressed by choosing the appropriate measurement equipment and most important the correct sampling time, since a sampling rate chosen to be too slow or too fast can cause well-known problems. The right choice concerning the measurement apparatus can only be made with the help of some *a priori* knowlege about the process in question, most commonly in the form of mathematical equations. If a mathematical structure (probably incomplete) of the underlying process is available it may still contain knowledge about the process that is not accessible through input-output data. For these reasons we believe the paradigms of learning systems and *a priori* mathematical-physical modelling must be combined to design controllers with enhanced capabilities. Both fields

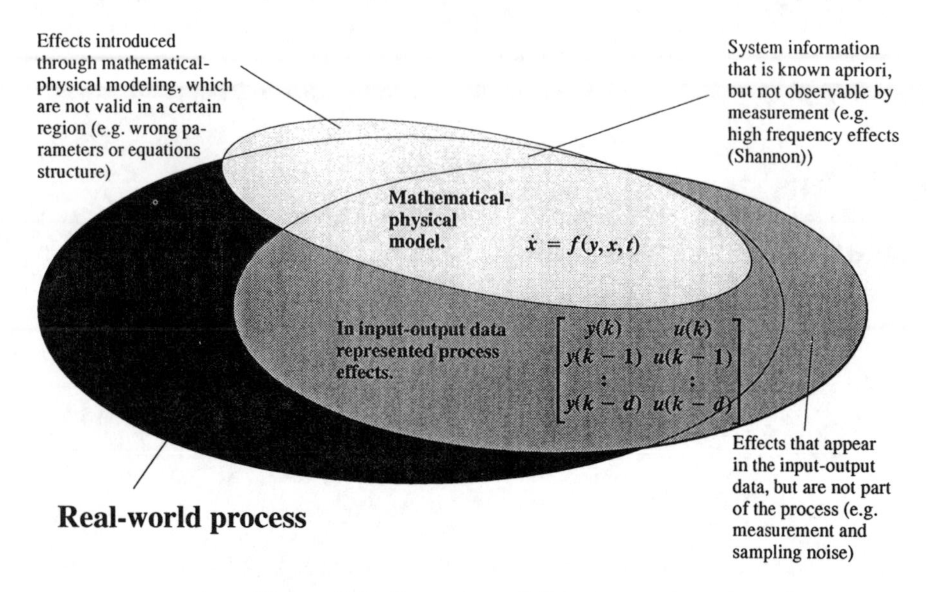

Figure 9.1 *Process modelling hierarchy*

contribute equally to this goal. We will use neural networks to fulfil the learning task posed in this special application.

As part of a research project for AEG [1,2] we developed a neural network control strategy for induction machines based on the model based predictive control concept [3]. This chapter will introduce the basic concepts, describe practical issues and then focus on aspects of the real-time implementation of a neural net controller and its application to an induction machine. The control algorithms have been implemented on a multi-transputer system, the neural network model has thus been parallelised in order to meet the real-time specifications. Details about the implementation on a transputer system will also be given.

9.2 Induction machine control concepts

The main driving force towards induction machines in industrial applications is the lack of required maintenance intervals as opposed to dc-drives. DC-drives can be very accurately modelled by a linear system and thus high performance linear control is available. Due to temperature changes, the linear system describing the dc-drive has to be identified on-line and the controller adequately adapted: this is not a trivial task, but solvable in the adaptive control framework. It is the goal of the approach described in this chapter to improve the existing induction machine

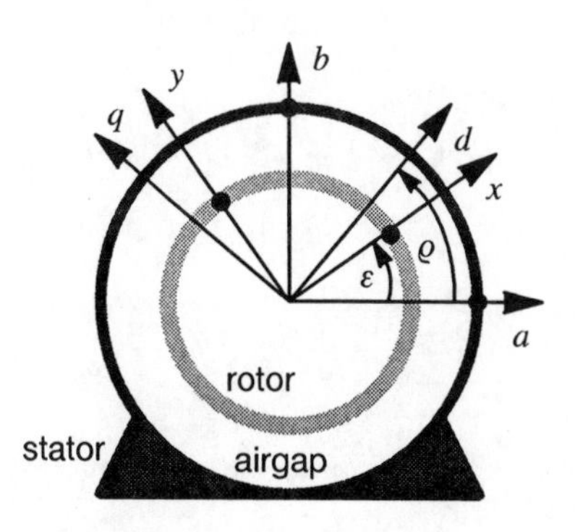

Figure 9.2 *Coordinate systems*

control techniques in order to replace dc-drives with ac-drives in high quality control applications.

Induction machines are in general nonlinear, multivariate systems, the mathematical equations are given as:

$$\vec{u}_s(t) = R_S\,\vec{i}_S(t) + \frac{d}{dt}\vec{\psi}_S(t) = R_S\,\vec{i}_S(t) + L_S\frac{d}{dt}\vec{i}_s(t) + L_H\frac{d}{dt}\left(\vec{i}_R(t)\,e^{j\varepsilon(t)}\right) \tag{9.1}$$

$$o = R_R\,\vec{i}_R(t)\,e^{j\varepsilon(t)} + \frac{d}{dt}\vec{\psi}_R(t) = R_R\,\vec{i}_R(t)\,e^{j\varepsilon(t)} + L_R\frac{d}{dt}\left(\vec{i}_R(t)\,e^{j\varepsilon(t)}\right) + L_H\frac{d}{dt}\vec{i}_S(t) \tag{9.2}$$

$$m_d(t) = \frac{2}{3}z_p L_H\,\mathrm{Im}\left(\vec{i}_S(t)\;\vec{i}_R(t)\,e^{j\varepsilon(t)\,*}\right) \tag{9.3}$$

$$\frac{d\omega\,(t)}{dt} = \frac{z_p}{J}\left(m_d(t) - m_L(t)\right) \tag{9.4}$$

$$\frac{d\epsilon\,(t)}{d} = \omega\,(t) \tag{9.5}$$

The conventional control method for induction machines, field-oriented control (FOC), is based on decoupling and linearising the multivariate system through feedback. A very insightful way to achieve this is to use a coordinate transform into a reference system rotating together with the rotor flux Ψ_R((d,q) as depicted in Figure 9.2) on eqns. 9.1 and 9.2. According to the notation given in [4], eqn. 9.3 can be rewritten as:

$$m_d = \frac{2}{3} z_p (1 - \sigma) L_S \left[(i_{Sq}(t)\ i_{mR}(t) \right] \tag{9.6}$$

where $i_{mR}(t)$ is a current vector defined by the rotor flux. The momentum term, eqn. 9.6, is now composed of two current components. The feedback terms mentioned above decouple and partially linearise the system equations 9.1 and 9.2 and enable us to design a fast linear controller for $i_{mR}(t)$ which represents the rotor flux. This method can only be applied since the stator currents and the rotor speed are measurable.

The task of the flux model (Figure 9.3) is to estimate the flux orientation, based on measurements of the drive stator currents and rotor speed, needed to carry out the reference frame transformation. If $i_{mR}(t)$ is quasi-constant with regard to the changes in $i_{Sq}(t)$ the control path from $i_{Sq}(t)$ to $m_d(t)$ can thus be seen as quasi-linear. Since eqn. 9.4 is linear, a linear controller can be designed for the speed path as well. The linear controllers in both paths are usually designed as PI-P cascades. The controller parameters can be found by various controller design techniques.

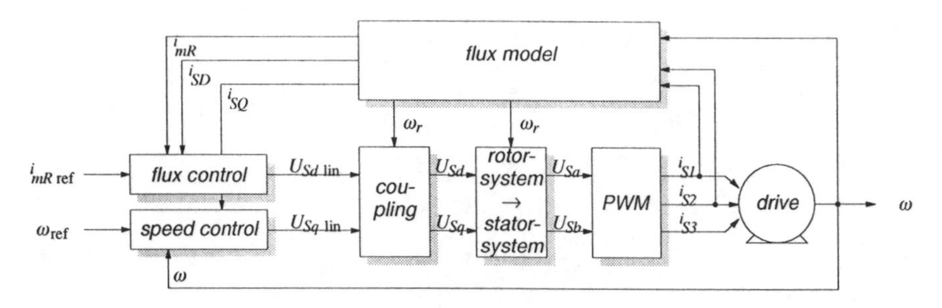

Figure 9.3 *Speed control using field oriented control*

The main disadvantage of the above approach is that unmodelled dynamics are neglected. This may lead to errors in the flux calculation and the coupling terms. As a result the quality of speed or momentum control is not sufficient for high end applications. As already mentioned above it is our aim to improve the quality of induction machine speed control. To achieve this goal we will replace the complete speed control path (see Figure 9.3) including the coupling term by a model based predictive controller (see Figure 9.4). To improve the modelling a neural net based input-output data learning strategy will be used to model the dynamics of the plant's speed path from $U_{Sq}(t)$ to $\omega(t)$.

9.3 Model predictive control

Model based predictive control [5] is often used for slow processes such as chemical plants [6]. A model of the process dynamics is the basis of a nonlinear optimisation procedure. The optimisation is carried out over a finite time horizon in order to determine a series of control values giving an optimal trajectory of the controlled variable. This procedure is computationally very expensive as it often requires iterative algorithms.

It was the major challenge of this particular project to adapt the idea for high speed control of servo drives. The desired sampling time was 2 ms.

Since an optimisation routine is the kernel of the model based predictive control procedure we will start by investigating the optimisation problem. We have

$$\min_{u \in R^m} \quad J \tag{9.7}$$

with J the functional to be minimised,

$$g_i = 0 \qquad i \in [1;n] \tag{9.8}$$

being the equality constraints and

$$g_j \geq 0 \qquad j \in [n+1;k] \tag{9.9}$$

the inequality constraints. For the problem in question we assume a global, minimal solution exists and the optimisation routine used will find this minimum. The performance index J is used to specify the desired control characteristics, e.g.

$$J(u) = \left(y_{ref} - y(u)\right)^2 \tag{9.10}$$

with y_{ref} being the reference signal and $y(u)$ the plant driven by the control signal u.

The equality constraints may be used to encode the process information, typically by means of a process model y_m

$$g_1(u) = 0 = y_m(u) - y(u) \tag{9.11}$$

Inequality constraints may be included in the optimisation problem, but are not necessarily needed. If bounds on control, state or output variables are to be taken into account, one can translate them into inequality constraints, e.g.

$$g_2(u) = |y(u)| - y_{max} \leq 0 \tag{9.12}$$

This is a convenient and straightforward method to introduce safety measures. Usually a maximum angular speed of the drive system has to be guaranteed in order to prevent the mechanical assembly from breaking or the bearings being damaged. Another safety sensitive part in the control loop is the actuator for which restriction to a maximum gain could be important. The optimisation problem for the induction machine application has a minimum and maximum bound on the control voltage U_{Sq}

$$g_2(U_{Sq}) = |U_{Sq}| - U_{Sq_{max}} \leq 0 \tag{9.13}$$

Considering practical, numerical issues it is often advantageous to include inequality constraints on the control variables in order to narrow down the search space to be evaluated by the optimisation routine.

The state-of-the-art technique to solve nonlinear, constrained optimisation problems is the sequential quadratic programming (SQP) procedure which can be found in standard optimisation software packages such as Matlab. We will mention another technique which is computationally less expensive and was thus used in the real-time application.

Now that the optimisation framework is given, we want to describe how the index J has been designed for this particular application. Reformulating eqn. 9.10 for discrete time, we have to introduce a time index k. Thus the prediction horizon has to be taken into account by means of weighted terms. We additionally augmented index J by an additional term penalising changes of the control value in order to smooth the control behaviour.

$$\begin{aligned} J(k) &= \sum_{i=T_d}^{T_h} \alpha_i \cdot \left(\omega_{ref}(k+i) - \omega\left[U_{Sq}(k+i)\right]\right)^2 \\ &+ \sum_{j=1}^{T_u} \beta_j \cdot \left[U_{Sq}(k+j) - U_{Sq}(k+j-1)\right]^2 \end{aligned} \tag{9.14}$$

The solution of the optimisation problem is a time sequence of control values to be applied. At time step k $U_{Sq}(k|k)$ (the control value for time step k calculated at k) is applied. Thus the optimisation has to be repeated at every time step. This procedure is computationally very expensive. Disturbances which have not been accounted for in the plant model, can only be compensated if the equality constraint eqn. 9.11 is properly extended:

$$g_1(U_{Sq}(k+1)) = 0 = \omega_m(U_{Sq}(k+1)) + e(k) - \omega(U_{Sq}(k+1)) \quad (9.15)$$

with $e(k)$ being the error between rotor speed measurement ω and model prediction ω_m. Substituting the future values of ω in eqn. 9.14 using constraint eqn. 9.15 gives

$$J(k) = \sum_{i=T_d}^{T_h} \alpha_i \cdot \left(\omega_{ref}(k+i) - \omega_m\left[U_{Sq}(k+i)\right] + e(k)\right)^2 + \sum_{j=1}^{T_u} \beta_j \cdot \left[U_{Sq}(k+j) - U_{Sq}(k+j-1)\right]^2 \quad (9.16)$$

In this form the index consists of two parts. The left sum in eqn. 9.16 weights the performance of tracking the future reference trajectory $\omega_{ref}(k+i)$ using the system model to predict the future plant behaviour $\omega_m(k+i)$ according to the chosen control variable trajectory. The starting point T_d can be chosen larger than one if a system deadtime is known while T_h is the control horizon.

The error term $e(k)$ is the only appearance of $\omega(k)$ in eqn. 9.16. Its importance lies in introducing feedback, we thus get a closed loop controller capable of compensating modelling mismatch and disturbances (compare Figure 9.4).

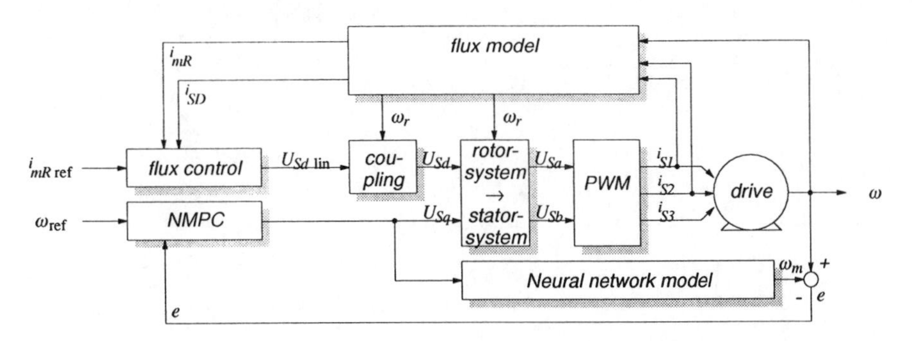

Figure 9.4 *Modifications for neural model predictive control*

The right sum in eqn. 9.16 weights the control effort. The β_j may be chosen large if the energy used to control the plant is to be kept small at the expense of slower output reference tracking. The special form of weighting the differences of the control variables (as seen in eqn. 9.16) can be used in drive control to prevent damage in the inverter as a consequence of too dynamic control actions.

In order to minimise the index J, a sequence of control values $U_{Sq}(k + i)$, $i \in \{1 .. T_h\}$ has to be calculated. Then the first control term $U_{Sq}(k + 1)$ is applied. In the following sampling period the optimisation takes new measurements and deviations between model and process of the previous sample into account. Obviously, the sampling period determines the maximum calculation time for the MPC optimisation. In the following subsection a computationally effective optimisation algorithm is described.

9.3.1 Interval search optimisation

In order to reduce the computational complexity of the optimisation task, several simplifications were made. Restricting the optimisation to the one control term to be applied next $U_{Sq}(k + 1)$ saves much of the computing power which would be needed to solve the multi-dimensional optimisation problem. This can be done assuming that subsequent time steps within the optimisation horizon use the same control value,

$$U_{Sq}(k + j) = U_{Sq}(k + 1), \quad j \in [2 .. T_h] \tag{9.17}$$

A second speed improvement stems from a mathematical simplification. Solving eqn. 9.16 leads to

$$\frac{\partial J}{\partial U_{Sq}(k + 1)} = 0$$
$$= 2 \cdot \sum_{i=1}^{T_h} \alpha_i \left(\omega_{ref}(k) - \omega_m\left[U_{Sq}(k + i)\right] + e(k)\right) \cdot \left(-\frac{\partial \omega_m}{\partial U_{Sq}(k + 1)}\right) \tag{9.18}$$

where the last brackets describe the partial derivative of the model with respect to the control variable. With the additional introduction of the constraints

$$\alpha_i = 0 \, , \quad i \in [1 .. T_h - 1] \tag{9.19}$$

and

$$\alpha_{Th} = 1 \tag{9.20}$$

a valid solution of eqn. 9.18 can be found by setting the term within the first brackets in eqn. 9.18 to zero, disregarding the partial derivatives.

$$0 = \omega_{ref}(k) - \omega_m\left[U_{Sq}(k+i)\right] + e(k) \tag{9.21}$$

The interval search algorithm (IS) is used to find the zero of eqn. 9.21. The algorithm stops at a limited number of iterations according to the sampling rate. The algorithm was found to work efficiently, even if the number of iterations is limited to three or four.

9.3.2 System model based on a radial basis function neural net

Radial basis function neural nets (RBFs) have been described in various papers e.g. [7,8]. The main advantages of RBF nets are good approximation capability at reasonable expense and good interpretability. As another benefit, basis function type networks offer good means for an implementation on parallel hardware. Figure 9.5 shows the RBF network structure where it can be seen that the input and output layer require almost no computational power. The intensive calculations have to be done in the hidden layer which can be scaled down to one hidden unit per processor on a transputer system. Computation speed reduction through interprocess communication is not a problem since the hidden units are not interconnected.

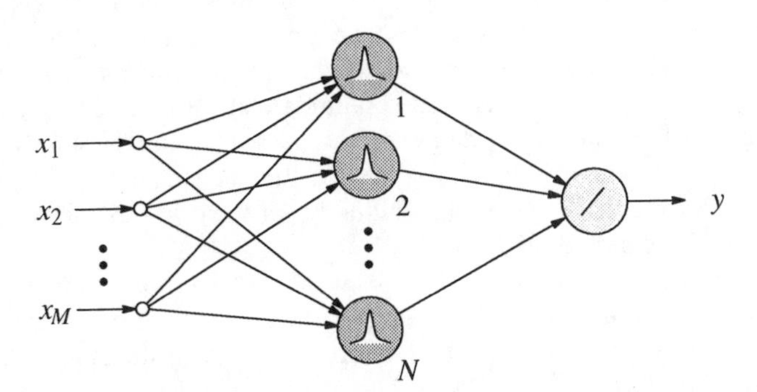

Figure 9.5 *Radial basis function network (RBF)*

The RBF network depicted in Figure 9.5 performs a static, nonlinear mapping of an input vector $\vec{x}$ to a scalar output y. The mapping is given by

$$y = \sum_{i=1}^{N} w_i\, \phi_i \tag{9.22}$$

with ω_i being weighting factors and ϕ_l the continuous, radial basis functions. Radial basis functions can be written in general as

$$\Phi\left(\left\| c - x \right\|\right) \tag{9.23}$$

where c is the centre of the basis function and the argument of ϕ is the Euclidian norm of the input vector to the centre. In our case the functions are described by the well-known Gaussians

$$\phi_i = e^{-\frac{1}{2}\left(\frac{\overline{x - c}}{\sigma_i}\right)^2} \tag{9.24}$$

with centres c_i and widths σ_i. The equations given here describe a MISO system. To obtain a MIMO system one has to add the same structure as in Figure 9.5 for every additional output. Since such a structure is not internally interconnected it can be parallelised just as easily. For this application, the nonlinear modelling task was to capture the induction machine's dynamics in terms of the voltage to speed relationship (see eqn. 9.25) represented in appropriate input-output data.

$$\omega(k+1) = f\left(\left[U_{Sq}(k),...,U_{Sq}(k-n)\right],\left[\omega(k),...,\omega(k-n)\right]\right) \tag{9.25}$$

The predictor-model structure in eqn. 9.25 will be designed to capture the dynamics of the system in question. A static structure as depicted in Figure 9.5 is thus not sufficient. Two different approaches to overcome this problem are shown in Figure 9.6, a straightforward solution already suggested by eqn. 9.25, and in Figure 9.7, the time delay neural network (TDNN) [8]. Both implementations are equally parallelisable and thus adequate for fast real-time applications. For the drive system modelling experiment in this chapter we will concentrate on these structures.

Choosing the inputs, outputs and their delays is a very critical step and in most cases done by exploiting *a priori* knowledge and experience during the modelling phase. However, we will now concentrate on the design of the static RBF. The learning procedure for RBFs can be split into three main parts. First the number of nodes N in the hidden layer and their location x_i must be chosen. We used an algorithm taken from [9] and described in detail in [8] which is very similar to the well known K-means procedure. The second step is to find the appropriate values for the bandwidths σ_i of the Gaussian basis functions. A straightforward way is to define

$$\sigma_i \sim \left\| x_i - x_{nearest\ neighbour} \right\| \tag{9.26}$$

where a proportional scaling factor may be included. Finally, the weights w_i have to be defined. This problem can be stated in a set of linear equations as follows

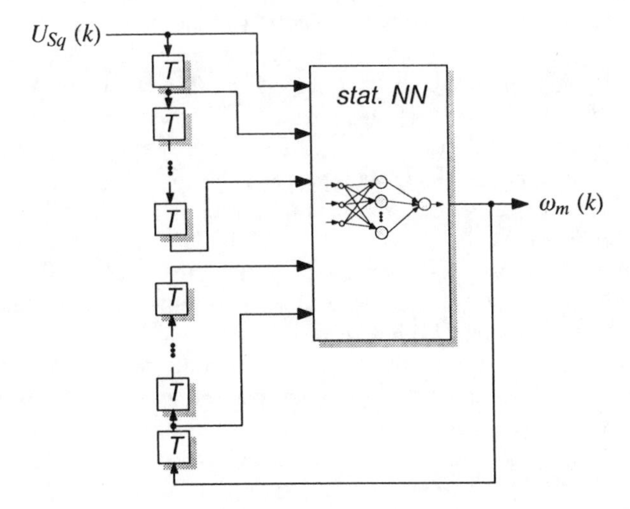

Figure 9.6 *A tapped delay solution to enable dynamic modelling*

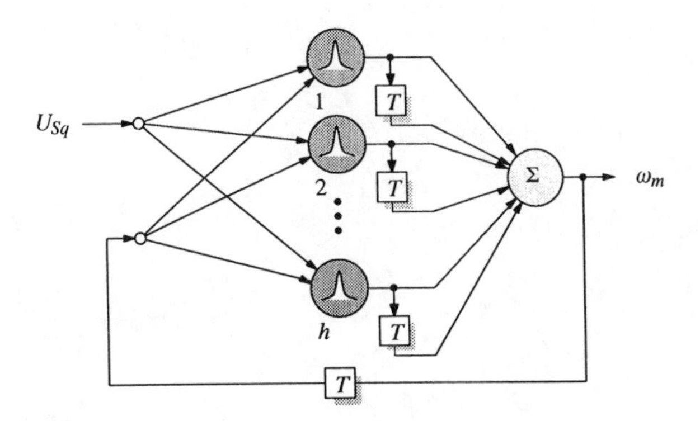

Figure 9.7 *Time delay neural network*

$$\begin{bmatrix} y_1 \\ \vdots \\ y_r \end{bmatrix} = \begin{bmatrix} \phi_{11} & \cdots & \phi_{1h} \\ \vdots & & \vdots \\ \phi_{r1} & \cdots & \phi_{rh} \end{bmatrix} \cdot \begin{bmatrix} w_1 \\ \vdots \\ w_h \end{bmatrix} \tag{9.27}$$

with the number of hidden nodes h and the number of training samples r. In identification problems it should be $h << r$ and thus the Φ-matrix is not regular and also not invertible. We chose the Moore-Penrose pseudo-inverse algorithm to solve this problem. Robust methods using singular value decomposition are available in many standard matrix calculation software packages.

As can be seen by the short description above, the procedure to train an RBF does not need time-consuming iterative algorithms as opposed to gradient methods and is thus much faster. It took approximately 5–15 minutes (according to the number of training samples used) on a SPARC 2 station to run a complete training cycle. This is also a big advantage compared to the multi-layer-perceptron (MLP) structure leaving the design engineer more time to evaluate various design parameter combinations.

It is a difficult, but essential task to set up an appropriate experiment design to receive input-output data that hold sufficient information about the system in question. We chose a ramp function as a basis for our excitation of the FOC speed control path to get a similar amount of training data from each operating region. Various noise signals were superimposed to excite the system. The exciting reference signal used for the described application is shown in Figure 9.8. The data aquisition was done with the available FOC structure, because this gave a guarantee of stable plant behaviour during the data aquisition procedure, a very important feature when real-world processes are to be modelled.

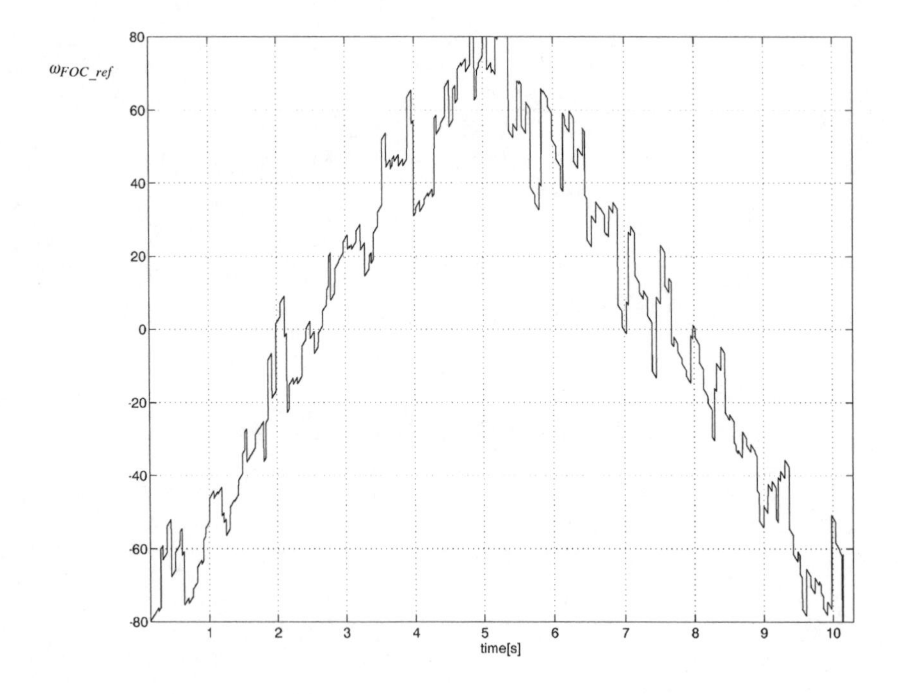

Figure 9.8 *Excitation of the FOC speed control path for input-output training data generation*

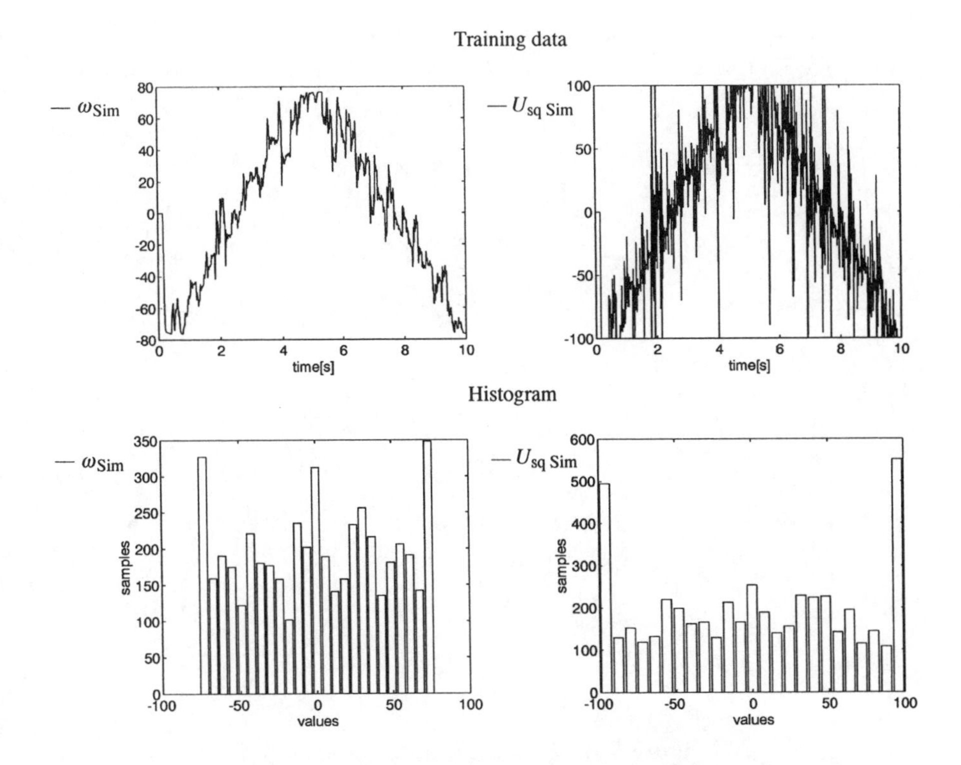

Figure 9.9 *Training data produced by simulation*

The training data collected by simulation using the excitation shown in Figure 9.8 are plotted in Figure 9.9. The histograms in the lower part of Figure 9.9 show that the training data are equally distributed in the operating range. A test data set was used to verify the model quality, note that the test set was not used during the training procedure. The excitation, rotor speed and control variable for the test set can be seen in Figure 9.10.

In Figure 9.11 the response of the real induction machine is plotted against the ac-drive simulation result (using eqns. 9.1 – 9.5) with the same nominal parameters as the real machine. As expected, the real machine showed a different output characteristic from the simulated machine when driven by the same input sequence. This is due to unmodelled effects, such as friction and disturbances, as well as uncertainties in the parameters. This observation explains why we engaged the learning system paradigm to model the ac-drive. We found in many experiments with the above training and test routines, that a RBF model with three delayed values of rotor speed ω and control variable U_{S_q} using approximately 100 hidden unit nodes meets our modelling specifications. To demonstrate this the one-step-ahead predicted behaviour of the neural net model is also given in Figure

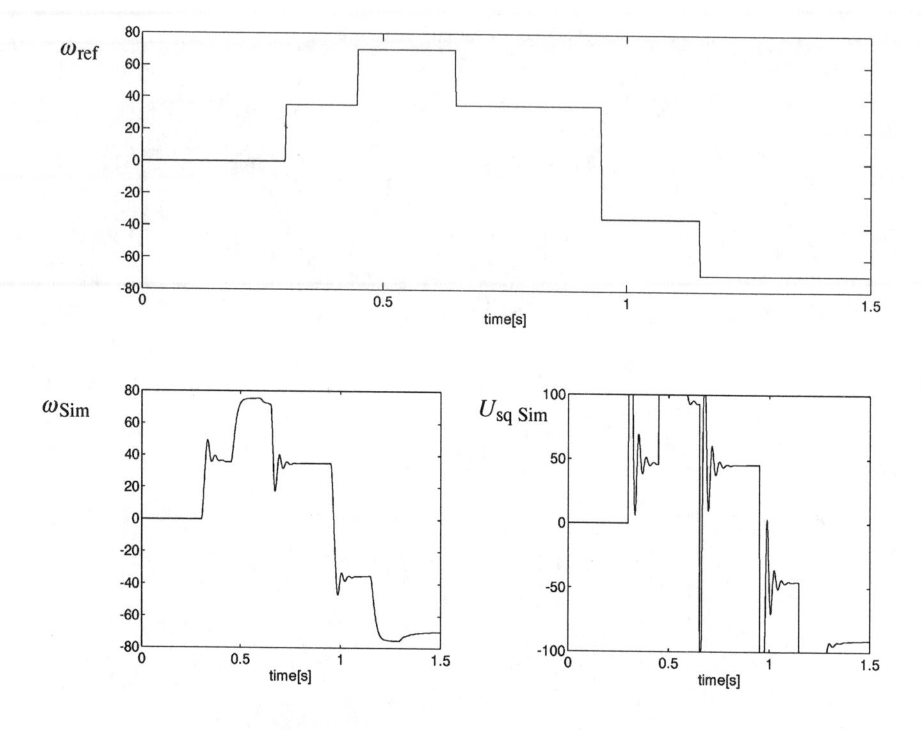

Figure 9.10 *Test data set for simulation*

9.11 and is found to give very good matching results. Since these results are obtained with an input sequence different from the training data, we can expect that the system behaviour is sufficiently captured by the neural networks for use in model predictive control.

Because of the different behaviour of the real-world plant and the simulation, two models had to be trained, one for the simulated and one for the real machine. The RBF model gained from the simulation was used to predesign the MPC controller and evaluate its capabilities before testing it on the real machine.

9.4 Simulation results

Before implementing the MPC controller using the RBF neural network model on the real machine, we estimated its performance via simulation. The simulation framework is the one depicted in Figure 9.4.

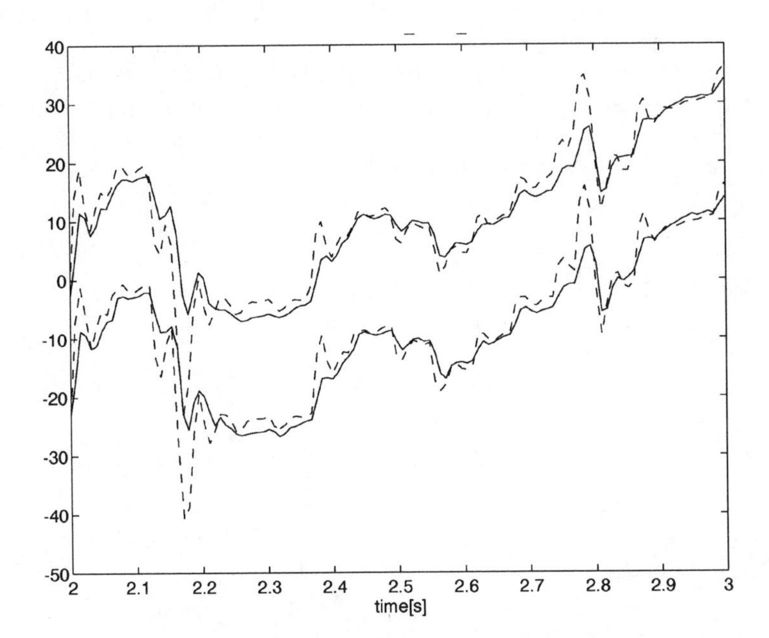

Figure 9.11 *Comparison: Simulation vs. real machine*
machine$_{(\text{shifted} + 10)}$: ——— ω_{real} - - - ω_{Sim}
RBF models$_{(\text{shifted} - 10)}$: ——— ω_{real} - - - ω_{Sim}

9.4.1 MPC using SQP

As mentioned in Section 9.3 the SQP algorithm (we used the version implemented in the Matlab optimisation toolbox) can be used to solve nonlinear, constrained minimisation problems. To demonstrate the controller performance we chose a simple step function as input reference ω_{refMPC} and a step disturbance (step load torque change). The best results were obtained with the following parameters for the index J:

$$T_d = 1,\ T_h = 5,\ \alpha = [5\ 5\ 5\ 5\ 5]$$

$$T_u = 3,\ \beta = [0.1\ 0.1\ 0.1]$$

There was also a bound of maximal 20 optimisation iteration steps, but usually much less were needed.

The step response using the above parameter settings is plotted in Figure 9.12. It can be seen that the controller is stable and the output converges fast towards the reference signal with almost no overshoot. The control signal is shown in the same figure on the bottom right. The oscillation near the steady state region is due to slight modelling errors plotted on the bottom left. In this special case it would be advantageous to neglect the modelling error, this would, however, reduce the disturbance rejection capabilities. An error filter can be added to compensate these effects, but was not investigated in our context. Using the modelling error e in the index J introduces feedback in the MPC control scheme and disturbance rejection can be achieved as demonstrated in Figure 9.13.

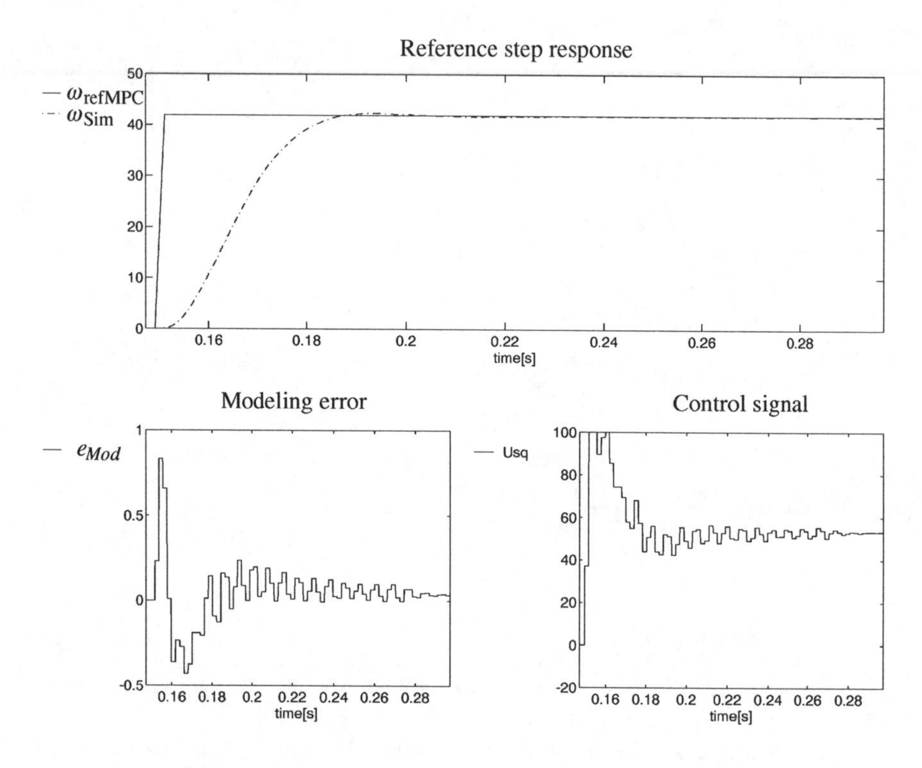

Figure 9.12 *MPC with SQP step input reference signal*

9.4.2 MPC using interval search

The real-time implementation requires a computationally less expensive optimisation routine than SQP. In Section 9.3.1 we introduced the interval search as a much simpler routine. Its main advantage is that given a certain interval range a solution within a known precision is found after a definite number of iterations. This enables us to precalculate the maximum computing time for a given accuracy, a very important feature for the real-time implementation.

The optimisation index parameters used are:

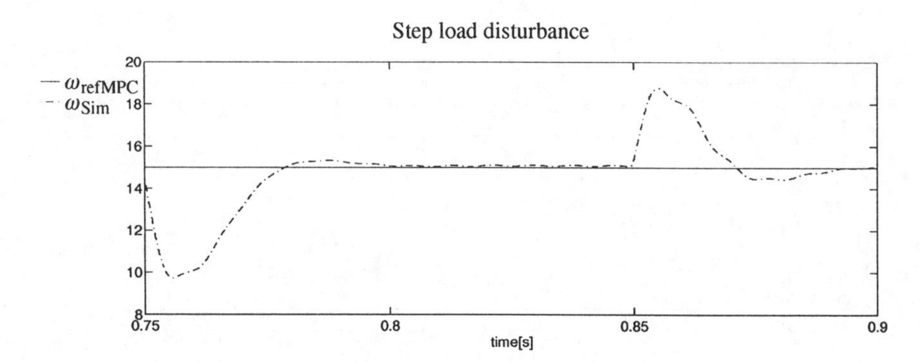

Figure 9.13 *MPC with SQP step disturbance signal*

$T_d = 1,\ T_h = 5,\ \alpha = [5\ 5\ 5\ 5\ 5]$

$T_u = 1,\ \beta = [1]$

Figure 9.14 shows the disturbance step response of this configuration. Comparing Figure 9.14 with Figure 9.13 one can observe that the amplitude of the control error is much bigger using the interval search than with SQP. This is a consequence of the fact that the interval search procedure optimises only the next control sequence value (the following are kept constant) and is thus more restrictive. Nevertheless, the disturbance rejection is satisfying. The next challenge is the real-time implementation described in the following section.

9.5 Real-time implementation using a transputer system

The neural model predictive control structure has been implemented on a multi-transputer hardware developed for various research projects, e.g. [10].

9.5.1 Hardware structure

Figure 9.15 shows the multi-processor structure. The transputer system included seven T800 transputers and one T222. The transputer's serial communication links can be soft-configured according to the process scheme in order to achieve a powerful system.

The acquisition of analogue and digital machine data is done by the T222-based I/O-board. It can be equipped with application specific I/O-modules. For this particular project the induction machine's speed is observed by an incremental encoder delivering a measure for the angular rotor position. By using a four-fold interpolation, a resolution of 10 000 increments per revolution is achieved. The

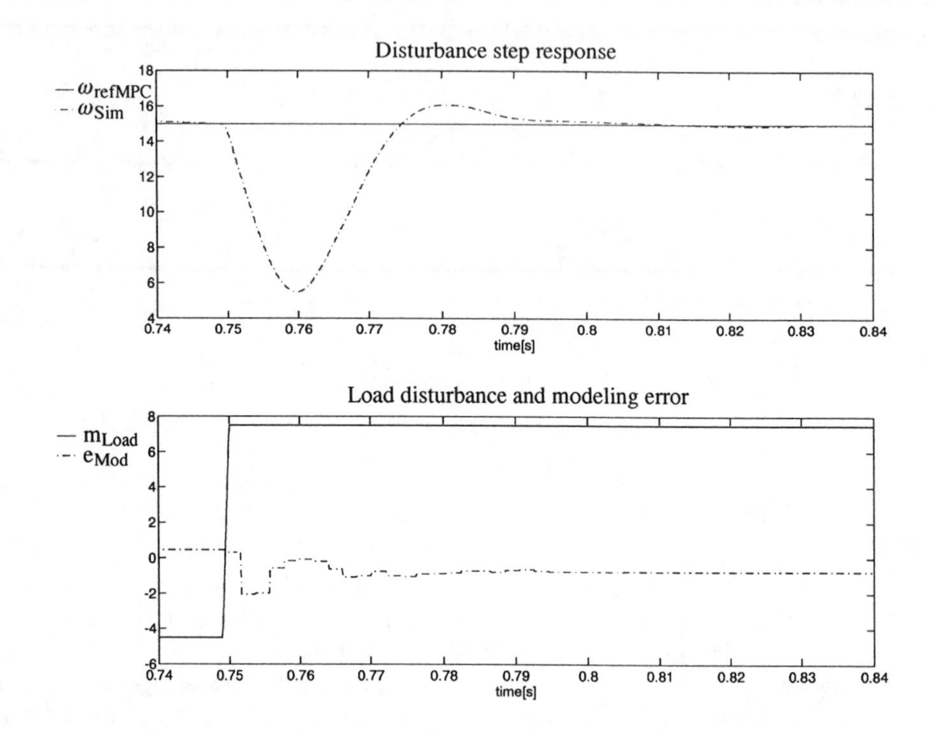

Figure 9.14 *Load disturbance rejection with MPC using interval search*

speed is then calculated by differentiating the position with respect to time. Two 12-bit A/D-converters with a conversion time of 5 μs are used to determine the two stator currents. The PWM signal is generated by two 16-bit timers and a programmable logic device. In order to achieve an almost sinusoidal shaped signal a PWM frequency of 15 625 kHz was used.

The voltage source inverter is realised with six MOSFETs. They form a full-bridge that is connected to the three windings of the star connected machine. The MOSFET inverter is controlled by a digital pulse pattern which is generated according to the value of U_{Sq} using a three space-phasor method.

The machine is a 1.5 kW three phase AC machine with four poles. The AC servo-drive is coupled with a DC machine to be able to apply loads. The machine is characterised by the following parameters:

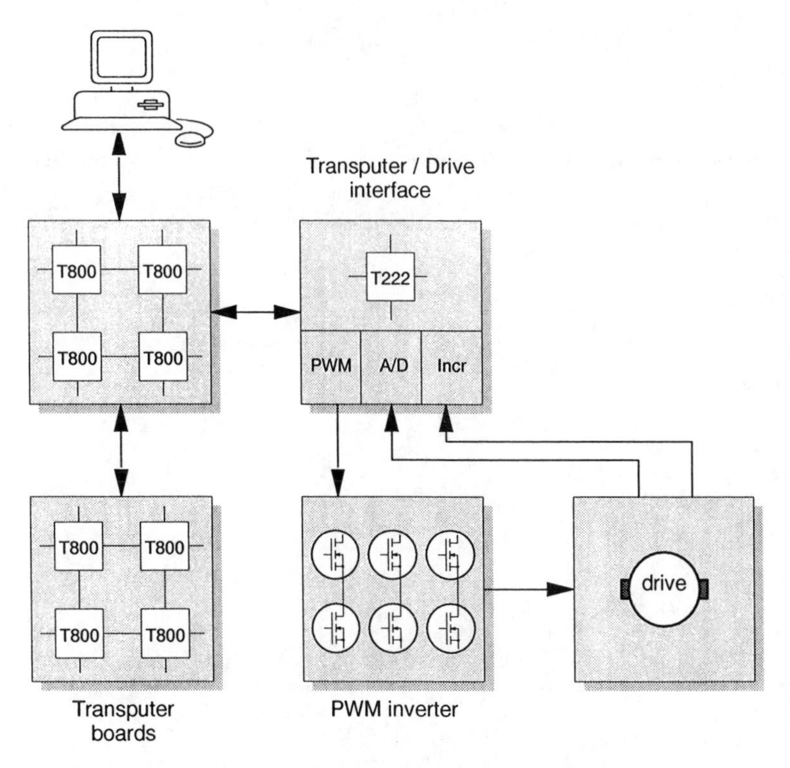

Figure 9.15 *The multi-transputer hardware*

Parameter	Symbol	Value
Nominal power	P_N	1.5 kW
Nominal voltage	U_N	380 V
Nominal current	I_N	3.7 A
Nominal torque	M_N	10 Nm
Number of pole pairs	z_p	2
Efficiency	cos Φ	0.82
Moment of inertia	J	0.009 Nms^2
Stator resistance	R_S	6.5 Ω
Rotor resistance	R_R	3.4 Ω
Stator inductance	L_S	0.34 H
Rotor inductance	L_R	0.34 H

9.5.2 *Software structure and parallel processing*

The control system software graph is shown in Figure 9.16. The system is booted and controlled by a PC. The 'Shell'-processor runs several user interface processes. The main process handling all the other parts of the software is called 'Control'. It is linked with the two control loops (Figure 9.4), named 'Flux' and 'MPC'. 'MPC' works together with four parallel 'RBF'-processes, each of them being a quarter of the total neural network.

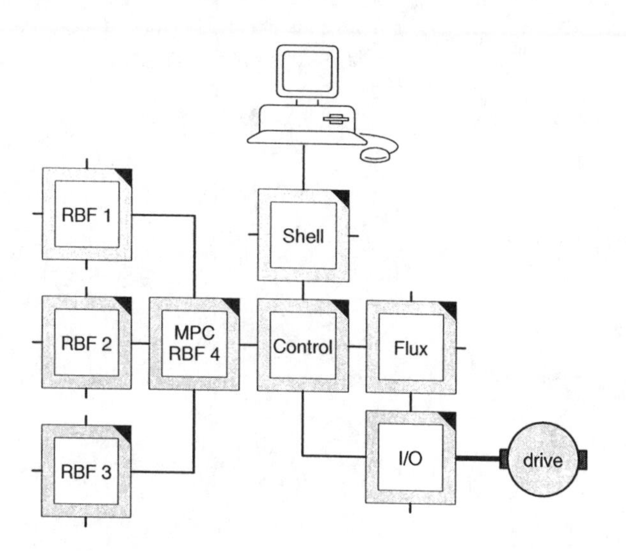

Figure 9.16 *Process graph showing the parallel computation of the RBF-network*

The flux control path and other maintenance tasks were already implemented [10]. The computation time for the RBF process on a single T800 transputer was approximately 2 ms. Different simulations showed that a prediction horizon of two to three time steps and at least four optimisation iterations would be necessary in order to achieve satisfying control results. Under these assumptions, the calculation of one predicted speed value would have to be repeated 12 times resulting in a sampling rate of 24 ms, which is far too long.

The problem was solved by utilising four transputers for the calculation of the exponential functions in the Gaussian layer. It is obvious that this parallelisation can be scaled down to one processor calculating one hidden unit output, since each hidden unit has the same input vector which can be supplied at the same time. The resulting calculation time would then be dominated by data exchange times.

In the implementation of the standard RBF-net, each of the four T800 processors had to calculate 30 hidden Gaussians. In addition, the MPC control process was executed by one of these processors. This implementation resulted in an execution time of 7.8 ms allowing for a speed path control rate of 8 ms.

A further improvement was achieved by using more powerful network architectures, namely the time-delay RBF (TDRBF) [8], and the local model network (LMN) [11]. The TDRBF offered a similar modelling accuracy with only 30 units. Using the LMN approach, the number of units was decreased to seven. This led to execution times below 4 ms and the LMN allowed 2 ms. The next section gives some results for the different network architectures.

9.6 Results

The results illustrate the strength of NMPC. A comparison of the neural control approach with a field-oriented control implementation on the basis of simulations is given in [3]. As mentioned earlier, the real-time implementation was targeted to achieve a control period of 2 ms. With a simple RBF network of 120 nodes the control period was limited to 8 ms. In addition, the maximum number of iterations allowed for the optimisation was limited.

We investigated two different values for the prediction horizon length T_h, in particular two and three steps. The two-step version NMPC (Figure 9.17) could interpolate up to six times. With a horizon of three steps only four iterations were possible.

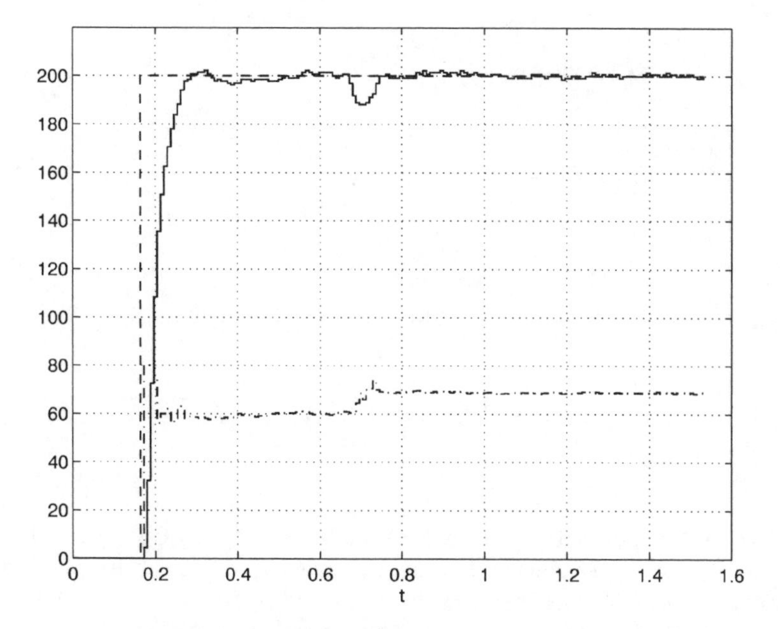

Figure 9.17 *MPC with RBF model, 8 ms, T_h = 2*

- - - n_{ref} ——— n - · - U_{Sq}

The plot above shows the result of real-time NMPC with $T_h = 2$. An almost perfect result without any overshoot can be seen. The predictive controller reduces the control value (dash-dotted line) as it recognises a potential overshoot in the future.

The deviation at about 0.7 s is due to a load change. The induction machine is coupled with a dc-drive. Since the dc-drive runs without electric load, the mechanical load is also zero. After the desired speed of 200 r.p.m. is reached, a resistor is switched into the dc-drive's anchor circuit. The characteristic of that load change varies for the different experiments since the switch is closed by hand.

The next two plots show the same experiment with different network achitectures. First, the result of NMPC with a TDRBF with 30 units is shown (Figure 9.18). Since the network size is less than the size of the RBF network, NMPC also worked with a sampling time of 4 ms. Due to the limitation of the number of iterations, the 4 ms result was not as satisfying as the former.

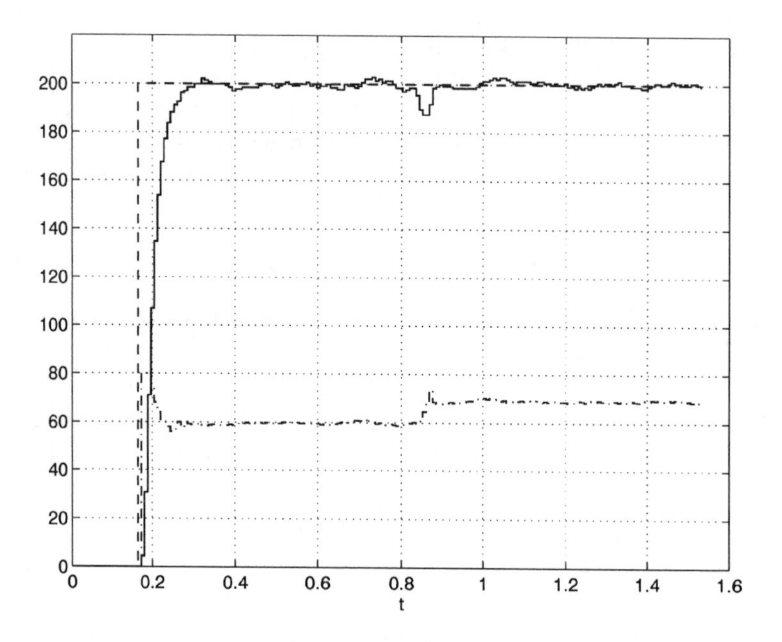

Figure 9.18 *MCP with time-delay RBF, 8 ms*

- - - n_{ref} ——— n - · · U_{Sq}

In contrast, Figure 9.19 shows a result of MPC at 4 ms. Obviously, the step response is much faster due to the higher dynamic capabilities of the optimisation procedure. The load response seems to have a similar behaviour. The LMN had just seven basis functions, so it was possible to run MPC at 2 ms. The results of the 2 ms experiment are worse due to limitations of the number of iterations and the reduced prediction horizon.

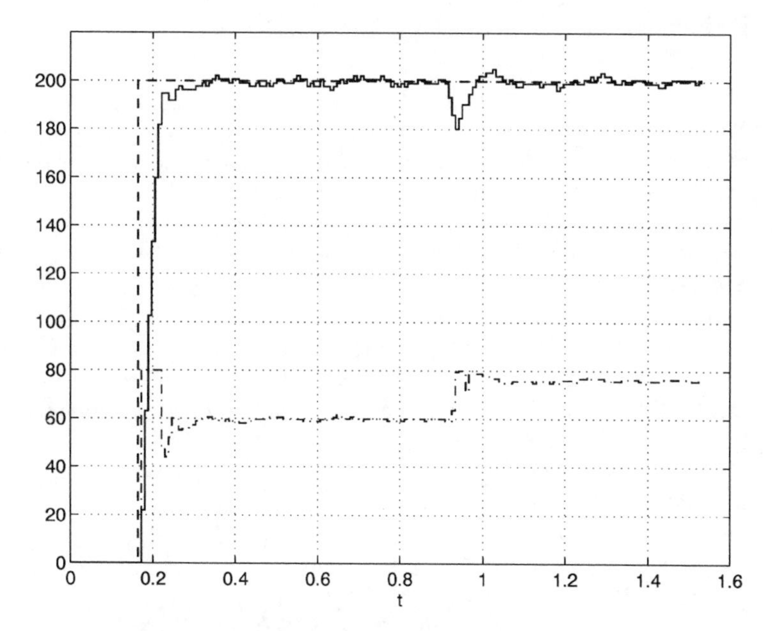

Figure 9.19 *MPC with local model network, 4 ms*
- - - n_{ref} —— n - · - U_{Sq}

9.7 Conclusions

Combining the techniques of neural network modelling with basis function neural nets and parallel processing hardware can provide fast, real-time nonlinear system models. This work shows that the control strategy MPC, which is usually used for slow control applications, can also be applied to highly dynamic systems.

The NMPC structure has been implemented and tested under real-world, real-time conditions controlling an induction servo drive with a dc-drive load. We showed results demonstrating the high potential of that nonlinear control structure. MPC is capable of combining an optimal rise time with minimised overshoot. Furthermore, the results of Figure 9.11 indicate that a neural network model trained on process data gives a more accurate description of the process behaviour than a mathematical model would do. Better representation of the process dynamics will improve the control performance as well.

The parallel computing hardware was fully exploited running parallelised versions of the different neural network architectures. With modern microprocessors, e.g. parallel DSPs, the proposed optimisation based on an interval search algorithm (IS) could be substituted by a more general method, the sequential quadratic programming algorithm (SQP). This state-of-the-art algorithm has been investigated in simulations. It needs at least four times the computing effort of the IS algorithm. With the next generation of control hardware these more sophisticated methods can also be applied.

9.8 References

[1] Murray-Smith, R., Neumerkel, D. and Sbarbaro-Hofer, D., 1992, Neural Networks for Modelling and Control of a Non-Linear Dynamic System, *Proceedings 7th IEEE International Symposium on Intelligent Control.*

[2] Sbarbaro-Hofer, D. Neumerkel, D. and Hunt, K., 1993, Neural Control of a Steel Rolling Mill, *IEEE Control Systems*, Vol. 13, No. 3.

[3] Neumerkel, D., Franz, J., Krüger, L. and Hidiroglu, A., 1994, Real-Time Application of Neural Predictive Control for an Induction Servo Drive, *Proceedings 3rd IEEE Conference on Control Applications*, p. 433.

[4] Leonhard, W., 1986, Microcomputer Control of High Dynamic Performance ac-Drives - A Survey, *Automatica*, Vol. 22, No. 1, pp. 1-19.

[5] García, C.E., Prett, D.M. and Morari, M., 1989, Model Predictive Control: Theory and Practice - a Survey, *Automatica*, Vol. 25, No. 3.

[6] Psichogios, D.C. and Ungar, L.H., 1989, Nonlinear Internal Model Control and Model Predictive Control Using Neural Networks, Appendix B in *Pattern Recognition and Neural Networks*, Adison-Wesley.

[7] Poggio, T. and Girosi, F., 1990, Networks for Approximation and Learning, *IEEE Proceedings*, Vol. 78, No. 9.

[8] Neumerkel, D., Murray-Smith, R. and Gollee, H., 1993, Modelling Dynamic Processes with Clustered Time-Delay Neurons, *Proceedings IJCNN-93-Nagoya International Joint Conference on Neural Networks.*

[9] Pao, Y., 1989, Unsupervised Learning Based on Discovery of Cluster Structure, *Proceedings 5th IEEE International Symposium on Intelligent Control*, Vol. 2.

[10] Beierke, S., 1992, Vergleichende Untersuchungen von unterschiedlichen feldorientierten Lagereglerstrukturen für Asynchron-Servomotoren mit einem Multi-Transputer-System, PhD thesis, Technical University of Berlin.

[11] Johansen, T.A. and Foss, B.A., 1992, A NARMAX Model Representation for Adaptive Control Based on Local Models, *Modeling, Identification and Control*, Vol. 13, No. 1.

Chapter 10

Fuzzy-neural control in intensive-care blood pressure management

D.A. Linkens

10.1 Introduction

Living systems, either at organ or whole body levels, offer major technological challenges to both dynamics understanding and manipulation. That such systems are dynamic is well-known and has long been recognised by physiologists and clinicians, with the phrase 'homeostasis' being coined by Cannon in 1932 to describe the inherent self-regulating feedback mechanisms of the human body. Having said that, there are many problems in eliciting and externally controlling these dynamics. The levels of complexity are such that vast interactions exist, leading to the need for multivariable intervention via therapeutic drug administration. In addition, the dynamics are frequently heavily nonlinear, often time-varying, and commonly involve time delays. Further, uncertainty often exists as to the precise structural nature of stimulus/response models, as well as massive parameter variability both intra-patient and inter-patient.

In spite of this daunting array of problems, successful feedback control of drug administration has been achieved in several areas of biomedicine [1]. In most cases this has employed classical 3-term controllers, and this has often proved inadequate to cater with the above-mentioned dynamics situation. To alleviate the patient variability problem, model based predictive control has been implemented in muscle relaxant anaesthesia [2], and extended to multivariable control to include the regulation of blood pressure (for depth of anaesthesia purposes) [3]. In simulation studies, the use of GPC (generalised predictive control) has been compared with SOFLC (self-organising fuzzy logic control) for this multivariable clinical application [4]. The conclusion reached was that GPC was superior in performance, but only if an accurate model of the patient could be elicited on-line and up-dated regularly.

The ability of fuzzy logic control to deal with nonlinearities and process uncertainties makes it an attractive possibility for biomedical applications. Indeed, in early work on muscle relaxant anaesthesia, simulation studies had shown its feasibility [5]. More recently, this has been validated in clinical trials [6] for single-variable regulation of drug-induced paralysis. A major bottleneck, however, is the well-known difficulty of eliciting an adequate rule base for the fuzzy inference engine. Thus, there has been much interest recently in the merging of fuzzy and neural technologies to provide synergetic advantages of good self-learning, explanation and robustness properties.

This chapter addresses three fundamental and important issues concerning the implementation of a knowledge based system, in particular a fuzzy rule based system, for use in intelligent control. These issues are knowledge acquisition, computational representation, and reasoning. Our primary objective is to develop systems which are capable of performing self-organising and self-learning functions in a real-time manner under multivariable system environments by utilising fuzzy logic, neural networks, and a combination of both paradigms with emphasis on the novel system architectures, algorithms, and applications to problems found in biomedical systems. Considerable effort has been devoted to making the proposed approaches as generic, simple and systematic as possible.

The research consists of two independent but closely correlated parts. The first part involves mainly the utilisation of fuzzy algorithm based schemes, whereas the second part deals primarily with the subjects of fuzzy-neural network based approaches. More specifically, in the first part a unified approximate reasoning model has been established suitable for various definitions of linguistic connectives and for handling possibilistic and probabilistic uncertainties.

The second (and major) part of the work has been devoted to the development of hybrid fuzzy-neural control systems. A key idea is first to map an algorithm based fuzzy system onto a neural network, which carries out the tasks of representation and reasoning, and then to develop the associated self-organising and self-learning schemes which perform the task of knowledge (rule) acquisition. Two mapping strategies, namely functional and structural mappings, are distinguished leading to distributed and localised network representation respectively. In the former case, by adopting a typical backpropagation neural network (BNN) as a reasoner, a multi-stage approach to constructing such a BNN based fuzzy controller has been investigated. Further, a hybrid fuzzy control system with the ability of self-organising and self-learning has been developed by adding a variable-structure Kohonen layer to a BNN network. In the latter case, we present a simple but efficient scheme which structurally maps a simplified fuzzy control algorithm onto a counter propagation network (CPN). The system is characterized by explicit representation, self-construction of rule-bases, and fast learning speed. More recently, this structural approach has been utilised to incorporate fuzzy concepts into neural networks based on radial basis functions (RBF) and the cerebellar model articulation controller (CMAC).

Throughout this research, it has been assumed that the controlled processes will be multi-input, multi-output systems with pure time delays in controls, that no mathematical model about the process exists, but some knowledge about the process is available. As a demonstrator application, a problem of dual-input/dual-output linear multivariable control for blood pressure management has been studied extensively with respect to each proposed structure and algorithm.

10.2 Fuzzy logic based reasoning

Assume that the system to be controlled has n inputs and m outputs denoted by $X_1, X_2, \ldots, X_n$ and Y_1, Y_2, ..., Y_m. Furthermore, it is assumed that the given L 'IF *situation* THEN *action*' rules are connected by ALSO, each of which has the form of

$$\begin{array}{l} \text{IF} \quad X_1 \text{ is } A_1^j \text{ AND } X_2 \text{ is } A_2^j \text{ AND } \cdots \text{ AND } X_n \text{ is } A_n^j \\ \text{THEN} \quad Y_1 \text{ is } B_1^j \text{ AND } Y_2 \text{ is } B_2^j \text{ AND } \cdots \text{ AND } Y_m \text{ is } B_m^j \end{array} \tag{10.1}$$

where A_i^j and B_k^j are fuzzy subsets which are defined on the corresponding universes, and represent some fuzzy concepts such as big, medium, small etc. Then the problems of concern are how to represent knowledge described by rules numerically and how to infer an approximate action in response to a novel situation. One of the solutions is to follow the inferencing procedures used by traditional data-driven reasoning systems and take the fuzzy variables into account [7,8]. The basic idea is first to treat the L rules one-by-one and then to combine the individual results to give a global output.

10.3 Functional fuzzy-neural control

The first type of fuzzy-neural control considers a *functional* mapping utilising standard backpropagation neural networks (BNN). Within this context, a crucial aim is the automatic knowledge acquisition of the fuzzy rules. Aiming at establishing an identical environment in both training and application modes without involving any converting procedures, here the linguistic labels are represented by fuzzy numbers, each of which is typically characterised by a central value and an interval around the centre. The central value is most representative of the fuzzy number and the width of the associated interval determines the degree of fuzziness. Thus, the control rule, eqn. 10.1, can be rewritten as

$$\text{IF } X \text{ is } (\hat{u}, \delta^{j}) \text{ THEN } Y \text{ is } (\hat{v}^{j}, \gamma^{j}) \tag{10.2}$$

where $\hat{u}^j$, $\hat{v}^j$, δ^j and γ^j are the input and output central value vectors and width vectors in the *j*th rule. Now it is possible to train the BNN with only the central value vectors, i.e ($\hat{u}^j$, $\hat{v}^j$) pairs while leaving the width vectors untreated explicitly. This can be justified by the fact that the BNN network inherently possesses some fuzziness which is exhibited in the form of interpolation over new situations [9]. This interpretation provides a basis for learning and extracting training rules directly from the controlled environment.

10.3.1 Neural network based fuzzy control

The principal objective in this section is to develop a BNN based fuzzy control system capable of constructing training rules automatically. The basic idea is first to derive required control signals and then to extract a set of training samples [9]. Thus the whole process consists of four stages: on-line learning, off-line extracting, off-line training, and on-line application. This cycle may be repeated if necessary.

10.3.1.1 Learning

Figure 10.1 shows a block diagram of the learning system consisting of a reference model, a learning mechanism, a controller input formation block (InF), a short term memory (STM), and the controlled process. By operating the learning mechanism iteratively, the desired control performances specified by the reference model subject to the command signals are gradually achieved and the desired control action denoted by v_d is derived.

At the same time, the controller input $\bar{u}$ together with the learned control v_d are stored in the STM as ordered pairs. Notice that there are two types of errors in Figure 10.1, namely, learning error e_L and control error e_C. The former is defined as the difference between the output of the reference model y_d and the output of the process y_p, whereas the latter is the discrepancy between the output of the command signal *r* and the output of the process y_p.

Suppose that the desired response vector y_d is designated by a diagonal transfer matrix, the goal of the learning system is to force the learning error $e_L(t)$ asymptotically to zero or to a pre-defined tolerant region ε within a time interval T of interest by repeatedly operating the system. More specifically, we require that $\|e_{Lk}(t)\| \to 0$ or $\|e_{Lk}(t)\| < \varepsilon$ uniformly in $t \in [0,T]$ as $k \to \infty$, where k denotes the iteration number. Whenever the convergence has occurred, the corresponding control action at that iteration is regarded as the learned control v_d.

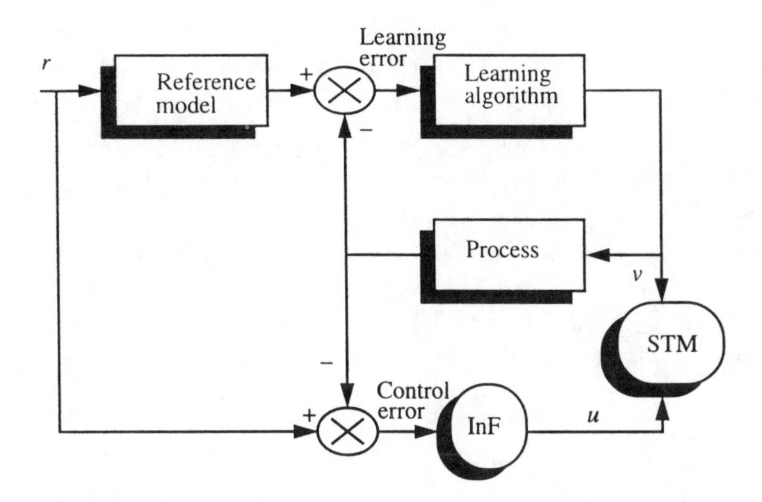

Figure 10.1 *A generalised fuzzy-neural learning scheme*

According to the learning goal described above, the corresponding learning law is given by [10]

$$v_{k+1}(t) = v_k(t) + P.e_{Lk}(t+\lambda) + Q.\dot{e}_{Lk}(t+\lambda) \tag{10.3}$$

where v_{k+1}, $v_k \in R^m$ are the learning control vector-valued functions at the kth and the $(k+1)$th iterations respectively, e_{Lk}, $\dot{e}_{Lk} \in R^m$ are the learning error and its derivative vector-valued functions, λ is an estimated time advance corresponding to the time delay of the process, and P, Q $\in R^{m\times m}$ are constant learning gain matrices.

10.3.1.2 Extracting

While the effects of δ^j and γ^j can be automatically and implicitly accommodated by the BNN, the primary effort for extracting training rules will be devoted to deriving the central value vectors $\hat{u}^j$ and $\hat{v}^j$. Suppose that, at the k_dth learning iteration, the desired control vector sequence $v_d(l)$ is obtained with which the response of the process is satisfied with $l \in [0, I_t]$, where I_t is the maximum sampling number. Meanwhile, the control error $e_c(l)$ is measured, expanded, and denoted by $\bar{u}(l)$. Combining $\bar{u}(l)$ with $v_d(l)$, a paired and ordered data set Γ_t: $\{\bar{u}(l), v_d(l)\}$ is created in the STM. Basically, to obtain a set of rules in the form of eqn. 10.2 based on the recorded data set is a matter of converting a time-based Γ_t into a value-based set Γ_v. Each data association in Γ_v is representative of several elements in Γ_t. Accordingly, the extracting process can be thought of as a grouping process during which I_t+1 data pairs are appropriately classified as I_v groups, each of which is represented by one and only one data pair. Thus two functions, grouping and extracting, are involved.

First, each component $\bar{u}(l)$ is scaled and quantised to the closest element of the corresponding universe, thereby forming a new data set Γ_t^*: $\{\tilde{u}(l), v_d(l)\}$. Next, the data pairs in Γ_t^* are divided into I_v groups by putting those pairs with the same value of u(l) into one group Γ_t^P ; that is,

$$\Gamma_t^P : (\tilde{u}^P, v_{d1}^P), (\tilde{u}^P, v_{d2}^P), \cdots, (\tilde{u}^P, v_{dQ}^P) \tag{10.4}$$

Note that, due to the pure time delay possessed by the controlled process and the clustering effects, the Q different v_d^P may be associated with the same $\tilde{u}^P$ in value, referred to as a conflict group. Therefore, the third step is to deal with conflict resolution. Here a simple but reasonable scheme is employed which takes the average of v_{d1}, v_{d2}, ... v_{dQ} as their representative denoted as v^P. Now after conflict resolution, I_v distinct data pairs are derived in a new data set Γ_v: $\{\tilde{u}^P, v^P\}$. By considering that $\tilde{u}^P$ and v^P are the central values corresponding to some fuzzy numbers, the above data pairs can be expressed in the form of 'IF $\tilde{u}^P$ THEN v^P' and can be used to train the BNN based controller.

The disadvantage of this structure is that it requires a 4-stage process for knowledge acquisition and implementation. The next section describes an alternative approach which is a single-stage paradigm.

10.3.2 Hybrid neural network fuzzy controller

Figure 10.2 shows a schematic of this system structure [11]. It consists of a basic feedback control loop containing a hybrid network based fuzzy controller, a controlled process, and a performance loop composed of reference models and learning laws.

Assuming that neither a control expert nor a mathematical model of the process is available, the objectives of the overall system are: (a) to minimise the tracking error between the desired output specified by the reference model and the actual output of the process in the whole time interval of interest by adjusting the connection weights of the networks, and meanwhile (b) to construct control rule bases dynamically, by observing, recording, and processing the input and output data associated with the net-controller. The whole system performs the two functions of control and learning simultaneously. Within each sampling period, a feedforward pass produces the suitable control input to the process and the backward step learns both control rules and connection weights. The function of each component in Figure 10.2 is described below.

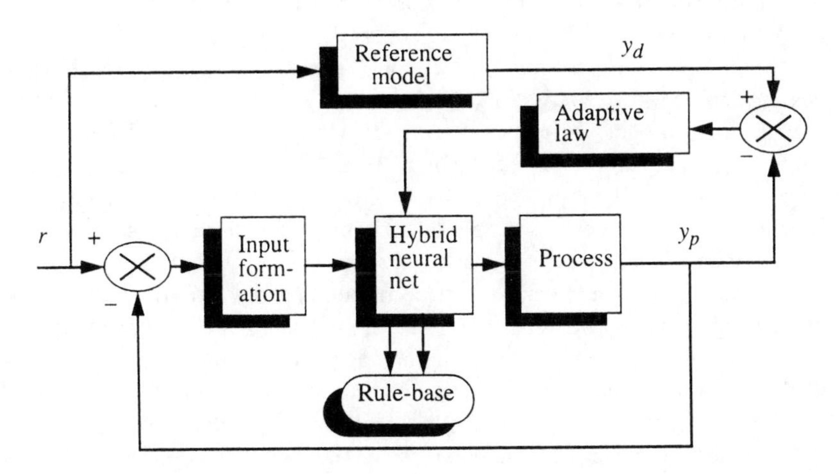

Figure 10.2 *A hybrid fuzzy-neural learning system*

Input formation. Corresponding to traditional PID controllers, the inputs to the network based controller are chosen to be various combinations of control error e_c, change-in-error c_c, and sum of error s_c. Three possible controller modes denoted by the vector u can be constructed by combining e_c with c_c, e_c with s_c, and e_c with c_c and s_c, being termed as EC, ES, and ECS type controllers respectively.

Reference model. The desired transient and state-steady performances are signified by the reference model which provides a prototype for use by the learning mechanism. For the sake of simplicity a diagonal transfer matrix is used as the reference model, with the elements being a second-order linear transfer function.

Hybrid neural networks. As a key component of the system, the hybrid neural network, as shown in Figure 10.3, is designed to stabilise the dynamical input vector u, compute the present control value, and provide the necessary information for the rule base formation module. To be more specific, the VSC network converts a time-based, on-line incoming sequence into value-based pattern vectors which are then broadcast to the BNN network, where the present control is calculated by the BNN using feedforward algorithms. Competitive or unsupervised learning can be used as a mechanism for adaptive (learning) vector quantisation (AVQ or LVQ) in which the system adaptively quantises the pattern space by discovering a set of representative prototypes. A basic idea regarding the AVQ is to categorise vectorial stochastic data into different groups by employing some metric measures with winner selection criteria. An AVQ network is usually composed of two fully connected layers, an input layer with n units and a competitive layer with p units. The input layer can be simply receiving the incoming input vector $u(t) \in R^n$ and forwarding it to the competitive layer through the connecting weight vectors $w_j(t) \in R^n$. In view of the above discussion, the operating procedures of the VSC competitive algorithm can be described as follows. With each iteration and each sampling time, the processed control error vector u is received. Then a competitive process takes place among the existing units in regard to the current u by using the winner selection criteria. If one of the existing units wins the competition, the corresponding w_j is modified; otherwise, a new unit is created with an appropriate initialisation scheme. Finally, the modified or initialised weight vector is broadcast to the input layer of the BNN net. Thus, the VSC can be regarded as a preprocessor or a pattern supplier for the BNN. From the knowledge processing point of view, the networks not only perform a self-organising function by which the required knowledge bases are constructed automatically, but also behave as a real-time fuzzy reasoner in the sense described previously.

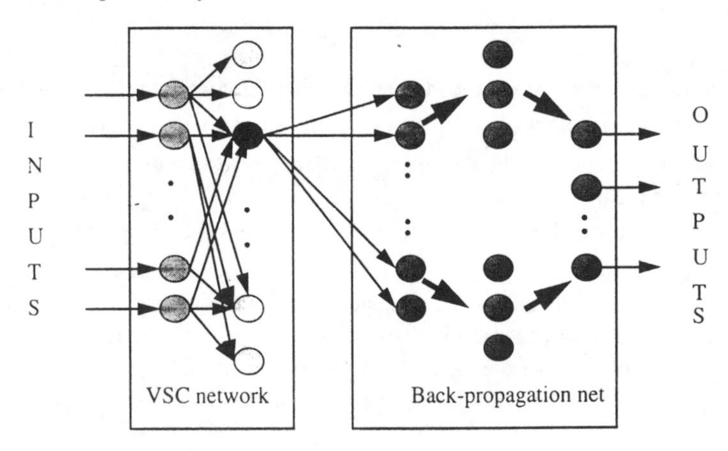

Figure 10.3 *Hybrid neural network structure*

Rule-base formation. By appropriately recording the input and output of the hybrid neural networks, the formed rule base, each rule having the format of IF *situation* THEN *action*, offers an explicit explanation of what the net based controller has done. This can be understood in the following way. Because the information used to guide the adjustment of the connection weights of the BNN is similar to that for constructing the IF-THEN rules, the mapping property of the resultant BNN must exhibit similar behaviour to that of the IF-THEN conditional statements. Again, from the knowledge representation viewpoint, both paradigms in effect represent the same body of knowledge using the different forms. The BNN represents it distributedly and implicitly, while IF-THEN models describe it concisely and explicitly.

Adaptive mechanism. While the VSC network can be self-organised by merely relying on the incoming information, the weights of the BNN network must be adjusted by backpropagating the predefined learning errors. Due to the unavailability of the teacher signals, a modified and simple adaptive law has been developed.

10.4 Structural fuzzy-neural control

In this section a number of approaches to *structural* mapping between fuzzy and neural approaches are given. In all of these, the initial basis is a simplified form of fuzzy control which is now described.

10.4.1 Simplified fuzzy control algorithm

The operation of a fuzzy controller typically involves three main stages within one sampling instant. However, by taking the nonfuzzy property regarding the input/output of the fuzzy controller into account, a very simple but efficient fuzzy control algorithm SFCA has been derived which consists of only two main steps, pattern matching and weighted average, and this is described below.

Assume that A_i^j and B_k^j in the rule eqns. 10.1 are normalised fuzzy subsets whose membership functions are defined uniquely as triangles, each of which is characterised only by two parameters, $M_{u,i}^j$ and $\delta_{u,i}^j$ or $M_{v,k}^j$ and $\delta_{v,k}^j$ with the understanding that $M_{u,i}^j$ ($M_{v,k}^j$) is the centre element of the support set of A_i^j (B_k^j), and $\delta_{u,i}^j$ ($\delta_{v,k}^j$) is the half width of the support set. Hence, the jth rule can be written as

$$\begin{aligned}&\text{IF } (M_{u,1}^j, \delta_{u,1}^j) \text{ AND } (M_{u,2}^j, \delta_{u,2}^j) \text{ AND} \ldots \text{AND } (M_{u,n}^j, \delta_{u,n}^j)\\ &\text{THEN } (M_{v,1}^j, \delta_{v,1}^j,) \text{ AND } (M_{v,2}^j, \delta_{v,2}^j) \text{ AND} \ldots (M_{v,m}^j, \delta_{v,m}^j)\end{aligned} \qquad (10.5)$$

Let $M_u^j = (M_{u,1}^j, M_{u,2}^j, \ldots M_{u,n}^j)$ and $\Delta_u^j = (\delta_{u1,}^j, \delta_{u,2}^j \ldots \delta_{u,n}^j)$ be two n-dimensional vectors. Then the *condition* part of the jth rule may be viewed as creating a subspace or a *rule pattern* whose centre and radius are $M_u{}^j$ and $\Delta_u{}^j$ respectively. Thus the *condition* part of the *j*th rule can be simplified further to 'IF $M\delta_u(j)$', where $M\Delta_u(j) = (M_u^j, \Delta_u^j)$. Similarly n current inputs $u_{0i} \in U_i (i = 1,2,\cdots n)$ with u_{0i} being a singleton, can also be represented as a n-dimensional vector u_0 or a input pattern.

The fuzzy control algorithm can be considered to be a process in which an appropriate control action is deduced from a current input and P rules according to some prespecified reasoning algorithms. The whole reasoning procedure is split into two phases: pattern matching and weighted average. The first operation deals with the IF part of all rules, whereas the second one involves an operation on the THEN part of the rules. From the pattern concept introduced above, the matching degrees between the current input pattern and each rule pattern need to be computed. Denote the current input by $u_0 = (u_{01}, u_{02}, \ldots u_{0n})$. Then the matching degree denoted by $S^j \in [0, 1]$ between u_0 and the *j*th rule pattern $M\Delta_u(j)$ can be measured by the complement of the corresponding relative distance given by

$$S^j = 1 - D^j (u_0, M\Delta_u(j)) \tag{10.6}$$

where $D^j(u_0, M\Delta_u(j)) \in [0, 1]$ denotes the relative distance from u_0 to $M\Delta_u(j)$. D^j can be specified in many ways. With the assumption of an identical width δ for all fuzzy sets $A_i{}^j$, the computational definition of D^j is given by

$$D^j = \begin{cases} \dfrac{\|M_u^j - u_0\|}{\delta} & \text{if } \|M_u^j - u_0\| \le \delta \\ 1 & \text{otherwise} \end{cases} \tag{10.7}$$

For a specific input u_0 and P rules, after the matching process is completed, the *k*th component of the deduced control action v_k is given by

$$v_k = \sum_{q=1}^{Q} S^q . M_{v,k}^q \Big/ \sum_{q=1}^{Q} S^q \tag{10.8}$$

where it is assumed that all membership functions $b_k{}^j(v)$ are symmetrical about their respective centres and have an identical width. Notice that, because only the centres of the THEN parts of the fired rules are utilised and they are the only element having the maximum membership grade 1 on the corresponding support sets, the algorithm can be understood as a modified maximum membership decision scheme in which the global centre is calculated by the *Centre Of Gravity* algorithm. Thus the rule form can be further simplified to

$$\text{IF } M\Delta_u(j) \text{ THEN } M_v^j \tag{10.9}$$

where $M_v^j = [M_{v,1}^j, M_{v,2}^j, \ldots M_{v,m}^j]$ is a centre value vector of the THEN part.

10.4.2 Counter propagation network (CPN) fuzzy controller

By combining a portion of the self-organising map of the Kohonen and the outstar structure of Grossberg, Hecht-Nielsen has developed a new type of neural networks named counter propagation (CPN) [12]. The CPN network consists of an input layer, a hidden Kohonen layer, and a Grossberg output layer with n, N, and m units respectively (Figure 10.4).

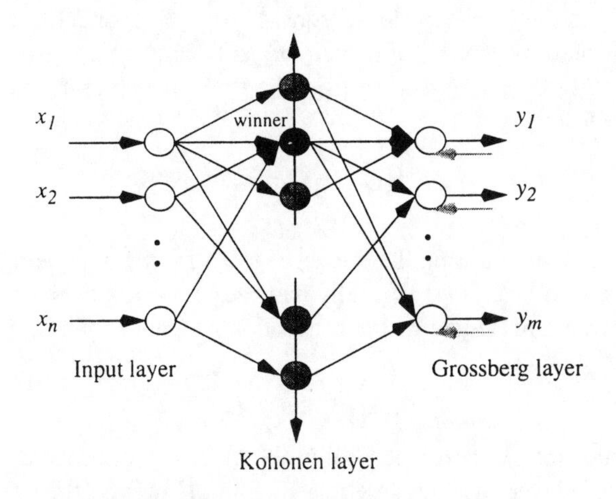

Figure 10.4 *Counter propagation network*

The aim of the CPN is to perform an approximate mapping $\phi: R^n \rightarrow R^m$ based on a set of examples (u^s, v^s) with u^s vectors being randomly drawn from R^n in accordance with a fixed probability density function ρ. The forward algorithms of the CPN in regard to a particular input u at the time instant t are outlined as follows:

a) Determine the winner unit I in the Kohonen layer competitively according to the distances of weight vector $w_i(t)$'s with respect to u

$$D(w_I(t), u) = \min_{i=1,N} \; D(w_i(t), u) \tag{10.10}$$

b) Calculate the outputs $z_i(t) \in \{0, 1\}$ of the Kohonen layer by a winner-take-all rule

$$z_i(t) = \begin{cases} 1 & \text{if } i = I \\ 0 & \text{otherwise} \end{cases} \tag{10.11}$$

c) Compute the outputs of the Grossberg layer by

$$v_j^* = \sum_{i=1}^{N} z_i(t).\pi_{ji}(t) \tag{10.12}$$

There exist some striking similarities between the fuzzy control algorithms eqns. 10.7 and 10.8 and the CPN algorithms eqns. 10.10, 10.11, and 10.12. By viewing the CPN as a hard version of the fuzzy or soft counterpart, the stable weight vectors connecting to and emanating from the ith Kohonen unit may be expressed as specifying a rule:

$$\text{IF } w_i \text{ THEN } \pi_i \tag{10.13}$$

In this perspective, by ignoring the effects of the Δ for the time being, each rule in the form of eqn. 10.9 can be structurally mapped into a Kohonen unit with IF and THEN parts being represented by the connection weights.

10.4.2.1 Self-constructing of rule base

So far the knowledge representation problem has been solved by structural mapping, provided that the rule base is available. However, the primary concern lies in building a rule base automatically from the controlled environment. In terms of CPN, this means that the number of units in the Kohonen layer must be self-organised and the associated weights must be self-learned.

Self-organising: IF part

Comparing the IF part of eqn. 10.9 with eqn. 10.13, there is a width vector Δ_q associated with the former. Δ_q can be roughly visualised as defining a neighbourhood for the ith rule centred at M_q. Q rules partition the input space into Q overlapping subspaces. This viewpoint, together with the concept of relative distance, provides some insight into finding a solution for the problem by using a modified CPN algorithm. By assigning each Kohonen unit a predefined width vector Δ_i , the winner I not only has a minimum distance among all the existing units in regard to the current input u as determined by eqn. 10.12, but also must satisfy the condition of u falling into the winner's neighbourhood as designated by

Δ_i. Thus if $D(w_I(t), u) \leq \Delta_I$ then unit I is considered to be the winner and the weight vectors in the Kohonen layer are adjusted by

$$w_i(t+1) = w_i(t) + \alpha_i(t) \,.\, [u - w_i(t)] \,.\, z_i \tag{10.14}$$

where I=1, N and $\alpha_i(t) \in [0,1]$ is a learning rate which decays monotonically with increasing time. On the other hand, if $D(w_I(t), u) > \Delta_I$, it indicates that no existing unit is adequate to assign u as its member and therefore a new unit should be created. It is clear that, starting from an empty state, the Kohonen layer can be dynamically self-organised in terms of the number of units and w weight associated with each unit, thereby establishing the IF part of the rule base.

Self-learning: THEN part

It is always the case that the lack of a teacher for any kind of neurocontroller, unless explicitly constructed, represents a formidable problem. Here, a simple but efficient scheme [10] is introduced which is capable of providing teacher signals in a real-time manner. By including a reference model specifying the desired response vector, the error vector, denoted by $\varepsilon \in R^m$ and defined as the difference between the desired and actual output of the process, is used as a basis for teacher learning. More specifically, a teacher vector $\gamma^k \in R^m$ at the kth iteration and at the time instant t is given by

$$\gamma^k(t) = \gamma^{k-1}(t) + G \,.\varepsilon^{k-1}(t) \tag{10.15}$$

where G is a learning gain matrix which can be diagonal in its simplest form.

Upon the γ^k being available, the weights π_{ji} at the Grossberg layer are adjusted by

$$\pi_{ji}(t) = \pi_{ji}(t-1) + \beta.[\gamma_j^k(t) - \pi_{ji}(t-1)] \,.\, z_i(t) \tag{10.16}$$

where i=1 to N, j=1 to m, and β is a constant learning rate within the range [0,1]. Denoting $\pi_i = [\pi_{1i}, \pi_{2i}, \ldots \pi_{mi}]$, we notice that π_i is a weight vector connecting the ith unit at the Kohonen layer to every unit at the output layer. Thus, by the representional convention described in the last section, the THEN part of the rule will be learned gradually. Further details can be found in [13].

10.4.3 Radial basis function (RBF) fuzzy control

Simply stated, a RBF network is intended to approximate a continuous mapping $f: R^n \rightarrow R^m$ by performing a nonlinear transformation at the hidden layer and subsequently a linear combination at the output layer (Figure 10.5). More specifically, this mapping is described by

$$\hat{f}_k(u) = \sum_{j=1}^{N} \pi_k^j . \phi^j\left(\left\|u - \omega^j\right\|\right) \tag{10.17}$$

where N is the number of the hidden units, $u \in R^n$ is an input vector, $\omega^j \in R^n$ is the centre of the jth hidden unit and can be regarded as a weight vector from the input layer to the jth hidden unit, ϕ^j is the jth radial basis function or response function, and π_k^j is the weight from the jth hidden unit to the kth output unit.

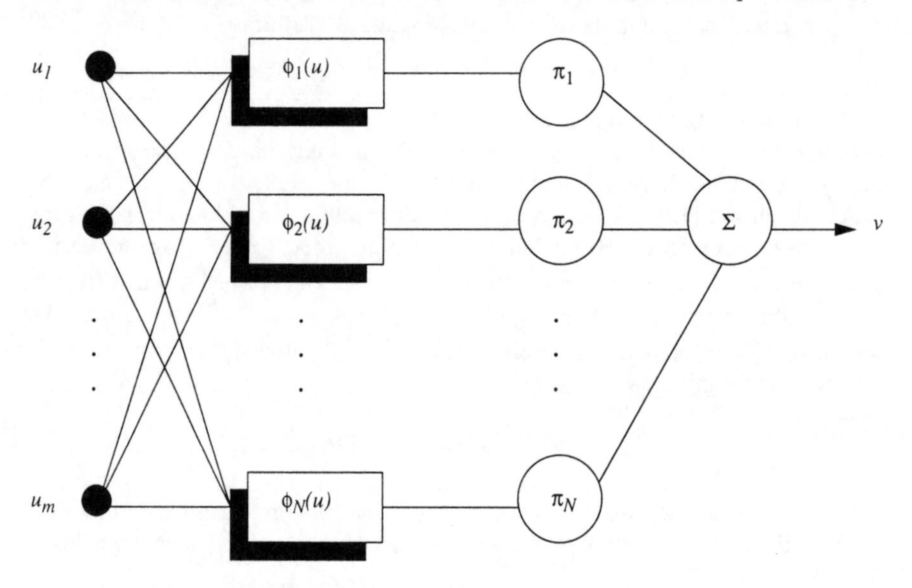

Figure 10.5 *Radial basis function (RBF) network structure*

Although there exist many possibilities for the choice of ϕ^j, as observed by Moody [14], Gaussian type functions given by

$$\phi^j(u) = \exp\left[\frac{\left\|u - \omega^j\right\|^2}{(\sigma^j)^2}\right] \tag{10.18}$$

offer a desirable property making the hidden units to be locally-tuned, where the locality of the ϕ^j is controlled by σ^j.

It is remarkable that the SFCA bears some intrinsic similarities with the RBF network, although they have originated from two apparently independent fields:

fuzzy logic theory and function approximation theory. From the knowledge representation viewpoint, the RBF network is essentially a net representation of IF-THEN rules. Each hidden unit reflects a rule: IF ω^j THEN π^j, where $\pi^j = (\pi_1{}^j, \ldots, \pi_m{}^j)$. The rationale behind fuzzy reasoning in the SFCA and interpolative approximation in the RBF seems to be the same: to create a similar action with respect to a similar situation or to produce a similar output in response to a similar input. By comparing eqn. 10.8 with eqn. 10.17, it is seen that the two systems have almost identical computational procedures i.e. matching degrees correspond to response function values and the resemblance of two approaches in the final combination step is evident. The following parameter correspondences are identified: $M\Delta_u(j)$ to ω^j, $M_v{}^j$ to π^j, δ to σ, and P to N.

Once a formal connection between these two paradigms is made, a hybrid system taking the advantages of both can be derived. One of the possibilities is to generalise the RBF network, by fuzzifying it, into a class of more general networks, referred to as fuzzified basis function networks, or FBFN for short. This can be done by simply replacing the radial basis function ϕ^j with matching degree S^j.

In what way can the FBFN be better than the RBF? Certainly, it would be of great benefit if the FBFN could be made to be self-constructed without relying on the control rules provided by human experts. In terms of FBFN, parameters ω^j, σ^j, π^j and N must preferably be self-determined. By pre-specifying N, Moody[14] developed a hybrid learning scheme to learn ω^j unsupervisedly, σ^j heuristically, and π^j supervisedly. Since the FBFN based controller must be operated in real-time, there are some difficulties in applying Moody's method. In particular, it is hard to specify N in advance due to the uncertain distribution of on-line incoming data, and more seriously, there are in general no teacher signals available to guide the learning of π^j. The problem is approached by using a fuzzified dynamical self-organising scheme and by incorporating an iterative learning control strategy to supply the teacher signals on-line.

With a prespecified identical width σ, the concern is how a specific region of the input space can be partitioned into subregions without knowing N in advance, each of which is represented by a centre vector ω^j. The competitive learning approach is adopted which adaptively quantises the input space by discovering a set of representative prototypes referred to as clusters or centroids. The original Kohonen algorithm is modified in the following ways. Instead of using absolute minimum distance as the winner selection criterion and a unique learning rate, a matching degree is employed to determine the winner and N local learning rates are used. If one of the existing hiddden units is able to win the competition, the corresponding ω^j is modified; otherwise, a new unit is created. In any case, the response vector ϕ is derived for computing the net output and modifying the corresponding weight vector π^j. In this manner, the required N and ω are dynamically learned in response to the incoming controller input u. For further details see [15].

10.4.4 Cerebellar model articulation controller (CMAC) fuzzy control

The CMAC is designed to represent approximately a multi-dimensional function by associating an input vector $u \in U\ R^n$ with a corresponding function vector $v \in V \subset R^m$. CMAC has a similar structure to a three-layered network with association cells playing the role of hidden layer units (Figure 10.6). Mathematically, CMAC may be described as consisting of a series of mappings: $U \to A \to V$, where A is a N-dimensional cell space.

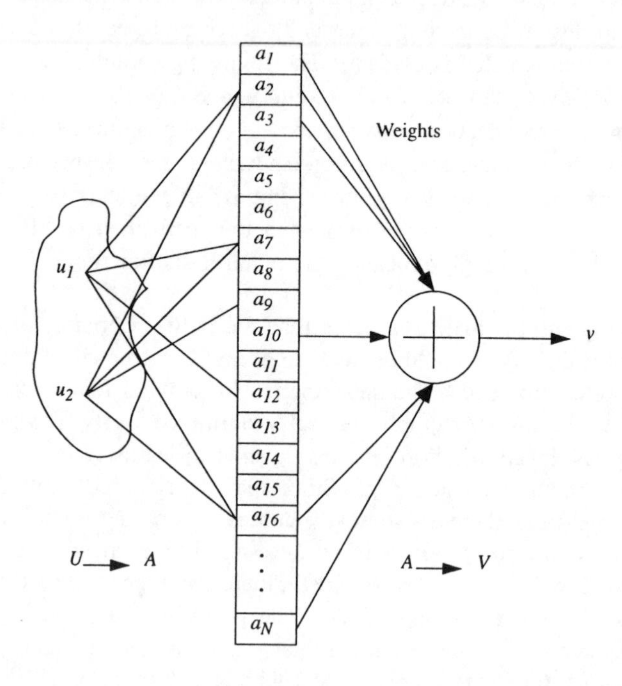

Figure 10.6 *Cerebellar model articulation controller (CMAC) network structure*

A fixed mapping $U \to A$ transforms each $u \in U$ into an N-dimensional binary associate vector $a(u)$ in which only N_L elements have the values of 1, where $N_L < N$ is referred to as the generalisation width. In other words, each u activates precisely N_L association cells or geometrically each u is associated with a neighbourhood in which N_L association cells are included. An important property of the CMAC is local generalisation derived from the fact that nearby input vectors u_i and u_j have some overlapping neighbourhood and therefore share some common association cells. The degree to which the neighbourhoods of u_i and u_j are overlapping depends on the Hamming distance H_{ij} of u_i and u_j. If H_{ij} is small, the intersection of u_i and u_j should be large and vice versa. At some values of H_{ij}

greater than N_L, the intersection becomes null indicating that no generalisation occurs.

According to the above principle, Albus [16] developed a mapping algorithm consisting of two sequential mappings: $U \rightarrow A \rightarrow V$. n components of u are first mapped into n N_L-dimensional vectors and these vectors are then concatenated into a binary association vector a with only N_L elements being 1. Albus also suggested an approach to further reducing the N association cells into a much smaller set by a hashing code.

The $A \rightarrow V$ mapping is simply a procedure of summing the weights of the association cells excited by the input vector u to produce the output. More specifically, each component v_k is given by

$$v_k = \sum_{j=1}^{N} a^j(u).\pi_k^j \tag{10.19}$$

where $\pi_k{}^j$ denotes the weight connecting the jth association cell to the kth output. Notice that only N_L association cells contribute to the output.

By carefully inspecting the SFCA and the CMAC, it is evident that there exist some striking similarities between these two systems. Functionally, both of them perform a function approximation in an interpolative look-up table manner with an underlying principle of generalisation and dichotomy: to produce similar outputs in response to similar input patterns and produce independent outputs to dissimilar input patterns. From the computational point of view, mapping $U \rightarrow A$ corresponds to the calculation of the matching degree in the SFCA, and mapping $A \rightarrow V$ corresponds to the weighted averaging procedure given by eqn. 10.8. While the latter similarity is apparent by comparing eqn. 10.8 with eqn. 10.19, where a^j, $\pi_k{}^j$, and N correspond to σ^j, $M_{v,k}{}^j$, and P, the former equivalence can be made clearer as follows.

Instead of saying that each input u is associated with a neighbourhood specified by N_L, it is equally valid to consider that each association cell Ψ^j is associated with a neighbourhood centred at, say, $\omega^j \in U$ referred to as the reference vector, with the width controlled by N_L. If a current input u is within the neighbourhood of Ψ^j, that cell is regarded as being active. In this view, the associate vector $a(u)$ can be derived by operating N neighbourhood functions $\psi^j(\omega^j, u)$ with respect to u, where

$$\psi^j(\omega^j, u) = \begin{cases} 1 & \text{if } u \in \Psi^j \\ 0 & \text{otherwise} \end{cases} \tag{10.20}$$

that is, $a(u) = (\psi^1, \psi^2, \ldots \psi^N)$. By appropriately selecting the ω^j, $a(u)$ can be made to contain only N_L '1's. Now it becomes evident that the associate vector a

is similar to the matching degree vector s except that the former uses the crisp neighbourhood function, whereas the latter adopts the graded one. In fact, by letting $s^{*j}=1$ for $s^j > 0$ and $s^{*j}= 0$ for other cases, the vector a will be precisely equal to the vector s^*, indicating that the former is a special case of the latter. Notice that a natural measurement of whether u belongs to Ψ^j in eqn. 10.20 is to use some distance metric relevant to the generalization width N_L. In fact, Albus himself [16] discussed the question of how the overlapping of u_i and u_j is related to the Hamming distance of u_i and u_j although he did not formulate explicitly this concept into the mapping process $U \rightarrow A$.

Now a position has been reached where a FCMAC can be implemented by replacing eqn. 10.19 with eqn. 10.8. Several advantages can be identified by this replacement. The concept of the graded matching degree not only provides a clear interpretation for $U \rightarrow A$ mapping, but also offers a much simpler and more systematic computing mechanism than that proposed by Albus where some very complicated addressing techniques are utilised and a further hashing code may be needed to reduce the storage. In addition, as noted by Moody [17] and Lane *et al.* [18], the graded neighbourhood functions overcome the problem of discontinuous responses over neighbourhood boundaries due to the crisp neighbourhood functions. However, in order to use FCMAC, reference vectors ω^j must be specified in advance. Fortunately, this requirement can be met by introducing a fuzzified Kohonen self-organising scheme as described in Sections 10.4.2 and 10.4.3. For further details see [19].

10.5 Multivariable modelling of the cardiovascular system

Over the years, a variety of mathematical models of the cardiovascular system have been developed [20], some of which are pulsatile and are suitable for investigating the phenomena that can change in a fraction of a heart beat. Models for studying the long-term effects are non-pulsatile. Blood circulation is described in terms of non-pulsatile pressures, flows and volumes. The model used in this study is a non-pulsatile model developed by Moller *et al.* [21]. A two-pump circulatory model was postulated and a basic relationship governing the physiologically closed cardiovascular system was derived. The two parts of the heart are represented by flow sources QL and QR for the left and for the right ventricle respectively. The systemic circulation is represented by the arterial systemic compliance CAS, the systemic peripheral resistance RA and the venous systemic distensible capacity CVS. Similarly, the pulmonary circulation consists of the arteriopulmonary compliance CAP, the pulmonary resistance RP and the venopulmonary distensible capacity CVP. The cardiovascular system dynamics governing the relationship between the blood pressures and the flow sources can be described by the following differential vector equation

$$\frac{dX}{dt} = AX + BV \tag{10.23}$$

where [PAS, PVS, PAP, PVP]^{T} with PAS, PVS, PAP and PVP being the systemic arterial pressure, systemic venous pressure, pulmonary arterial pressure and pulmonary venous pressure respectively; $\mathbf{V}=[QL,QR]^{T}$. And

$$A = \begin{bmatrix} (-CAS.RA)^{-1} & (CAS.RA)^{-1} & 0 & 0 \\ (CVS.RA)^{-1} & (-CVS.RA)^{-1} & 0 & 0 \\ 0 & 0 & (-CAP.RP)^{-1} & (CAP.RP)^{-1} \\ 0 & 0 & (CVP.RP)^{-1} & (-CVP.RP)^{-1} \end{bmatrix}$$

$$B = \begin{bmatrix} \frac{1}{CAS} & 0 \\ 0 & \frac{-1}{CVS} \\ 0 & \frac{1}{CAP} \\ \frac{-1}{CVP} & 0 \end{bmatrix}$$

Equation 10.23 is a nonlinear vector equation because the resistances and the compliances are nonlinear functions of the pressure. Furthermore, the inputs QL and QR are also pressure dependent, that is,

$$QL=SV_{L}.HR \qquad QR=SV_{R}.HR \tag{10.24}$$

where HR stands for the heart rate, and SV_L and SV_R for the stroke volumes of the left and right ventricles respectively. SV can be related to the arterial and venous pressures by a complicated nonlinear algebraic function [21].

If the compliances and the resistances are treated as pressure independent, eqn. 10.23 represents a linear state-space model with QL and QR as independent system inputs. By selecting the heart rate HR and resistance RA as the system inputs, Moller *et al.* derived a linear model near the stationary state PAS_0=117.5 mmHg, PVS_0=7.15 mmHg, PAP_0=17.18 mmHg and PVP_0=10.87 mmHg, which is given by

$$\frac{d\Delta\mathbf{X}}{dt} = A_l\Delta\mathbf{X} + B_l\Delta\mathbf{U} \tag{10.25}$$

where $\Delta\,\mathbf{X}=[\Delta\,\mathrm{PAS}, \Delta\,\mathrm{PVS}, \Delta\,\mathrm{PAP}, \Delta\,\mathrm{PVP}]^{\mathrm{T}}$, $\Delta\,\mathbf{U}=[\Delta\,\mathrm{HR}, \Delta\,\mathrm{RA}]^{\mathrm{T}}$ and

$$A_l = \begin{bmatrix} -3.4370 & 1.8475 & 0.0 & 18.7584 \\ 0.01834 & -0.3015 & 0.06855 & 0.0 \\ 0.0 & 7.3514 & -10.1131 & 8.3333 \\ 0.2049 & 0.0 & 4.1667 & -6.5846 \end{bmatrix}$$

$$B_l = \begin{bmatrix} 125.8 & 194.3 \\ -0.5048 & -1.929 \\ 13.1058 & 0.0 \\ -16.2125 & 0.0 \end{bmatrix}$$

Equation 10.25 indicates that the processes can be controlled by manipulating heart rate and systemic resistance with a fast response time. It should be noted that the activation of HR is currently feasible through direct electrical stimulation of the heart, but is not yet available directly for RA.

The simultaneous regulation of blood pressure and cardiac output (CO) is needed in some clinical situations, for instance congestive heart failure. It is desirable to maintain or increase CO and, at the same time, to decrease the blood pressure. This goal can be achieved by simultaneous infusions of a positive inotropic agent, which increases the heart's contractility and cardiac output, and a vasodilator which dilates the vasoculature and lowers the arterial pressure. Two frequently used drugs in clinical practice are the inotropic drug Dopamine (DOP) and the vasoactive drug Sodium NitroPrusside (SNP). It is worth noting that the inputs are interacting with respect to controlled variables CO and mean arterial pressure (MAP). The inotropic agent increases CO and thus MAP, whereas the vasoactive agent decreases MAP and increases CO.

An accurate dynamical model associating cardiac output and mean arterial pressure with DOP and SNP is not available to date. However, Serna *et al.* [22] derived a first-order model in which different time constants and time delays in each loop were used. The steady-state gains in the model were obtained from Miller's study [23]. The dynamics in the s-domain are given by

$$\begin{bmatrix} \Delta\mathrm{CO}_d \\ \Delta\mathrm{RA}_d \end{bmatrix} = \begin{bmatrix} \dfrac{K_{11}e^{-\tau_1 s}}{sT_1+1} & \dfrac{K_{12}e^{-\tau_1 s}}{sT_1+1} \\ \dfrac{K_{21}e^{-\tau_2 s}}{sT_2+1} & \dfrac{K_{22}e^{-\tau_2 s}}{sT_2+1} \end{bmatrix} \begin{bmatrix} I_1 \\ I_2 \end{bmatrix} \qquad (10.26)$$

where ΔCO_d (ml/s) is the change in cardiac output due to I_1 and I_2; ΔRA_d (mmHg.s/ml) is the change in systemic resistance due to I_1 and I_2; I_1 (mg/Kg/min) is the increment in infusion rate of DOP; I_2 (ml/h) is the increment in infusion rate of SNP; K_{11}, K_{12}, K_{21} and K_{22} are steady-state gains with typical values of 8.44, 5.275, -0.09 and -0.15 respectively; τ_1 and τ_2 represent two time delays with typical values τ_1=60 s and τ_2=30 s; T_1 and T_2 are time constants typified by the values of 84.1 s and 58.75 s respectively. The model parameters presented above can be varied during simulations in order to evaluate the robustness of the proposed controller.

Because the accessible measurable variables are MAP and CO, a model which relates the ΔCO_d and ΔRA_d due to drug infusions is needed. Moller's cardiovascular model can be used for this purpose. Note that the cardiovascular dynamics is much faster than the drug dynamics. Consequently, it is reasonable to neglect the cardiovascular dynamics and only retain the steady state gains in the CVS model. By this consideration, Mansour *et al.* [24] derived a simulation model from Moller's CVS model and the drug dynamics, which is given by

$$\begin{bmatrix} \Delta CO \\ \Delta MAP \end{bmatrix} = \begin{bmatrix} 1.0 & -24.76 \\ 0.6636 & 76.38 \end{bmatrix} \begin{bmatrix} \dfrac{K_{11}e^{-\tau_1 s}}{sT_1+1} & \dfrac{K_{12}e^{-\tau_1 s}}{sT_1+1} \\ \dfrac{K_{21}e^{-\tau_2 s}}{sT_2+1} & \dfrac{K_{22}e^{-\tau_2 s}}{sT_2+1} \end{bmatrix} \begin{bmatrix} I_1 \\ I_2 \end{bmatrix} \tag{10.27}$$

10.6 Experimental results

All of the fuzzy-neural techniques described in the previous sections have been validated using the multivariable cardiovascular model summarised in eqn. 10.27. A few sample results are given here for the CPN self-organising method, which typifies the performance of the other single-stage approaches. Using a range of first- and second-order reference models, it was found that the system converged to low errors within three iterations of the learning loop. The resulting controlled structure gave good performance under either noise-free (Figure 10.7) or noise-contaminated (Figure 10.8) environments. Sensitivity studies showed that the system was robust to changes in the process model parameters, as shown in Figure 10.9 for 10% variations in the gains and time constants.

The convergence properties of the overall self-organising structure depend on a number of parameters. In the case of CPN the most important is the learning matrix P_L. Large values of P_L speed up the convergence process, but at the risk of overall instability. Lower values slow down the convergence, as can be seen in Figure 10.10. The Grossberg learning rate β controls the updating of the π weighting coefficients, but the system is not sensitive to its choice, with a value of

0.5 being suitable. The valid radius δ controls the self-organising structures and is related to the number of Kohonen units or control rules which are self-elicited. Convergence of the system was not largely sensitive to δ, although the number of rules was. The type of distance metric, e.g. Hamming, Euclidean or maximum, likewise did not affect radically the performance of the architecture. It was concluded, therefore, that suitable design parameters are not difficult to select, with nominal ones being quite easy to obtain followed by tuning if necessary.

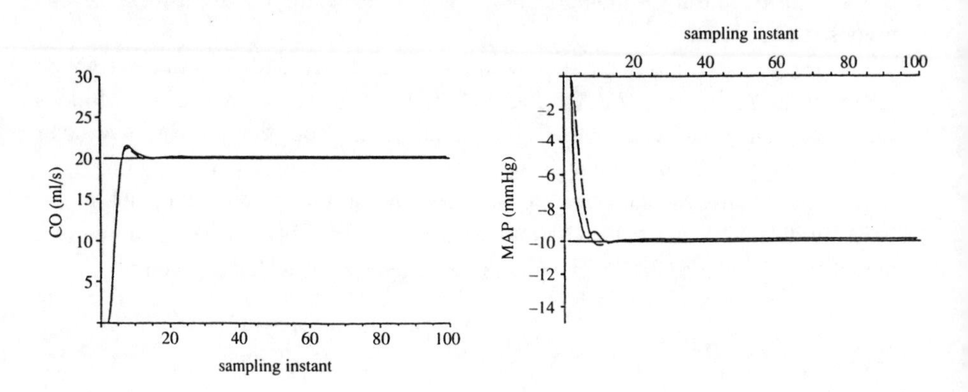

Figure 10.7 *Adapted response for CPN control of multivariable cardiovascular process*

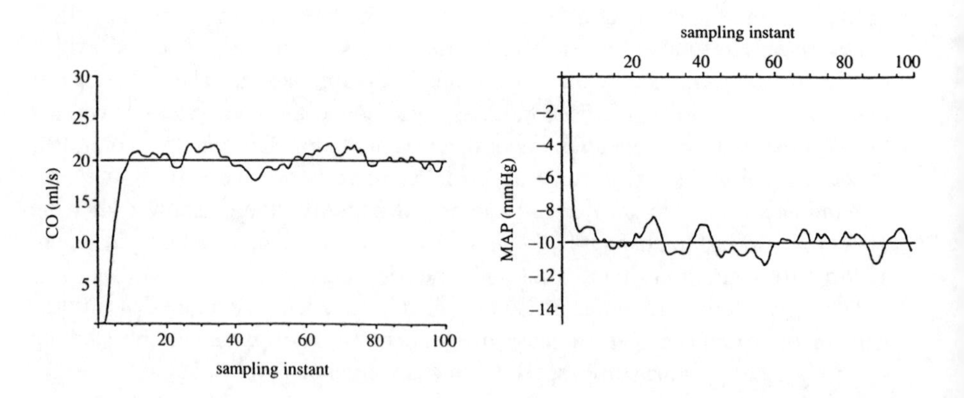

Figure 10.8 *Cardiovascular outputs with noise-contaminated measurements under CPN control*

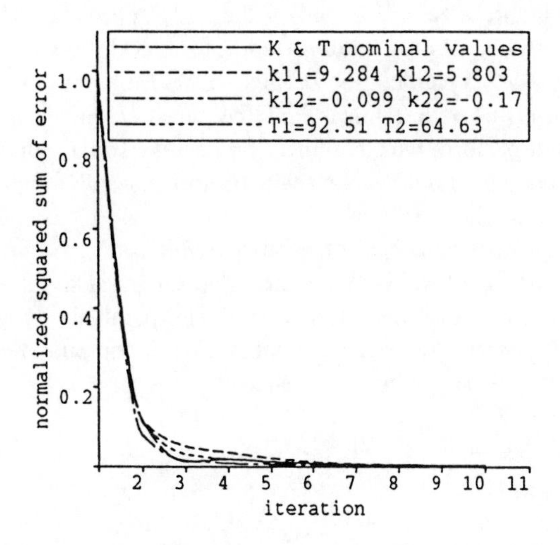

Figure 10.9 *Adaptive capability of CPN for different cardiovascular process parameters*

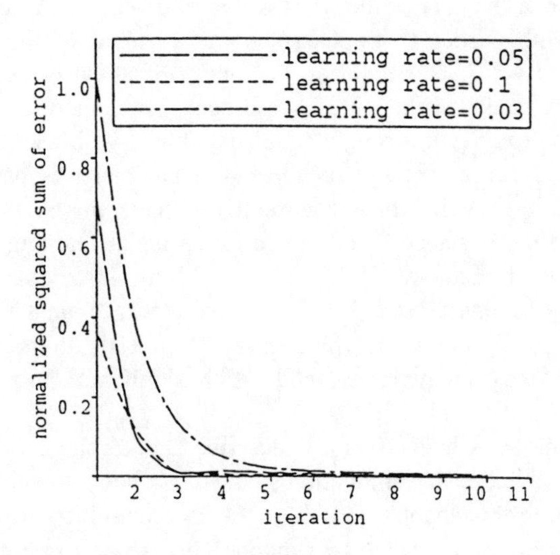

Figure 10.10 *Convergence properties of CPN-based system for different learning rate parameter values*

Similar remarks apply to the other methods of FBFN and FCMAC, which have a very similar basic architecture to that for the CPN approach. The system used for validation, being that of drug administration for blood pressure management in intensive care therapy, is challenging because of its multivariable and large time delay nature. It is, however, linear and the question arises as to how such architectures would perform with nonlinear processes. In fact, the CPN approach has also been tried with nonlinear anaesthetic models and found to perform in equivalent manner to the cardiovascular system. In this case, the outputs are depth of unconciousness and level of paralysis which are measured inferentially by blood pressure and evoked EMG signals. Control is via simultaneous infusion of two drugs, using propofol and atracurium. The paralysis dynamics has very severe nonlinearity, with the pharmacodynamics (i.e. output effect from blood concentration) being highly sigmoidal in shape.

10.7 Conclusions

In recent years much progress has been made in linking the fuzzy and neural paradigms in a synergetic manner. The advantages of learning via neural networks and the transparent explanation via fuzzy logic have been coupled in a unified framework. In the work described in this chapter, an additional technique related to model reference adaptive control (MRAC) has also been incorporated to provide the necessary on-line learning signals and also an architecture which allows the system to develop its own parsimonious model based structure.

The application at the core of the work is that of on-line drug administration in intensive care, with particular emphasis on management of the cardiovascular system post-operatively. This requires the control of blood pressure and cardiac output using simultaneous infusion of two drugs. The patient dynamics for this application are multivariable, strongly coupled and has significant time delays. In addition, there is large uncertainty in the individual patient's parameters. The neuro-fuzzy techniques which have been explored have all shown characteristics of good transient performance, good convergence under learning, and relatively low sensitivity to parameter variability. Coupled with this, it was found that the necessary design parameters are largely insensitive and can usually be set to nominal values for initial prototyping. Initial investigations with nonlinear systems support these viewpoints, and offer significant promise for these techniques in the future.

All of the techniques which are briefly described in this chapter are discussed in detail, together with extensive simulation studies, in a book devoted to the subject [25]. In concluding this chapter, I wish to acknowledge fully the intensive work and extensive research undertaken by Junhong Nie, which forms the basis of the work presented and which is fully expounded in his augmented Ph.D. thesis published as [25].

10.8 References

[1] Linkens, D.A. and Hacisalihzade, S.S., 1990, Computer control systems and pharmacological drug administration: a survey, *J. Med. Eng. & Tech.*, Vol. 14, pp. 41-54.

[2] Mahfouf, M., Linkens, D.A., Asbury, A.J., Gray, W.M. and Peacock, J.E., 1992, Generalised predictive control (GPC) in the operating theatre, *Proc. IEE, Pt.D*, Vol. 139, pp. 404-420.

[3] Linkens, D.A. and Mahfouf, M., 1992, Generalised predictive control with feedforward (GPCF) for multivariable anaesthesia, *Int. J. Contr*, Vol. 56, pp. 1039-1057.

[4] Linkens, D.A., Mahfouf, M. and Abbod, M.F., 1992, Self-adaptive and self-organising control applied to non-linear multivariable anaesthesia: a comparative model-based study, *Proc. IEE, Pt.D*, Vol. 139, pp. 381-394.

[5] Linkens, D.A. and Mahfouf, M., 1988, Fuzzy logic knowledge-based control for muscle relaxant anaesthesia, *IFAC Symp. on Modelling and Control in Biomedical Systems*, Venice, Pergamon, pp. 185-190.

[6] Mason, D.G., Linkens, D.A., Abbod, M.F., Edwards, N.D. and Reilly, C.S., 1994, Automated drug delivery in muscle relaxation anaesthesia using self-organising fuzzy logic control, *IEEE Eng. in Med. & Biol.*, Vol. 13, No. 5, pp. 678-686.

[7] Linkens, D.A. and Nie, J., 1992, A unified real time approximate reasoning approach, Part 1, *Int. J. Contr.*, Vol. 56, pp. 347-393.

[8] Linkens, D.A. and Nie, J., 1992, A unified real time approximate reasoning approach, Part 2, *Int. J. Contr.*, Vol. 56, pp. 365-393.

[9] Nie, J. and Linkens, D.A., 1992, Neural Network-based Approximate Reasoning, *Int. J. Contr.*, Vol. 56, pp. 394-413.

[10] Linkens, D.A. and Nie, J., 1993, Constructing rule-bases for multivariable fuzzy control by self-learning, *Int. J. Syst. Sci.*, Vol. 24, pp. 111-128.

[11] Nie, J. and Linkens, D.A., 1994, A hybrid neural network-based self-organising fuzzy controller, *Int. J. Contr.*, Vol. 60, pp. 197-222.

[12] Hecht-Nielsen, R., 1987, Counterpropagation networks, *Applied Optics*, Vol. 26, pp. 4979-4984.

[13] Nie, J. and Linkens, D.A., 1994, Fast self-learning multivariable fuzzy controller constructed from a modified CPN network, *Int. J. Contr.*, Vol. 60, pp. 369-393.

[14] Moody, J. and Darken, C., 1989, Fast-learning in networks of locally-tuned processing units, *Neural Comput.*, Vol. 1, pp. 281-294.

[15] Nie, J. and Linkens, D.A., 1994, Learning Control using fuzzified self-organising Radial Basis Function network, *IEEE Trans. Fuzzy Sys.*, Vol. 1, pp. 280-287.

[16] Albus, J.S., 1975, A new approach to manipulator control: the Cerebellar Model Articulation Controller (CMAC), *J. Dyn. Sys. Meas. Contr.*, Vol. 97, pp. 220-227.

[17] Moody, J., 1989, Fast-learning in multi-resolution hierarchies, *Advances in Neural Information Processing Systems*, Vol. 1 (Towetzky, D., Ed.), Morgan Kaufmann.

[18] Lane, S.H., Handelman, D.A and Gelfard, J.J., 1992, Theory and development of high-order CMAC neural networks, *IEEE Contr. Syst. Mag.*, Vol. 12, pp. 23-30.

[19] Nie, J. and Linkens, D.A., 1994, FCMAC: A fuzzified Cerebellar Model Articulation Controller with self-organising capacity, *Automatica*, Vol. 30, pp. 655-664.

[20] Mansour, N-E., 1988, Adaptive control of blood pressure, Ph.D. thesis, University of Sheffield.

[21] Moller *et al.,* 1983, Modelling, simulation and parameter estimation of the human cardiovascular system, *Advances in control systems and signal processing*, p. 4.

[22] Serna *et al.*, 1993, Adaptive control of multiple drug infusion, *Proc. JACC*, p. 22.

[23] Miller, R.R. *et al.*, 1977, Combined dopamine and nitroprusside therapy in congestive heart failure, *Circulation*, Vol. 55, p. 881.

[24] Mansour, N-E, and Linkens, D.A., 1990, Self-tuning pole-placement multivariable control of blood pressure for post-operative patients: a model-based study, *Proc. IEE, Pt.D*, Vol. 137, pp. 13-29.

[25] Nie J. and Linkens, D.A., 1995, *Fuzzy-Neural Control: principles, algorithms and applications*, Prentice-Hall.

Chapter 11

Neural networks and system identification

S. A. Billings and S. Chen

11.1 Introduction

Neural networks have become a very fashionable area of research with a range of potential applications that spans AI, engineering and science. All the applications are dependent upon training the network with illustrative examples and this involves adjusting the weights which define the strength of connection between the neurons in the network. This can often be interpreted as a system identification problem with the advantage that many of the ideas and results from estimation theory can be applied to provide insight into the neural network problem irrespective of the specific application.

Feedforward neural networks, where the input feeds forward through the layers to the output, have been applied to system identification and signal processing problems by several authors (e.g. [1–13]) and the present study continues this theme. Three network architectures, the multi-layered perceptron, the radial basis function network and the functional link network, are discussed and several network training algorithms are introduced. A recursive prediction error algorithm is described as an alternative to backpropagation for the multi-layered perceptron. Two learning algorithms for radial basis function networks, which incorporate procedures for selecting the basis function centres, are discussed and the extension of these ideas to the functional link network is described.

Feedforward neural networks are of course just another functional expansion which relates input variables to output variables and this interpretation means that most of the results from system identification [14,15] can usefully be applied to measure, interpret and improve network performance. These concepts and ideas are discussed in an attempt to answer questions at least partially such as: how to assign input nodes, does network performance improve with increasing network complexity, is model validation useful, will noisy measurements affect network performance, is it possible to detect poor network performance, can we judge when the generalisation properties of a network will be good or bad, and so on.

While almost all of the results are applicable for alternative learning algorithms and should apply to the range of problems for which neural networks have been considered, a discussion that relates specifically to the identification of nonlinear systems is also included.

Throughout, the algorithms are compared and illustrated using examples based on data from both simulated and real systems.

11.2 Problem formulation

Consider the nonlinear relationship

$$y(t) = f(y(t-1),\ldots, y(t-n_y), u(t-1),\ldots, u(t-n_u)) + e(t) \tag{11.1}$$

where

$$y(t) = [y_1(t),\ldots, y_m(t)]^{\mathrm{T}},\ u(t) = [u_1(t),\ldots, u_r(t)]^{\mathrm{T}},\ e(t) = [e_1(t),\ldots, e_m(t)]^{\mathrm{T}} \tag{11.2}$$

are the system output, input and noise vectors respectively and $f(.)$ is some vector valued nonlinear function. In the realistic case where the output is corrupted by noise, lagged noise terms have to be included within $f(.)$ and this defines the NARMAX model (nonlinear autoregressive moving average model with exogenous inputs) introduced by Billings and Leontaritis [16,17] and studied extensively in nonlinear system identification [18]. In an attempt to keep the concepts as simple as possible most of the current analysis will relate to the model of eqn. 11.1 with the aim of approximating the underlying dynamics $f(.)$ using neural networks. Throughout the network input vector will be defined as

$$x(t) = [y^{\mathrm{T}}(t-1),\ldots, y^{\mathrm{T}}(t-n_y)\ u^{\mathrm{T}}(t-1),\ldots, u^{\mathrm{T}}(t-n_u)]^{\mathrm{T}} \tag{11.3}$$

with dimension $n_I = m \times n_y + r \times n_u$, where the one-step-ahead predicted output is expressed as

$$\hat{y}(t) = \hat{f}(x(t)) \tag{11.4}$$

and the model predicted output as

$$\hat{y}_d(t) = \hat{f}(\hat{y}_d(t-1),\ldots,\hat{y}(t-n_y),u(t-1),\ldots,u(t-n_u)) \tag{11.5}$$

11.3 Learning algorithms for multi-layered neural networks

11.3.1 The multi-layered perceptron

A neural network is a massively parallel interconnected network of elementary units called neurons. The inputs to each neuron are combined and the neuron produces an output if the sum of inputs exceeds a threshold value. A feedforward neural network is made up of layers of neurons between the input and output layers called hidden layers with connections between neurons of intermediate layers. The general topology of a multi-layered neural network is illustrated in Figure 11.1.

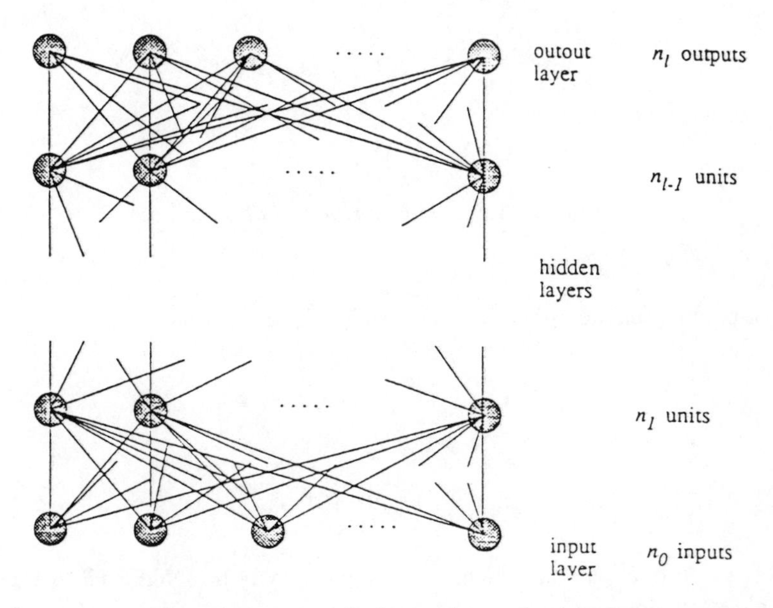

Figure 11.1 *A multi-layered neural network*

The input layer acts as an input data holder which distributes the inputs to the first layer. The outputs from the first layer nodes then become inputs to the second layer and so on. The last layer acts as the network output layer and all the other layers below it are called hidden layers. A basic element of the network, the *i*th neutron in the *k*th layer, illustrated in Figure 11.2, incorporates the combining and activation functions.

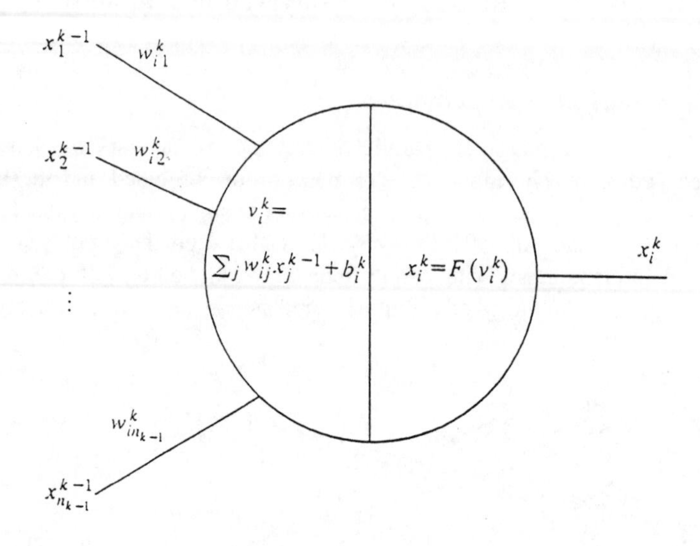

Figure 11.2 *A hidden neuron i of layer k*

The combining function produces an activation for the neuron

$$v_i^k(t) = \sum_{j=1}^{n_{k-1}} w_{ij}^k \, x_j^{k-1}(t) + b_i^k \tag{11.6}$$

$$x_i^k(t) = F(v_i^k(t)) \tag{11.7}$$

where w_{ij}^k and b^k are the connection weights and threshold and $F(.)$ is the activation function.

In applications to modelling nonlinear dynamic systems of the form of eqn. 11.1, the network input is given by $x(t)$ (eqn. 11.3) with $n_0 = n_I$ and the number of output nodes is $n_l = m$. The activation function of the hidden nodes is typically chosen as

$$F(v(t)) = \frac{1}{1+\exp(-v(t))} \tag{11.8}$$

or

$$F(v(t)) = \frac{1-\exp(-2v(t))}{1+\exp(-2v(t))} \tag{11.9}$$

The output nodes usually do not contain a threshold parameter and the activation functions are linear to give

$$\hat{y}_i\ (t) = \sum_{j=1}^{n_{l-1}} w_{ij}^l\ x_j^{l-1}\ (t) \tag{11.10}$$

Cybenko [19] and Funahashi [20] have proved that the multi-layered perceptron is a general function approximator and that one hidden layer networks will always be sufficient to approximate any continuous function. To simplify the notation therefore only one hidden layer networks and only single-input single-output (SISO) systems (m = 1, r = 1) will be considered in the present study. It is important to emphasise however that all the results are applicable to networks with several hidden layers and multi-input multi-output (MIMO) systems. The SISO restriction means that only one output neuron, that is $n_2 = 1$ is required and the index i in eqn. 11.10 can be dropped. With these simplifications the output of the network is

$$\hat{y}\ (t) = \sum_{i=1}^{n_l} w_i^2 x_i^l\ (t) = \sum_{i=1}^{n_l} w_i^2\ F\Big(\sum_{j=1}^{n_0} w_{ij}{}^l x_j(t)\ + b_i{}^l\Big) \tag{11.11}$$

The weights w_i and thresholds b_j are unknown and can be represented by the parameter vector $\Theta=[\theta_1,\ \theta_2,\ldots,\theta_{n_\theta}]^{\mathrm{T}}$. The objective of training the neural network is to determine Θ such that the discrepancies defined as

$$\varepsilon\,(t) = y(t) - \hat{y}\ (t) \tag{11.12}$$

called the prediction errors or residuals are as small as possible according to a defined cost function.

11.3.2 Backpropagation

The backpropagation method of training neural networks was initially introduced by Werbos [21] and later developed by Rumelhart and McClelland [22]. Backpropagation is just a steepest descent type algorithm where the weight connection between the jth neuron of the (k-1)th layer and the ith neuron of the kth layer and the threshold of the ith neuron of the kth layer are respectively updated according to

$$w_{ij}^{k}(t) = w_{ij}^{k}(-1) + \Delta w_{ij}^{k}(t)$$
$$b_{i}^{k}(t) = b_{i}^{k}(t-1) + \Delta b_{i}^{k}(t) \tag{11.13}$$

with the increment $\Delta w_{ij}^{k}(t)$ and $\Delta b_{i}^{k}(t)$ given by

$$\Delta w_{ij}^{k}(t) = \eta_{w}\, \rho_{i}^{k}(t) x_{j}^{k-1}(t) + \alpha_{w}\, \Delta w_{ij}^{k}(t-1)$$
$$\Delta b_{i}^{k}(t) = \eta_{b}\, \rho_{i}^{k}(t) + \alpha_{b}\, \Delta b_{i}^{k}(t-1) \tag{11.14}$$

where the subscripts w and b represent the weight and threshold respectively, α_w and α_b are momentum constants which determine the influence of past parameter changes on the current direction of movement in the parameter space, η_w and η_b represent the learning rates and $\rho_{i}^{k}(t)$ is the error signal of the ith neuron of the kth layer which is backpropagated in the network. Because the activation function of the output neuron is linear, the error signal at the output nodes is

$$\rho_{i}^{l}(t) = y_{i}(t) - \hat{y}_{i}(t) \tag{11.15}$$

and for the neurons in the hidden layer

$$\rho_{i}^{k}(t) = F'(v_{i}^{k}(t)) \sum_{j} \rho_{j}^{k+1}(t)\, w_{ji}^{k+1}(t-1) \qquad k=l-1,\ldots,2,1 \tag{11.16}$$

where $F'(v)$ is the first derivative of $F(v)$ with respect to v.

Similar to other steepest descent type algorithms, backpropagation is often slow to converge, it may become trapped at local minima and it can be sensitive to user selectable parameters [4].

11.3.3 Prediction error learning algorithms

Estimation of parameters in nonlinear models is a widely studied topic in system identification [14,23] and by adapting these ideas to the neural network case a class of learning algorithms known as prediction error methods can be derived for the multi-layered perceptron. The new recursive prediction error method for neural networks was originally introduced by Chen *et al.* [6,7] and Billings *et al.* [4] as an alternative to backpropagation. The full recursive prediction error (RPE) algorithm is given by

$$
\begin{aligned}
&\Delta(t) = \alpha_m \Delta(t-1) + \alpha_g \Psi(t)\, \varepsilon(t) \\
&P(t) = [P(t-1) - P(t-1)\Psi(t)\,(\lambda I + \Psi^T(t)\, P(t-1)\, \Psi(t))^{-1}\, \Psi^T(t)\, P(t-1)\,]/\lambda \\
&\hat{\Theta}(t) = \hat{\Theta}(t-1) + P(t)\Delta(t) \\
&\varepsilon(t) = y(t) - \hat{y}(t)
\end{aligned}
\tag{11.17}
$$

where α_g and a_m are the adaptive gain and momentum respectively, λ is the forgetting factor, Θ is the estimate of the parameter vector and Ψ^T represents the gradient of the one-step-ahead predicted output with respect to the model parameters

$$
\Psi(t,\Theta) = \left[\frac{d\hat{y}\;(t,\Theta)}{d\Theta}\right] \tag{11.18}
$$

By partitioning the Hessian matrix into $q \times q$ sub-matrices, for a network with q nodes, a parallel version of the above algorithm can be derived as

$$
\left.
\begin{aligned}
&\Delta_i(t) = \alpha_m \Delta_i(t-1) + \alpha_g \Psi_i(t)\, \varepsilon(t) \\
&\hat{\Theta}_i(t) = \hat{\Theta}_i(t-1) + P_i(t)\Delta_i(t)
\end{aligned}
\right\} \quad 1 \le i \le q
\tag{11.19}
$$

where the formula for updating $P_i(t)$ is identical to that used for $P(t)$ in eqn. 11.17.

A rigorous derivation of these algorithms together with implementation details, properties and comparisons with backpropagation on both simulated and real data sets is available in the literature [4,5,6,7].

Both recursive prediction error algorithms given above converge significantly faster than backpropagation at the expense of increased algorithmic complexity. Full RPE (eqn. 11.17) violates the principle of distributed computing because it is a centralised learning procedure. However, the parallel prediction error algorithm eqn. 11.19, which is a simple extension to full RPE, consists of many subalgorithms, each one associated with a node in the network [7]. The parallel recursive prediction error algorithm is therefore computationally much simpler than the full version and like backpropagation learning is distributed to each individual node in the network. Although parallel RPE is still algorithmically more complex than backpropagation the decrease in computational speed that this imposes per iteration is often offset by a very fast convergence rate so that overall the RPE algorithm is computationally far more efficient.

Comparisons of backpropagation and parallel RPE are available in the literature [4,11] and in Section 11.4.2.

11.4 Radial basis function networks

Radial basis function (RBF) networks [8,9,24,25] consist of just two layers and provide an alternative to the multi-layered perceptron architecture. The hidden layer in a RBF network consists of an array of nodes and each node contains a parameter vector called a centre. The node calculates the Euclidean distance between the centre and the network input vector and the result is passed through a nonlinear function. The output layer is just a set of linear combiners. A typical RBF network is illustrated in Figure 11.3.

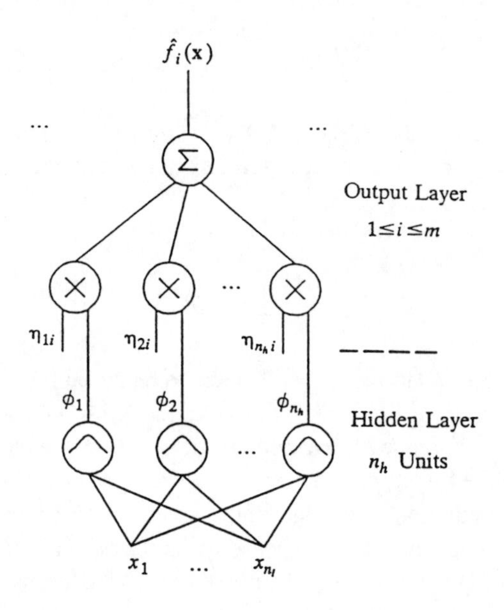

Figure 11.3 *Schematic of radial basis function network*

The response of the RBF network can be expressed as

$$\hat{f}_i(x) = \sum_{j=1}^{n_h} \eta_{ji}\phi_j = \sum_{j=1}^{n_h} \eta_{ji}\phi(\|x - c_j\|, \rho_j) \qquad 1 \le i \le m \tag{11.20}$$

where η_{ji} are the weights of the linear combiners, $\|.\|$ denotes the Euclidean norm, c_j are called RBF centres, ρ_i are positive scalars called widths, n_h is the number of nodes in the hidden layer, m=1 for SISO systems, and $\phi(\ .\ ,\ \rho)$ are functions typically chosen as

$$\phi(z,1) = z^2 \log(z) \tag{11.21}$$

the thin plate spline function or

$$\phi(z, \rho) = \sqrt{z^2 + \rho^2} \tag{11.22}$$

the multi-quadratic function etc. Clearly the topology of the RBF network is very similar to the two layer perceptron with the main difference being the characteristics of the hidden nodes. Park and Sandberg [26] have shown that any continuous function can be uniformly approximated to an arbitrary accuracy by a RBF network and Poggio and Girosi [25] have shown that they exhibit the best approximation property confirming the importance of these architectures.

11.4.1 Learning algorithms for radial basis function networks

If a block of data is available for processing the parameters of a RBF network can be estimated using a prediction error method. This would however result in a nonlinear learning rule. Significant advantages can be gained by selecting some data points as RBF centres and developing learning rules based on linear regression methods [24]. The disadvantage of this is the difficulty in selecting appropriate centres from the large number of candidate choices since an inappropriate selection will often lead to unsatisfactory results.

However, by interpreting the RBF network in terms of a NARMAX model and exploiting the extended model set idea and associated estimation algorithm of Billings and Chen [27] the RBF centres can be optimally positioned within a linear regression framework [8,9]. This means computationally cheap learning algorithms, fast adaption and avoidance of problems associated with local minima because there is only one global minimum.

The algorithm which consists of interpreting the RBF network (eqn. 11.20) in terms of the linear regression model

$$y(t) = \sum_{j=1}^{M} \phi_j(t)\theta_{ji} + e_i(t) \quad 1 \le i \le m, \tag{11.23}$$

where $\phi_j(t)$ are known as regressors and θ_{ji} are the parameters to be estimated, is available in the literature [8,9]. The orthogonal least squares training algorithm has recently been enhanced by combining regularisation with the parsimonious principle to produce a regularised orthogonal training procedure. The new algorithm is computationaly efficient and tends to produce parsimonious RBF networks with excellent generalisation properties [28].

For on-line identification applications using the RBF network a recursive algorithm which updates the centres and weights at each sample time is required. Moody and Darken [29] showed that the n-means clustering technique of pattern classification can be used to update RBF centres. Chen *et al.* [10] adapted this procedure to derive a new hybrid learning algorithm for RBF networks which consists of a recursive n-means clustering subalgorithm for adjusting the centres and a recursive least squares subalgorithm for updating the weights. Although the new hybrid algorithm can be implemented in batch form the main advantage of the hybrid method is that it can naturally be implemented in recursive form. Space limitations preclude a full description but details are available in the literature [10]. Modifications of this algorithm include the use of different clustering schemes including a cluster variation-weighted measure [30], and fuzzy clustering to determine the RBF centres in a manner which overcomes the sensitivity to the initial centre positions and the inclusion of direct linear links [31].

11.4.2 Backpropagation vs RPE vs the hybrid clustering algorithm

Several examples which illustrate the performance of these three algorithms have been described in the literature cited above. To illustrate the typical results that are obtained consider the identification of a nonlinear liquid level system based on sampled signals of the voltage to a pump motor which feeds a set of tanks one of which is conical. This is a SISO system $r = m = 1$. A multi-layered perceptron with one hidden layer defined by $n_I = n_y + n_u = 3+5$, $n_1 = 5$, $n_2 = 1$ giving $n_\theta = 50$ was used to model this process. The hidden node activation function of eqn. 11.8 was selected and the initial weights and thresholds were set randomly between ± 0.3. After several trial runs it was found that $\eta_w = \eta_b = 0.01$ and $\alpha_w = \alpha_b = 0.8$ were appropriate for the backpropagation algorithm eqns. 11.13 and 11.14.

For the parallel recursive error algorithm eqn. 11.19 a constant trace technique was used to update $P_i(t)$ and the parameters of the algorithm were set to $\alpha_g = 1.0$, $\alpha_m = 0.0$, $\lambda = 0.99$, $k_o = 60.0$ and $P_i(0) = 1000.0I$ where k_o is a parameter in the constant trace technique.

The structure of the RBF network employed to identify this system was defined by $n_I = n_y + n_u = 3+5$ giving $n_h = 40$ and ϕ (.) was chosen as the thin plate spline function eqn. 11.21. The set-up parameters for the hybrid identification algorithm were selected as $P(0) = 1000.0I$, $\lambda_o = 0.99$, $\lambda(0) = 0.95$ and $\alpha_c(0) = 0.6$, and the initial centres were randomly set. $P(t)$ is the usual matrix associated with RLS, λ_o and $\lambda(0)$ define the initial forgetting factor, and $\alpha_c(t)$ is a slowly decreasing learning rate for the RBF centres.

The evolution of the mean squared error in dBs for all three algorithms is illustrated in Figure 11.4. The results typify the performance of the three algorithms observed over several examples and clearly show that the parallel RPE provides a significant improvement compared with backpropagation and the hybrid algorithm for RBF networks is better again.

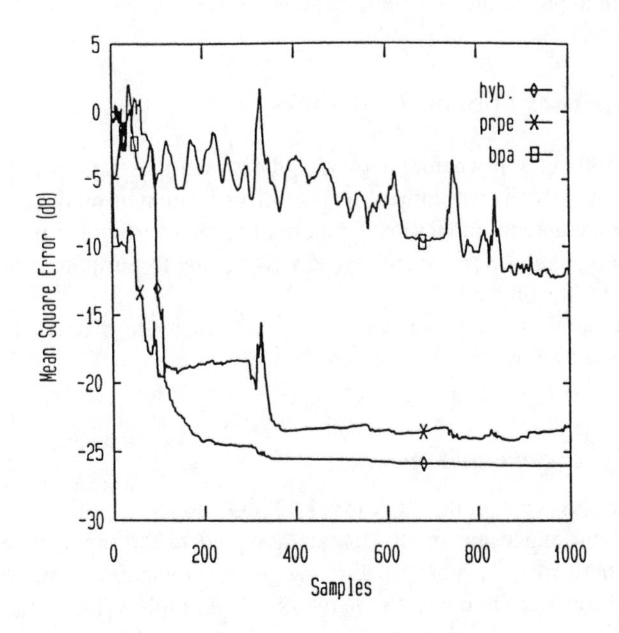

Figure 11.4 *Evolution of mean square error*

11.5 The functional link or extended model set network

The main disadvantage of the MLP is that the nonlinear activation functions create problems when training the network. Nonlinear learning methods have to be applied, the learning rate can be unacceptably slow and in some applications local minima cause problems. One way to avoid nonlinear learning and hence the associated problems is to perform some nonlinear transformation or expansion of the network inputs initially and then linearly to combine the resulting terms. This is similar to the concept of RBF networks except that now there are no free parameters like the widths and centres associated with eqn. 11.20; instead functions such as absolute value $|x_i - x_j|$, hyperbolic tanh (x_i), exponential $\exp(-x_i^{\ 2})x_j$ etc. can be employed. The potential list of functions is endless. This concept was introduced by Billings and Chen [27] and referred to as an extended model set (EMS). Pao [32] discussed a similar idea and called it the functional link network (FLN). A related idea has been used by Brown and Harris [3] to formulate the B-spline network.

Networks based on these concepts are very flexible and can be used to represent a rich class of systems with the major advantage that the learning algorithms are linear. Optimal procedures for training such networks are given in [27].

11.6 Properties of neural networks

It is easy blindly to apply neural networks to data sets and to get what appear to be credible results. But as with other forms of identification or model fitting it is also easy to be misled and it is therefore important to be aware of the properties of the networks and the algorithms which are employed and to formulate methods which validate the results obtained.

Some properties of multi-layered networks which are discussed in [4,5] are briefly reviewed below.

11.6.1 Network expansions

It is easy to see from eqns. 11.6 to 11.10 that classical feedforward networks provide a static expansion of the signals assigned to the input nodes. In other words the networks do not generate lagged or dynamic components of the variables assigned as inputs to the network. This implies that if the model of a system involves an expansion of $u(t-1)$, $y(t-6)$ and $y(t-18)$ say then the network model of this system is likely to be poor, irrespective of which training algorithm is employed, unless network input nodes are assigned to these variables. Input node assignment can therefore be crucial to network performance. This is not an easy problem to solve because in reality the appropriate lagged *u*s and *y*s may be distributed over a very wide range and attempts to over-specify the network input nodes simply leads to increased problems of dimensionality and slow training.

11.6.2 Model validation

Model validity tests are procedures designed to detect the inadequacy of a fitted model. Irrespective of the particular discrepancy of the model including in the neural network context incorrect input node assignment, insufficient hidden nodes, noisy data or a network that has not converged model validity tests should detect that the model is in error.

Well known tests for linear systems [14] are inappropriate for the present application but Billings and Voon [33,34] have shown that, for the NARMAX model, the following conditions should hold if the fitted nonlinear model is adequate

$$
\begin{aligned}
&\Phi_{\varepsilon\varepsilon}(\tau) = E[\,\varepsilon(t-\tau)\,\varepsilon(t)] = \delta(\tau) && \forall\tau \\
&\Phi_{u\varepsilon}(\tau) = E[u(t-\tau)\,\varepsilon(t)] = 0 && \forall\tau \\
&\Phi_{u^{2'}\varepsilon}(\tau) = E[(u^2(t-\tau) - E[u^2(t)])\,\varepsilon(t)] = 0 && \forall\tau \\
&\Phi_{u^{2'}\varepsilon^2}(\tau) = E[(u^2(t-\tau) - E[u^2(t)])\,\varepsilon^2(t)] = 0 && \forall\tau \\
&\Phi_{\varepsilon(\varepsilon u)}(\tau) = E[\,\varepsilon(t)\,\varepsilon(t-1-\tau)\,u(t-1-\tau)] = 0 && \tau \geq 0
\end{aligned}
\tag{11.24}
$$

Whilst it is very difficult to prove definitively for neural networks, which are often highly nonlinear in the parameters, that the tests of eqn. 11.24 will detect every possible model deficiency, the tests have been shown to provide excellent results in practice. Normalisation to give all the tests a range of plus or minus one and approximate 95% confidence bands at $1.96/\sqrt{N}$ make the tests independent of the signal amplitudes and easy to interpret.

Consider the problem of incorrect input node assignment to illustrate the power of the model validity concept. The system defined by

$$
\begin{aligned}
\underline{y}(t) &= \frac{0.4}{1 + e^{-(0.3u(t-1)+0.7u(t-2)+0.5\underline{y}(t-1)+0.1)}} \\
n(t) &= e(t) + 0.6e(t\text{-}1) \\
y(t) &= \underline{y}(t) + n(t)
\end{aligned}
\tag{11.25}
$$

was simulated with a zero mean uniformly distributed white noise input $u(t)$. Clearly this system can be represented exactly by a MLP network with just one hidden neuron and input node assignment $u(t-1), u(t-2), y(t-1)$. Initially the network was trained with the input vector incorrectly assigned as

$$x(t) = [u(t-1)\; y(t-1)]^{\mathrm{T}}$$

The model validity tests for this case are illustrated in Figure 11.5.

The tests $\Phi_{u\varepsilon}(\tau)$ and $\Phi_{u^{2'}\varepsilon^2}(\tau)$ are well outside the confidence bands at lag two suggesting that term of this lag has been omitted. When the network was retained with a correct input node assignment

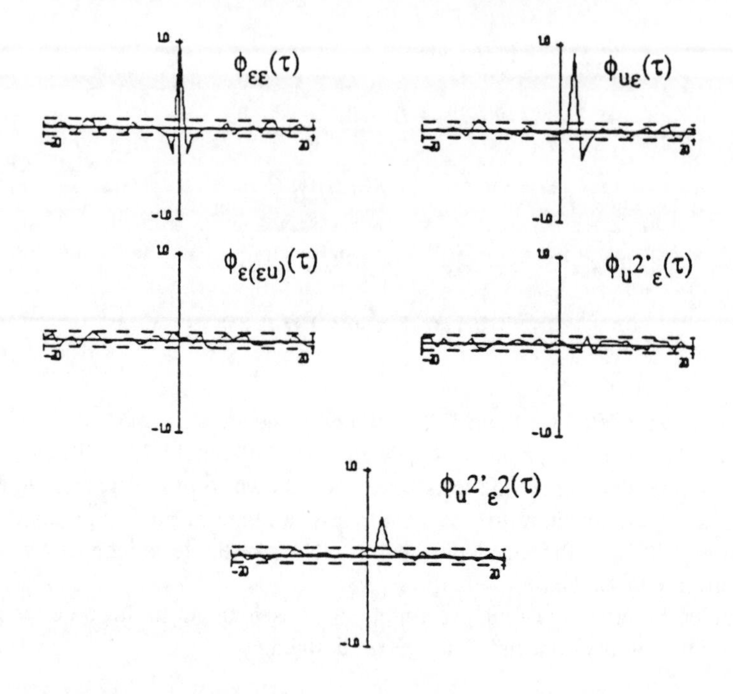

Figure 11.5 *Model validation showing incorrect node assignment*

$$x(t) = [u(t-1)\ u(t-2)\ y(t-1)]^{\mathrm{T}}$$

$\Phi_{u\varepsilon}(\tau)$ and $\Phi_{u^{2'}\varepsilon^{2}}(\tau)$ were inside the bands indicating that the input node selection was now sufficient. The autocorrelation of the residuals $\Phi_{\varepsilon\varepsilon}(\tau)$ does not satisfy eqn. 11.24 because the simulation eqn. 11.25 included coloured noise and $\Phi_{\varepsilon\varepsilon}(\tau)$ detects this condition which will induce bias. This is discussed in the next section.

11.6.3 Noise and bias

Almost every single application of neural networks to the identification and modelling of nonlinear systems ignores the fact that noise will always be present if the data have been recorded from a real system. Unless the effects of the noise are understood and the noise is accommodated appropriately then incorrect or biased models will result. Bias is a well known concept in estimation theory and can often only be eliminated by fitting a noise model. If the system which is to be modelled is nonlinear then there is no reason to assume the noise will be purely linear either. The noise is likely to enter in some nonlinear way and even if it can be assumed to be additive at the system output (highly unlikely unless the system is linear), the noise on the model will in general involve cross products between the inputs and outputs [17].

Bias is often difficult to detect because even if a biased network model has been obtained, when this network is used to predict over the data set that was used to determine the weights and thresholds, a good prediction will often be obtained. This is to be expected because the network has been trained to minimise a function of the squared prediction errors, to curve fit to the data, but it does not mean the network provides a model of the underlying mechanism.

The importance of the bias problem can be illustrated by using the example defined by

$$\underline{y}(t) = \frac{0.6}{1+\mathrm{e}^{(-(0.5u(t-1)+0.4\underline{y}(t-1)+0.1))}}$$
$$y(t) = \underline{y}(t) + n(t) \tag{11.26}$$

where the input $u(t)$ was a zero mean uniformly distributed white noise sequence and $n(t)$ represents additive noise. The form of eqn. 11.26 has been specifically chosen to match the activation function eqn. 11.8 so that the estimated weights can be associated with the true model parameters. Initially $n(t)$ was set to zero and the network was trained and produced the model

$$\underline{y}(t) = \frac{0.6}{1+\mathrm{e}^{(-(0.50u(t-1)+0.398\underline{y}(t-1)+0.102))}} \tag{11.27}$$

A comparison of eqns. 11.26 and 11.27 shows that the network is a good representation of the system.

The system was simulated again but this time with $n(t)$ defined as coloured noise

$$u(t) = e(t) + 0.6\, e(t-1)$$

where $e(t)$ was Gaussian white $N(0,0.01)$. The network was retrained and produced the model

$$\underline{y}(t) = \frac{0.578}{1+e^{(-(0.519u(t-1)+0.652\underline{y}(t-1)+0.091))}} \tag{11.28}$$

This model is biased because the coefficient of $y(t-1)$ is 0.652 when it should be approximately 0.4. The bias in this example only affects one parameter, in general it will alter all the weights. It is easy to detect in this example because we know the real answer.

Fortunately, the model validity tests of eqn. 11.24, or in the time series case the tests of Billings and Tao [35], can be used to detect these problems. The tests for the biased network of eqn. 11.28 are illustrated in Figure 11.6.

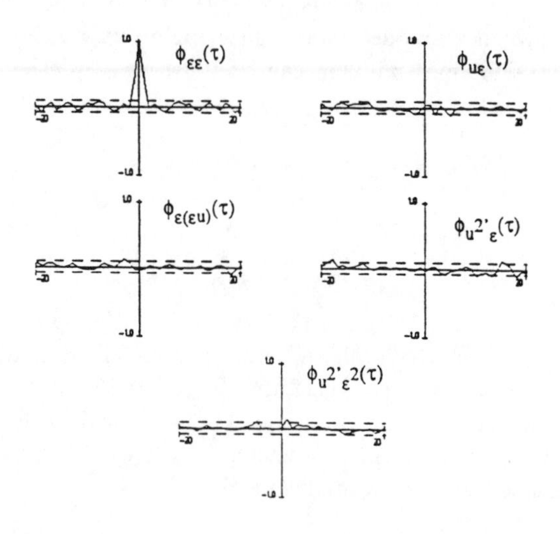

Figure 11.6 *Model validity tests for the network defined by eqn. 11.28 showing bias*

All the correlation functions except $\Phi_{\varepsilon\varepsilon}(\tau)$ are satisfied. This indicates that the network structure is correct but that a noise model will be required to reduce the prediction errors to an unpredictable sequence and eliminate bias.

One way to avoid bias is to fit a noise model and a network architecture which can accommodate this was proposed in [5]. This consisted of augmenting the network by adding a linear mapping from lagged prediction error terms $e(t-1)$, $e(t-n_e)$, computed at the previous iteration, onto the usual network architecture. Alternatively a much more general solution could be obtained by simply using more input nodes in an MLP assigned to lagged prediction error terms [8].

11.6.4 Network node assignment

The effects of bias on the network will be influenced by the assignment of network nodes.

In general a complex nonlinear system can be represented by the NARMAX model of eqn. 11.1. This maps past inputs and outputs into the current output according to the nonlinear expansion or, in our case, the network architecture. The advantage of this type of model is that it provides a very concise representation of the system. In neural network terms we would assign input nodes to represent both past inputs $u(t-1)$, $u(t-2)$... and past outputs $y(t-1)$,$y(t-2)$... This minimises the number of input nodes required and hence implies faster training. The disadvantage is that the parsimonious property is offset by the need to fit noise models to accommodate bias as discussed in the previous section.

The alternative is only to allow the model or the network to be a function of past inputs. So in eqn. 11.1 all the past outputs are eliminated from the rhs to give

$$y(t) = f^{1}(u(t-1), u(t-2), \ldots, u(t-n_{u'})) + e(t) \tag{11.29}$$

The advantage of this is that since the noise will almost always be independent of the input even if it is coloured, the estimates will be unbiased even without a noise model, provided the noise is purely additive. However, this property is obtained at the expense of a considerably larger network $n_{u'} >> n_u + n_y$, increased complexity and consequently much slower training. This follows because if lagged *y*s are disallowed the expansion eqn. 11.29 will involve a large, possibly infinite number, of past *u*s. In terms of neural networks this means considerably more input nodes in the network.

This analysis may be rather simplistic because the elimination of bias in the latter case will only be correct if the noise is additive at the output. If the system being modelled is nonlinear and if the noise occurs at internal points within the system, multiplicative noise terms will be induced into the system model and biased models will be obtained, unless this effect is accommodated even if the network is just an expansion in terms of past *u*s. When considering real systems therefore the choice of network node assignment is not clear cut. Again the model validity tests can be used to try and detect these effects.

Many authors have discussed how to include temporal information into networks. These include subset selection methods, recurrent network concepts, backpropogation through time and other related ideas [36,37,38].

11.6.5 Network complexity

Over-fitting by over-specifying the input node assignments and number of hidden nodes is likely to produce misleading results. Over-fitting, like bias, is a subtle effect [5]. Increasing the complexity of the network is likely to improve the prediction capabilities of the network because at each stage additional terms are added which further reduce the mean squared error. However, taking the extreme case, the prediction error could be reduced to zero simply by employing a look-up table which associates the input at each time instant to the recorded output. It is clear that such a table or model is virtually useless because it only represents one data set.

It is well known in linear estimation theory that model over-fitting should be avoided. The usual practice is to fit models of increasing order and to select the model of minimal complexity which just satisfies the model validity tests. Two approaches are available to overcome over-fitting for neural networks, namely pruning and regularisation. The first approach consists of fitting a large network and then pruning by weight elimination to reduce the size of the network. Several methods are available including the well known optimal brain damage method. Regularisation improves generalisation by adding an extra penalty function to the network cost function to penalise over-fitting.

11.6.6 Metrics of network performance

The common measure of predictive accuracy of network performance is the one-step-ahead prediction of the system output. This was defined in eqn. 11.4 to be

$$\hat{y}(t) = \hat{f}(y(t-1),\ldots,y(t-n_y),u(t-1),\ldots,u(t-n_u)) \qquad (11.30)$$

Inspection of eqn. 11.30 suggests that this may not be a good metric to use because at each step the model is effectively reset by inserting the appropriate values in the rhs of eqn. 11.30. Any errors in the prediction are therefore reset at each step and consequently even a very poor model tends to produce reasonable one-step-ahead predictions. In reality we may wish to use the model in simulation to predict the system response to different inputs and in this situation the required output values, the *y*s on the rhs of eqn. 11.30, will be unknown. It is for these reasons that we recommend that the one-step-ahead prediction should always be augmented by computing the model predicted output defined by eqn. 11.5.

$$\hat{y}_d(t) = \hat{f}(\hat{y}_d(t-1),\ldots,\hat{y}_d(t-n_y),u(t-1),\ldots,u(t-n_u)) \qquad (11.31)$$

Now all the output terms on the rhs are predicted not measured values, $\hat{y}_d(t)$ can be computed for any $u(t)$ and any errors in the prediction should now become apparent because they will quickly accumulate.

11.7 System identification

Whilst the discussions in previous sections are relevant for all applications of neural networks this section will be devoted specifically to the use of neural networks and other procedures for modelling or identifying nonlinear dynamic systems.

Neural networks provide a new and exciting alternative to existing methods of system identification. The advantages and disadvantages of these two approaches are briefly discussed below.

11.7.1 Neural network models

Neural networks are just another way of curve fitting to data. They have several advantages; they are conceptually simple, easy to use and have excellent approximation properties, the concept of local and parallel processing is important and this provides integrity and good fault tolerant behaviour, and they produce impressive results especially for ill defined fuzzy problems such as speech and vision. The disadvantage is that most of the current procedures ignore all the analysis of parameter estimation and they destroy the structure of the system. Some of these points will not be relevant to all applications but they certainly are for dynamic modelling. To take an extreme example to illustrate the point, consider the situation where we have input/output data from a system which we wish to identify. The data were recorded from a linear system but this will not be apparent from an inspection of the data records and so an MLP network is used to model the system. The model obtained will probably predict the system output quite well but it will not reveal that the underlying system is a simple first-order lag with one time constant and a gain. The neural network model destroys and does not reveal the simplicity of the underlying mechanism which produced the data. This information is distributed and encoded in the network weights and thresholds. Consequently it is difficult to analyse the fitted model to determine the range of system stability, often critical in control systems design, or the location of resonances, the sensitivity of the process output to the model time constant and gain, or to relate the model parameters to the physical components of the system.

More traditional methods of system identification overcome some of these limitations.

11.7.2 The NARMAX methodology

Traditional methods of system identification for nonlinear systems have been based upon functional series methods [39], block structured system models [40] and parameter estimation procedures [27].

Parameter estimation is of course very straightforward if the form of the model is known *a priori*. This information is rarely available and the NARMAX methodology [41] is based around the philosophy that determining the model structure, or which terms to include in the model, is a critical part of the identification process. So that if the system is linear with a simple gain and time constant this should be revealed during the identification. Adopting this approach ensures the model is as simple as possible. Complex behaviour does not necessarily equate to complex models. If we do not try and determine the model structure then many different phenomena from several systems which are indeed described by the same law will appear as different laws. We will not see the simplicity and the generality of the underlying mechanism, and surely this is one of the main objectives of system identification.

The NARMAX methodology attempts to break the problem down into the following steps:

nonlinear detection: is the system linear or nonlinear?
structure detection: which terms are in the model?
parameter estimation: what are the values of the unknown coefficients?
model validation: is the model the correct model?
prediction: what is the output at some future time?
analysis: what are the properties of the system?

These components form an estimation tool-kit which allows the user to build a concise mathematical description of a system that can be used as a basis for analysis and design. Both polynomial, rational and extended model set terms can be accommodated as required. The disadvantage of this approach is that the algorithms are much more complex than simple backpropagation for example, the user needs to have an understanding of parameter estimation theory and needs to interact more during the identification process.

Both the neural network and the parameter estimation approach therefore have strengths and weaknesses.

Users need to be aware of both approaches [13,15] but the choice between them will often depend on the purpose of the identification. Algorithmic computing and neurocomputing therefore tend to complement each other. The former is ideal for modelling engineering systems and the like. The latter is ideal for pattern recognition, fuzzy knowledge processing, speech and vision. But there is a large area in between which will benefit from a fusion of ideas from both approaches.

11.8 Conclusions

Neural networks have opened up a whole new area of opportunity which even at this early stage of development have shown very impressive results. But we should not ignore algorithmic methods and estimation theory which offer alternative approaches and suggest solutions to many of the current problems in neural network research. A marriage of the best of both these approaches should provide a powerful tool-kit for analysing an enormous range of systems and stimulate research for many years to come.

11.9 Acknowledgments

The authors gratefully acknowledge support for part of this work from the UK Engineering and Physical Sciences Research Council.

11.10 References

[1] Lapedes, A. and Farber, R., 1988, How neural nets work, in *Neural Information Processing Systems*, Anderson, D.Z., (Ed.), New York, American Institute of Physics, pp. 442-456.

[2] Narendra, K.S. and Parthasarathy, K., 1990, Identification and control of dynamical systems using neural networks, *IEEE Trans. Neural Networks*, Vol. 1, pp. 4-27.

[3] Brown, M. and Harris, C. J., 1992, The B-spline neuro-controller, in *Parallel Processing for Control*, Rogers, E. (Ed.) Prentice-Hall.

[4] Billings, S.A., Jamaluddin, H.B. and Chen, S., 1991, A comparison of the backpropagation and recursive prediction error algorithms for training neural networks, *Mechanical Systems and Signal Processing*, Vol. **5**, pp. 233-255.

[5] Billings, S.A., Jamaluddin, H.B. and Chen, S., 1992, Properties of neural networks with applications to modelling nonlinear dynamical systems, *Int. J. Control*, Vol. 55, pp. 193-224.

[6] Chen, S., Billings, S.A. and Grant, P.M., 1990a, Non-linear systems identification using neural networks, *Int. J. Control*, Vol. 51, pp. 1191-1214.

[7] Chen, S., Cowan, C.F.N., Billings, S.A. and Grant, P.M., 1990b, Parallel recursive prediction error algorithm for training layered neural networks, *Int. J Control*, Vol. 51, pp. 1215-1228.

[8] Chen, S., Billings, S.A., Cowan, C.F.N. and Grant, P.M., 1990c, Practical identification of NARMAX models using radial basis functions, *Int. J. Control*, Vol. 52, pp. 1327-1350.

[9] Chen, S., Billings, S.A., Cowan, C.F.N. and Grant, P.M., 1990d, Nonlinear systems identification using radial basis functions, *Int. J. Systems Science*, Vol. 21, pp. 2513-2539.

[10] Chen, S., Billings, S.A. and Grant, P.M., 1992a, Recursive hybrid algorithm for nonlinear system identification using radial basis function networks, *Int. J. Control*, Vol. 55, pp. 1051-1070.

[11] Chen, S. and Billings, S.A., 1992b, Neural networks for nonlinear dynamic system modelling and identification, *Int. J. Control*, Vol. 56, pp. 319-346.

[12] Willis, M.J., Montague, G.A., Morris, A.J. and Tham, M.J., 1991, Artificial neural networks - a panacea to modelling problems, in *Proc American Control Conf*, pp. 2337 - 2342.

[13] Montague, G.A., Morris, J. and Willis, M.J., 1994, Artificial neural networks: methodologies and applications in process control, *Chapter 7 this volume*.

[14] Ljung, L. and Soderstrom, T., 1983, *Theory and Practice of Recursive Identification*, Cambridge, MIT Press.

[15] Sjorberg, J., Hjalmarsson, H. and Ljung, L., 1994, Neural networks in system identification, *Research Report LiTH-ISY-R-1622,* Linkoping University.

[16] Billings, S.A. and Leontaritis, I.J., 1981, Identification of nonlinear systems using parameter estimation techniques, in *Proc. IEE Conf. Control and Its Applications*, Warwick, U.K., pp. 183-187.

[17] Leontaritis, I.J. and Billings, S.A., 1985, Input-output parametric models for non-linear systems - Part 1: deterministic non-linear systems; Part 2: stochastic non-linear systems, *Int. J. Control*, Vol. 41, pp. 303-344.

[18] Chen, S. and Billings, S.A., 1989, Representation of non-linear systems: the NARMAX model, *Int. J. Control*, Vol. 49, pp. 1013-1032.

[19] Cybenko, G., 1989, Approximations by superpositions of a sigmoidal function, *Mathematics of Control, Signals and Systems*, Vol. 2, pp. 303-314.

[20] Funahashi, K., 1989, On the approximate realization of continuous mappings by neural networks, *Neural Networks*, Vol. 2, pp. 183-192.

[21] Werbos, P.I., 1974, Beyond regression: new tools for prediction and analysis in the behaviour sciences. Ph.D. thesis, Harvard University, Cambridge MA.

[22] Rumelhart, D.E. and McClelland, J.L. (Eds.), 1986, *Parallel Distributed Processing: Explorations in the Microstructure of Cognition,* Vol. 1: Foundations, MIT Press.

[23] Billings, S.A. and Chen, S., 1989a, Identification of non-linear rational systems using a prediction-error estimation algorithm, *Int. J. Systems Sci.*, Vol. 20, pp. 467-494.

[24] Broomhead, D.S. and Lowe, D., 1988, Multivariable functional interpolation and adaptive networks, *Complex Systems*, Vol. 2, pp. 321-355.

[25] Poggio, T. and Girosi, F., 1990, Networks for approximation and learning, *Proc IEEE*, Vol. 78, pp. 1481-1497.

[26] Park, J. and Sandberg, I.W., 1991, Universal approximation using radial-basis-function networks, *Neural Computation*, Vol. 3, pp. 246-257.

[27] Billings, S.A. and Chen, S., 1989b, Extended model set, global data and threshold model identification of severely non-linear systems, *Int. J. Control*, Vol. 50, pp. 1897-1923.

[28] Chen, S., Chng, E. S. and Alkadhimi, K., 1994, Regularised orthogonal least squares algorithm for constructing radial basis function networks. Submitted for publication.

[29] Moody, J. and Darken, C., 1989, Fast-learning in networks of locally-tuned processing units, *Neural Computation*, Vol. 1, pp. 281-294.

[30] Chinrungrueng, C. and Sequin, C.H., 1994, Enhanced k-means techniques for partitioning the input domain in function approximation, *J Applied Science and Computations*, Vol. 1.

[31] Zheng. G. and Billings, S.A., 1994, Network training using a fuzzy clustering scheme. Submitted for publication.

[32] Pao, Yoh-Han, 1989, *Adaptive Pattern Recognition and Neural Networks*, Reading, Addison-Wesley.

[33] Billings, S.A. and Voon, W.S.F., 1983, Structure detection and model validity tests in the identification of nonlinear systems, *Proc. IEE Pt. D*, Vol. 130, pp. 193-199.

[34] Billings, S.A. and Voon, W.S.F., 1986, Correlation based model validity tests for non-linear models, *Int. J. Control*, Vol. 44, No. 1, pp. 235-244.

[35] Billings, S.A. and Tao, Q.M., 1991, Model validity tests for nonlinear signal processing applications, *Int. J. Control*, Vol. 54, pp. 157-194.

[36] Marren, A., Harston, C. and Pap, R., 1990, *Handbook of Neural Computing Applications*, Academic Press.

[37] Miller, W. T., Sutton, R. S. and Werbos, P. J., 1990, *Neural Networks for Control,* MIT Press.

[38] Haykin, S., 1994, *Neural Networks,* Prentice Hall.

[39] Billings, S.A., 1980, Identification of nonlinear systems - a survey, *Proc. IEE, Part D*, Vol. 127, pp. 272-285.

[40] Billings, S.A. and Fakhouri, S.Y., 1982, Identification of systems composed of linear dynamic and static nonlinear elements, *Automatica*, Vol. 18, pp. 15-26.

[41] Billings, S.A., 1986, An introduction to nonlinear systems analysis and identification, in Godfrey, K. and Jones, P. (Eds) *Signal Processing for Control*, Springer Verlag, pp. 263-294.

Chapter 12

Neurofuzzy adaptive modelling and construction of nonlinear dynamical processes

K.M. Bossley, M. Brown and C.J. Harris

12.1 Introduction

System modelling has become an important task in many different areas of research, none more so than in the control community where it is termed *system identification*. Many modern control techniques rely on accurate, reliable models of the controlled process, sensors and environment to synthesise robust controllers. Generally the controller has the same complexity and structure as that of the process model, therefore it is critical to generate parsimonious models utilising any input/output data. This chapter addresses a range of neurofuzzy algorithms that automatically construct parsimonious models of nonlinear dynamical processes. The process (sensors and environment) dynamics are typically unknown and complex (i.e. multivariate, nonlinear and time varying) making the generation of accurate models by conventional methods, such as linear and nonlinear regressions impractical. In these instances more sophisticated (intelligent) modelling techniques are required. Typically the task of modelling entails *robustly* approximating the continuous input-output map-

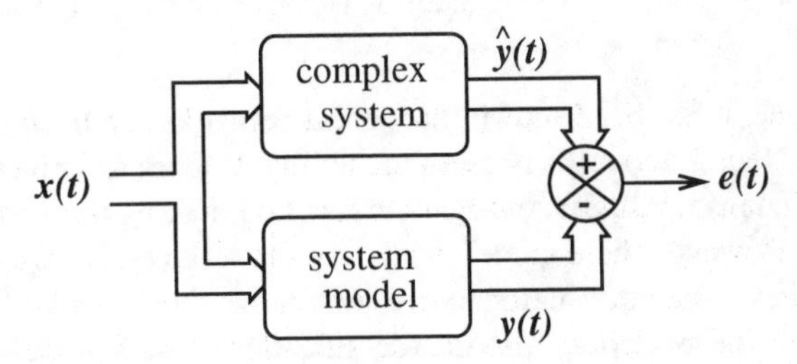

Figure 12.1 *Illustration of the modelling task; $e(t)$ should be sufficiently close to zero*

ping performed by the system, as illustrated in Figure 12.1. This task can be treated as a *learning* task where the model is taught such that the output error, $e(t)$, is sufficiently close to zero. The internal representation of these models is generally fixed, except for a series of adjustable parameters, called *weights*, which adapt the continuous function mapping of the model. Weight identification, known as *learning*, is achieved by optimising the weights with respect to some error criteria across a set of input-output pairs. This set is known as a *teaching set* and must adequately represent the systems dynamic behaviour. Typically this type of modelling is termed *black box* modelling where the internal representation does not reflect the behaviour of the physical system. The ability of a model to approximate the system output accurately, is not the only requirement and irrespective of which modelling technique is employed it should include the following desirable properties:

- **Accuracy:** the model should accurately approximate the desired system, across the training set;

- **Generalisation:** the model should generalise, this is the ability to model the desired system accurately for unseen inputs;

- **Interpretability:** often the understanding of the underlying system is limited and it is beneficial to the modeller if the model provides knowledge about the underlying physical process, i.e. qualitative physics. Such knowledge is also useful in validating the behaviour of the model.

- **Ability to encode *a priori* knowledge:** *a priori* knowledge is often available describing certain aspects of the systems operation. Any such knowledge should be utilised, hence a modelling technique should be capable of incorparating such knowledge;

- **Efficient implementation:** the model technique must use computational resources (i.e. speed and memory) efficiently as in many applications these are limited;

- **Fast adaption:** this property is particularly attractive when the model is employed on-line, where on-line adaption is performed as new system knowledge is acquired.

In the last decade, it has been shown that neural networks can be applied to system identification. Neural networks possess the ability to *learn* to universally approximate any continuous nonlinear multivariate function making them ideal candidates for modelling. However, these models tend to be black boxes, consisting of complex *opaque* structures. The information stored within the neural network cannot easily be interpreted by the modeller. Also the identification of such models is still a black art, where the choice of neural network class, optimisation technique and model structure has to be determined by the modeller.

An alternative approach to modelling is the use of fuzzy logic systems. Unlike neural networks, fuzzy logic systems reason in a seemingly natural manner, by the application of a series of linguistic rules. Typically, a fuzzy rule would be of the following form:

IF (*error is small*) THEN (*output is large*).

As the information in these fuzzy systems is stored as a set of interpretable rules, they are said to be *transparent*. However, these fuzzy systems are generally identified from heuristics and limited empirical knowledge. This approach tends to produce inadequate models to which no standard system identification mathematical analysis can be applied. Neurofuzzy systems, a relatively new modelling paradigm [4] are a particular type of fuzzy system. The fundamental components of the fuzzy system are constrained, enforcing a mathematical structure. The result is a fuzzy system defined on a neural network type structure providing a fuzzy system to which thorough mathematical analysis can be applied. Neurofuzzy systems have become an attractive powerful new modelling technique, combining the well established learning techniques of associative memory networks (AMNs) with the transparency of fuzzy systems. This technique can be termed *grey box* modelling as a qualitative insight into the behaviour of the system is given in terms of fuzzy rules. However, the modelling capabilities of any model technique are heavily dependent on their structure and to make modelling possible an appropriate structure must be identified. When little *a priori* knowledge of the system is known, model construction algorithms are required to identify neurofuzzy models. Also, neurofuzzy systems like many modelling techniques suffer from the *curse of dimensionality* i.e. the size of the rule base is an exponential function of the input dimension. Due to the exponentially increasing memory requirement and size of the teaching set required for a given output resolution, the use of neurofuzzy models on high dimensional problems (> 4 inputs) is impractical. To solve high-dimensional modelling problems some form of model complexity reduction must be performed, producing parsimonious neurofuzzy models. Hence, during model construction the following fundamental principles should be employed:

- **Principle of data reduction:** the smallest number of input variables should be used to explain a maximum amount of information;
- **Principle of network parsimony:** the best models are obtained using the simplest possible, acceptable structures that contain the smallest number of adjustable parameters.

12.2 Conventional neurofuzzy systems

A fuzzy system is a modelling technique that reasons by the application of vague fuzzy production rules of the form:

IF (*error is small*) THEN (*output is small*)

where a complete fuzzy system consists of a collection of these linguistic rules, known as a *rule base*. An inference engine uses this rule base to map fuzzy inputs to fuzzy outputs. Each fuzzy rule is given a *confidence* and these are also inferred by the inference engine, the learning mechanism in the system. Real-world inputs (*crisp*) are converted to vague fuzzy variables by the *fuzzifier*, these are then presented to the rule base. The rule base produces a collection of fuzzy output variables from the consequences of the rules, and the *defuzzifier* converts these to real valued outputs. There are various different types of fuzzy systems, but they all conform to the basic system outlined in Figure 12.2.

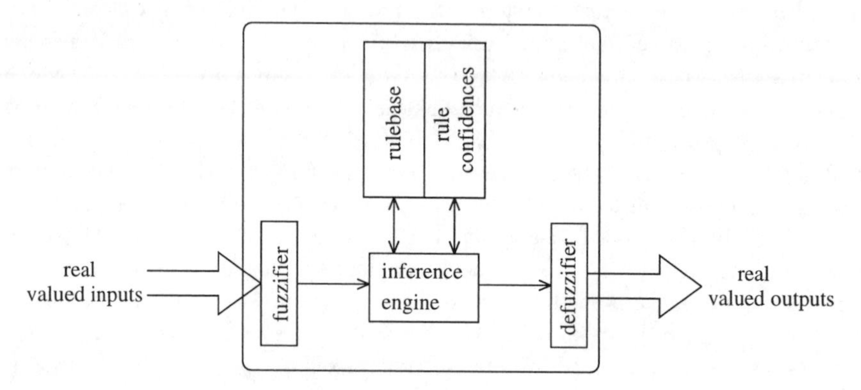

Figure 12.2 *The basic components of a fuzzy system*

A fuzzy system is defined by the underlying fuzzy membership functions and fuzzy operators used. Neurofuzzy systems [4, 20] are a particular type of fuzzy system in which the specialised structure gives the resulting models a well structured mathematical foundation. Fuzzy systems have previously been criticised as they lack mathematical rigour, preventing standard model identification algorithms and tests being applied. Neurofuzzy algorithms disperse this criticism as an in-depth mathematical analysis can be applied to the underlying mathematical structure [4]. In neurofuzzy systems the fuzzy membership functions and the fuzzy operators are defined in such a manner that there exists a direct invertible relationship between the fuzzy rule base and an associative memory network (AMN). All the well understood behaviour and established theorems of AMNs are inherited by neurofuzzy systems, with the added invaluable advantage of transparency. This transparency enables interpretability of the final model, providing the modeller with an insight into the behaviour of the system and a means by which the model can be validated. Also, *a priori* knowledge can be encoded into the model as a set of fuzzy rules. Neurofuzzy systems circumvent the drawbacks of both AMNs, with their *opaque* structure, and conventional fuzzy systems. B-spline networks belong to this class of AMNs and have been shown to exhibit many desirable properties when applied to the approximation of nonlinear functions. In Brown and Harris [4], B-spline networks have been shown to be ideal for the generation of neurofuzzy systems with a host of desirable properties. These neurofuzzy systems are adopted in this chapter, and this Section describes their structure by initially describing B-spline modelling. Single output models are considered throughout, but all the theory can easily be generalised to a multiple output models.

12.2.1 B-spline modelling

The fundamental structure of a conventional neurofuzzy system is the B-spline network. B-spline networks are commonly used as surface fitting algorithms in graphical applications, a task comparable to modelling a continuous input-output mapping. These networks are a type of associative memory network and hence form the output by a weighted sum of multi-dimensional basis functions, given by:

$$y(x) = \sum_{i=1}^{p} a_i(x) w_i = a^T w$$

where $y(x)$ is the output, a is the vector of the multi-dimensional basis function outputs $(a_0(x), \ldots, a_p(x))$ when excited by the present input $x = (x_1, \ldots, x_n)$ and w is the vector of the associated weights. The modelling capability of AMNs is determined by the nonlinear mapping performed by the p n-dimensional basis functions. The fundamental nonlinear function produced by the model is controlled by a set of adjustable weights, w, which must be appropriately optimised. The structure of this type of model is illustrated in Figure 12.3.

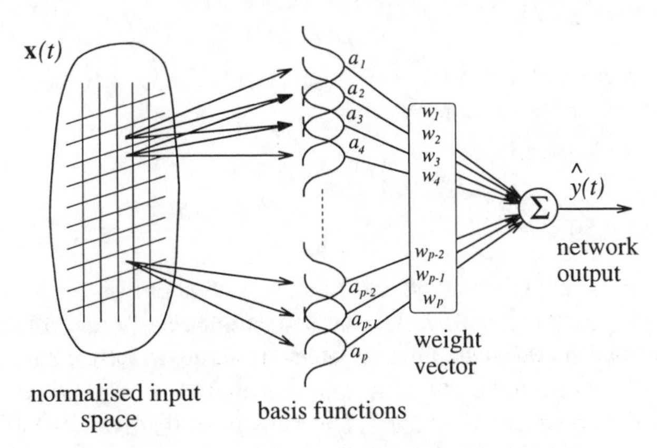

Figure 12.3 *The basic structure of an AMN*

Unsurprisingly, B-spline networks employ B-spline basis functions producing a continuous piecewise polynomial output, ideal for function approximation. Each multi-dimensional B-spline is composed of several univariate ones. A series of these univariate basis functions are defined, on the complete input space of each input variable. It is the shape and position of these basis functions that determines the accuracy and flexibility of the resulting model. The shape of a univariate B-spline is defined by its *order*, B-splines of different orders are shown in Figure 12.4

respectively giving piecewise constant, linear and quadratic approximations. A

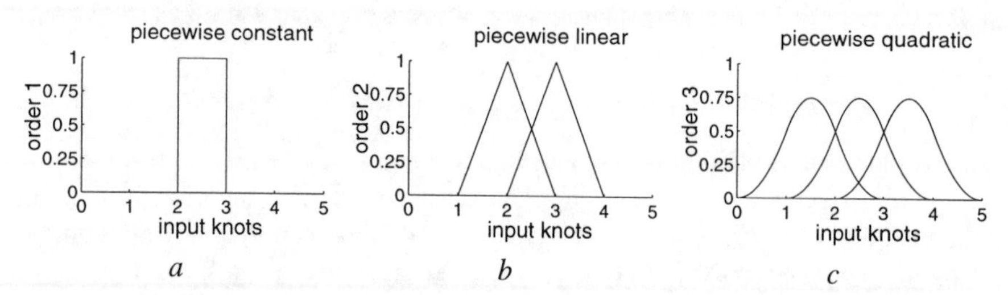

Figure 12.4 *Examples of univariate B-splines of different order: (a) shows order* 1, *piecewise constant, (b) shows order* 2, *piecewise linear and (c) shows order* 3, *piecewise quadratic*

set of univariate B-splines are defined by a knot vector, $\lambda = (\lambda_0, \lambda_1, \ldots, \lambda_{r+2k})^T$ where k is the order of the splines, and r is the number of univariate basis functions defined on this axis. The input domain of a set of univariate basis functions is given by $[\lambda_k, \lambda_r]$ giving a total of $(r - 2k)$ interior knots, λ_k and λ_r are known as the boundary knots. Univariate B-spline outputs are evaluated by a simple, stable recursive relationship:

$$a_{j,k}(x) = \frac{x - \lambda_j}{\lambda_{j+k} + \lambda_j} a_{j,k-1}(x) + \frac{\lambda_{j+k+1} - x}{\lambda_{j+k+1} - \lambda_{j+1}} a_{j+1,k-1}(x)$$

where $a_{j,k}$ represents the jth basis function with order k. For basis functions of order one, i.e. $(k = 1)$, the output is given by:

$$a_{j,1}(x) = \begin{cases} 1 & \lambda_j \le x < \lambda_{j+1} \\ 0 & \text{otherwise.} \end{cases}$$

A series of univariate B-splines are defined on each input variable. Multivariate basis functions are constructed by tensor multiplication of these univariates. Every univariate and consequent multivariate basis function is *tensor* multiplied with every other one defined on the remaining variables, resulting in lattice partitioned input space. In each partition the output is approximated by a polynomial specified by the order of the univariate B-splines. The number of multivariate basis functions produced from this operation is an exponential function of the number of inputs:

$$p = \prod_{i=1}^{n} (r_i)$$

where r_i is the number of basis functions on each axis. This property is known as the *curse of dimensionality* [1] and is common to many modelling techniques. As a result the cost of implementing a network and obtaining an output increases exponentionally as the input space grows linearly. This is the one major disadvantage of B-spline networks, limiting the use of such networks to problems of low dimensions (i.e. $n < 4$). This is unfortunate since B-spline networks possess many desired properties: efficient implementation; partition of unity [21] maintaining smooth (well interpolated) outputs; linear in the weights and hence simple, well understood, linear learning techniques can be applied, this also allows on-line adaption via instantaneous learning rules; the outputs are continuously differentiable which can be used to impose differential constraints on the output. These desirable modelling properties are enhanced by the development of neurofuzzy systems, which make these previously opaque networks transparent, as is explained in the following Section.

12.2.2 *Neurofuzzy interpretation*

A standard n-dimensional neurofuzzy system consists of a rule base composed of production rules of the following form:

$$\text{IF } \overbrace{(x_1 \; is \; A_1^i) \text{ AND } \cdots \text{ AND } (x_n \; is \; A_n^i)}^{\text{antecedent}} \text{ THEN } \overbrace{(y \; is \; B^j)}^{\text{consequent}} \; (c_{ij})$$

where A_k^i, $(k = 1, .., n)$ and B^j are linguistic variables that represent vague descriptions such as *small*, *medium* or *large*. These linguistic variables are represented by univariate fuzzy membership functions, and the linguistic statements (x_j is A_j^i) are evaluated by the intersection of the jth input variable with the A_j^i fuzzy membership function. This process can be viewed as *fuzzification*. The antecedent (also referred to as a rule premise) of each input is formed by the intersection of n input variables with n univariate fuzzy membership functions. An individual rule maps an antecedent to a consequence which describes the output in terms of a linguistic variables, an output univariate fuzzy membership function. Each rule has a rule confidence, c_{ij}, which lies in the interval [0, 1] and represents the confidence in a particular rule being true. From the complete fuzzy rule base a series of linguistic variables statements describing the output are found and these are *defuzzified* to produce a single output.

The structure of a neurofuzzy system is determined by the functions used to represent the linguistic fuzzy variables, the fuzzy logic operators and the fuzzification and defuzzification strategies employed. If the centre of gravity defuzzification algorithm is used, and algebraic operators are used to implement the fuzzy logical operators AND, IF(.) THEN(.) and OR, it can be shown that the systems output is smooth. Also, the individual rule antecedents can be treated as multivariate fuzzy membership functions, produced from the tensor product of the univariate ones. It is shown in Brown and Harris [4] that when B-splines are used to implement these fuzzy membership functions, the output of a neurofuzzy system is given by:

$$y(x) = \sum_{i=1}^{p} \mu_{A^i}(x) w_i \tag{12.1}$$

where $\mu_{A^i}(x)$ is the ith multivariate fuzzy membership function. This representation is identical to a B-spline network, where the multivariate basis functions, $a_i(x)$, are the multivariate fuzzy membership functions. The advantage of this direct equivalence is very significant, since all the mathematical tools used to teach and analyse B-spline networks can be applied to neurofuzzy systems, with the added advantage that the behaviour of model can be represented as a set of fuzzy rules. To make this representation transparent it must can be converted to a fuzzy rule base. Each multivariate fuzzy membership function $\mu_{A^i}(x)$ represents a rule antecedent. The consequence and rule confidence of this rule is represented by the weight, w_i. If the fuzzy output membership functions are chosen as symmetrical B-splines of order $(k) \leq 2$, the following relationship holds:

$$w_i = \sum_{j=1}^{q} c_{ij} y_j^c \tag{12.2}$$

where q is the number of fuzzy output sets, and y_j^c is the centre of the jth output set. To make this relationship possible the confidence vectors, $c_i = (c_{i1}, \ldots, c_{iq})^T$ are normalised, in which only k elements are non-zero. This unique invertiable relationship allows the transformation from a network representation to the fuzzy rule base, and *vice versa* with no loss of information. A full explanation and proof of this direct equivalence resulting in a neurofuzzy system is given in Brown and Harris [4].

Using B-spline basis functions to represent the univariate fuzzy membership functions, provides the neurofuzzy system with the flexibility to incorporate the common fuzzy sets. The shape of the fuzzy set is defined by the B-spline order, k, as illustrated in Figure 12.4 where binary, triangular and smoother sets can be represented. Also, standard sets such as trapezoidal and Π fuzzy sets can be created by the sum of several B-splines, also shown by the dashed lines in Figure 12.4, while still maintaining the partition of unity and other desirable properties. The antecedents of the rules represent multivariate membership functions, formed from the tensor multiplication of n univariate sets. These antecedents result in a lattice based partition of the input space, where every possible combination of the univariate membership functions is taken. If a 2-dimensional example is considered where the standard seven triangular fuzzy sets (order 2 B-splines) are defined on each axis, the partition of the input space described in Figure 12.5 would be produced. In this figure the multivariate membership function shown represents the antecedent ((x_1 is AZ) AND (x_2 is AZ)). Each antecedent appears k times in the complete rule base if B-splines of order k are used to represent the output fuzzy sets, as only k rule confidences are non-zero in each rule confidence vector. The weight associated with the multivariate membership functions are used to evaluate the rule confidences by the intersection of the weights with the output membership functions. The rules produced by the neurofuzzy system shown in Figure 12.5 would be of the form:

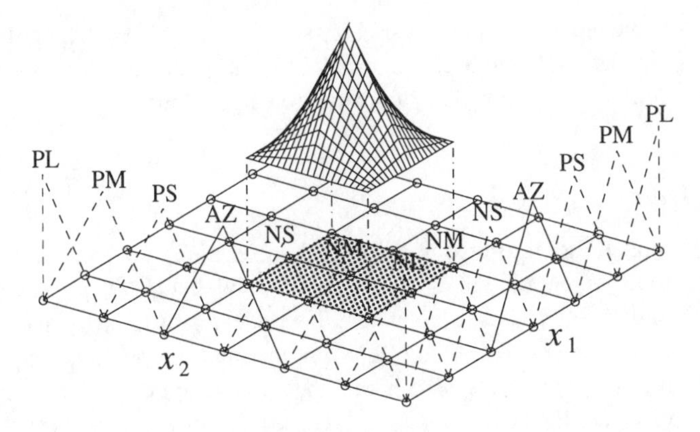

Figure 12.5 *An example of a 2-dimensional partition of the input space produced by a neurofuzzy system. Seven triangular fuzzy sets are defined on each axis, where the letters NL mean negative large, NM negative medium, NS negative small, AZ almost zero, PS positive small, PM positive medium and PL positive large. The multivariate membership function shown represents the antecedent (x_1 is AZ) AND (x_2 is AZ)*

	IF x_1 is NL AND x_2 is NL THEN	y is AZ	$(c_{1,4})$
OR	IF x_1 is NM AND x_2 is NL THEN	y is PS	$(c_{1,5})$
	IF x_1 is NM AND x_2 is NL THEN	y is NS	$(c_{2,3})$
OR	IF x_1 is NM AND x_2 is NL THEN	y is AZ	$(c_{2,4})$
⋮	⋮		
OR	IF x_1 is PL AND x_2 is PL THEN	y is PM	$(c_{49,4})$

Here it is assumed that the standard seven triangular fuzzy sets (2nd order B-splines) are defined on the output, giving 2 non-zero rule confidences for each antecedent.

Unfortunately, conventional fuzzy systems suffer from the *curse of dimensionality*. The number of rules in the rule base grows exponentionally as the dimension of the input space grows:

$$\text{number of rules} \propto \prod_{i=1}^{n} (r_i)$$

This effectively restricts these systems to problems involving less than 4 input variables. This drawback is a direct consequence of employing the lattice based partitioning strategy used by B-spline networks. Another problem, also encountered by most modelling techniques is model structure identification. The resulting model's accuracy and capability are determined by the structure of the model. When compared with B-spline modelling this task is made easier as now *a priori* knowledge can easily be encoded into the system. Often such *a priori* knowledge is unavailable and hence the structure has to be determined by the modeller. In the case of a conventional neurofuzzy system this is controlled by the choice of the shape (order) and the position of the individual univariate fuzzy sets (B-splines). Both these problems, the curse of dimensionality and model structure identification, are addressed in this chapter by using off-line neurofuzzy construction algorithms.

12.2.3 Training

The task of modelling requires that the model successfully approximates the continuous input-output mapping of the system being modelled, $\hat{y}(t) : \Re^n \longmapsto \Re^m$. This task is commonly treated as learning, where the model learns to produce this desired mapping. It can be seen in equation (12.1) that the input-output mapping is dependent on the weight vector, w, and it is this that is adjusted during training so that the model reproduces the system response. These weights are directly related to the rule confidences, so by adapting the weights the strength with which each rule fires is modified.

Training is usually performed off-line, where weights are optimised across a training set. A training set is a collection of observed system input-output pairs, given by $\{x(t), \hat{y}(t)\}_{t=1}^{T}$ where $\hat{y}(t)$ is the desired output given the input $x(t)$. This training set should adequately represent the systems behaviour throughout its working envelope. There are several methods used to improve the quality of the training set by ensuring the system is adequately excited, but often obtaining data from the system is expensive and hence the modeller has little control over its size and contents. Normally supervised training algorithms are employed which use the present error between the model output and the desired output to update the weights. The output of conventional neurofuzzy models is linear with respect to the weights, as shown in equation (12.1), and hence well established training algorithms can be employed. This ability is an important property of conventional neurofuzzy systems. Training algorithms optimise the weights of the model to improve its performance. The mean square output error (MSE) is adopted to measure model performance producing simple well behaved training algorithms. This MSE error criterion is given by:

$$J = \frac{1}{T}\sum_{t=1}^{T}(\hat{y}(t) - y(t))^2$$

This criterion can be expanded to the non-negative, quadratic function of the weight vector given by:

$$J = w^T R w - 2w^T p + \frac{y^T y}{T}$$

where $R = \frac{1}{T}\sum_{t=1}^{T} a(t)^T a(t)$ is the autocorrelation matrix, $p = \frac{1}{T}\sum_{t=1}^{T} y(t)^T a(t)$ is the crosscorrelation matrix, and $y(t) = (y(1), \ldots, y(T))^T$ is the vector of desired outputs. This quadratic function has a stationary point at w^* when $\partial J/\partial w_i$ is identically zero $\forall\, i$,

$$\frac{\partial J}{\partial w} = 2Rw^* - 2p = 0$$

Hence a stationary point satisfies the linear system of normal equations

$$Rw^* = p$$

The act of identifying a weight vector **w** such that the MSE of the system is minimised, can be treated as finding $w^* = R^{-1}p$. The autocorrelation matrix is by definition a real, symmetric, positive semi-definite matrix, and the identification of the weight vector relies on the availability of its inverse. If **R** is non-singular, it can be inverted, the normal equations have a unique solution and the MSE has a global minimum in weight space. By contrast if **R** is singular there exists an infinite number of solutions and the problem is undetermined. Often the existence of a singular autocorrelation matrix is the result of a poorly distributed training set. This highlights the importance of the task of obtaining a training set.

There are two different methods for solving linear equations, *direct* and *indirect*. Direct methods solve the linear system of normal equations and after a known number of steps the solution will be found. The autocorrelation matrix **R** of B-spline models will be sparse and only a small proportion of its elements will be non-zero. This property makes B-spline networks ideal candidates for the conjugate gradient method, a direct method well suited to the solution of large, sparse linear systems which will converge after N iterations. For details of this algorithm the reader is referred to Shewchuk [17], and it has been applied to neurofuzzy systems in Mills *et al* [12]. The conjugate gradient method is often implemented as an iterative (indirect) method, as typically a solution can be found in less than N iterations [17]. Indirect methods iteratively estimate a solution to the quadratic function. The change in the current weight vector is formulated as a function of the present error. Gradient descent adaption rules are probably the best known supervised learning rule, producing a technique ideally suited to on-line instantaneous learning. The reason for this is that the resulting techniques have low memory requirements and low computational cost. Also a large number of theoretical results have been derived which guarantee stability and convergence. Instantaneous training rules use information provided in a single training example, and hence a model can be updated on-line as more training pairs are made available. Gradient descent learning algorithms adapt the weight vector in the direction of the negative gradient of the error function, which is expressed by:

$$\Delta w = \delta(\hat{y}(t) - y(t))a(t)$$

where δ is the learning rate determining the stability and performance of the learning rule. Complete descriptions of these methods can be found in the literature [4, 8].

12.3 Iterative construction algorithm

Traditionally model identification is performed by human experts, where limited subjective knowledge about the system is used. An *ad hoc* approach of this kind often produces conservative models relying on the availability of often limited linguistic domain knowledge. These models lack a mathematical foundation, where no attempt is made to avoid over-fitting and under-fitting of the data. These deficiencies are highly prominent in all facets of the modelling community, i.e. regression analysis, neural networks, and fuzzy logic, and have hence motivated the development of model construction algorithms which generate their model from a training set.

These algorithms are typically iterative (stepwise) ones, where modelling is initiated with a simple model which can encompass any *a priori* knowledge. Several refinements are performed on this model to enhance its modelling capabilities, producing a series of new models. The new model that performs optimally across the training set, with respect to some predefined criteria, is retained and used as the basis for further model development. This process is continued until the model of the optimal complexity is produced. One of the earliest iterative construction algorithms is the group method of data handling (GMDH) algorithm [6], developed by A. G. Ivakhnenko in 1966. GMDH automatically builds up a parsimonious regression model from the information in a training set. In the literature iterative construction algorithms have been applied to many different modelling techniques all based on this simple method. In this chapter neurofuzzy construction algorithms based on this algorithm are discussed, and as an overview to such a technique a general, one-step-ahead, neurofuzzy, iterative construction algorithm is described. This construction and model selection algorithm, called COSMOS, can be applied to the different neurofuzzy model representations introduced in Section 12.4.

1. **Initiate the algorithm with a simple model.** As a neurofuzzy model can be described as a series of linguistic fuzzy rules, any *a priori* knowledge of this form can be encoded into this initial model. This model is then trained across the training set as described in Section 12.2.3.

2. **Perform step refinements.** A series of step refinements are performed on the current model, each producing a new model. Each step refinement changes the complexity and flexibility of the existing model in a search for improvements to its modelling capabilities. There are two distinct types of step refinement, *building* and *pruning*. Usually construction algorithms are concerned with building refinements that increase the existing model complexity, by the introduction of new parameters. As the algorithm should identify a parsimonious structure the complexity of the model must be minimised. Hence it is intuitively appealing to allow for pruning refinements which reduce the model complexity, giving the ability to *delete* superfluous rules. Such destructive steps are equally as important as building steps, and to allow the construction of truly parsimonious models every building refinement must have an opposite pruning refinement. This is particularly important in one-step ahead construction, to allow for the correction of erroneous refinements.

3. **Train the set of refined models.** Each of the candidate models produced

from the refinements are trained on the training set. The complexity of this task depends on the neurofuzzy representation used, but conventionally this is a simple linear optimisation task, as described in Section 12.2.3.

4. **Model selection.** From the set of candidate models one must be chosen to update the current model. This selection must achieve a compromise between the accuracy and the complexity of the approximation which can be viewed as the *bias/variance* dilemma. Reducing the bias can be achieved by increasing model complexity, which tends to produce high variance such that the expected error of the model across all possible training sets increases. There are a number of techniques that try to address this problem, all of which try to minimise the expected performance of a model when presented with new observations, known as the *prediction risk* [14]. This prediction risk can be estimated using

$$P = E\left\{\frac{1}{T'}\sum_{t=1}^{T'}\left(\widehat{y}'(t) - y(\mathbf{x}'(t))\right)^2\right\}$$

where $\{\mathbf{x}'(t), \widehat{y}'(t)\}_{t=1}^{T'}$ are new observations independent from the training set. There are several techniques that try to strike this balance between the bias and variance of the final model:

   (i) Validation is the commonest method for this, where a separate independent *test set* is generated. This new test set is used to provide an accurate estimate of the prediction risk. In many cases the availability of data is limited, and hence a test set cannot realistically be generated.

   (ii) Statistical significance measures derived from information theory heuristics measure network parsimony. These combine the mean square error (MSE):

$$MSE = \frac{1}{T}\sum_{t=1}^{T}(\hat{y}(t) - y(x(t)))^2$$

   across the training set with some measure of network complexity. There are several statistical significance measures including: Akaike information criterion (AIC); Bayesian statistical significance measure (BSSM); and structural risk minimisation (SRM). As an example, the Bayesian measure is given by:

$$BSSM = T\ln(MSE) + n_p\ln(T)$$

   where n_p is the number of free parameters. The Bayesian measure is the only statistical significance measure that does not rely on an arbitrary constant that has to be determined by the modeller.

(iii) Regularisation techniques [2, 11, 18] are standard statistical methods used to avoid over-fitting when teaching neural networks. Over-fitting occurs when the *error variance* becomes too large, and the model fits the noise in the training data and hence captures the underlying function poorly. This inability to generalise is typically due to the use of over-parameterised models, a property we are trying to avoid by the construction of parsimonious models. Regularisation involves redefining the model cost function by adding a constraint to the MSE to produced a new criterion

$$J(\theta) = \overbrace{E(y(t) - \hat{y}(w, x))^2}^{\text{MSE}} + \overbrace{\lambda p_r}^{\text{regularisation penalty}}$$

where λ is the regularisation parameter and controls how much the regularisation penalty influences the final model. This constraint is typically a measure of the smoothness of the output of the resulting model, hence producing a model which generalises well. Smoothness of the model is either calculated from the second derivative of the model (if available), i.e. $p_r = E(\hat{y}''(w, x))$ or estimated by the changes in the weight vector, i.e. $p_r = w^T w$. Regularisation is a good method for controlling the ability of models to generalise, and could be implemented into this construction algorithm when training the models. The choice of candidate model could be based on the new criteria, $J(\theta)$. The drawback with this method is the choice of the parameter λ and the penalty measure used, to which the success of this method is sensitive. These choices are typically governed by *a priori* knowledge of the function being approximated.

12.4 Parsimonious neurofuzzy models

This Section discusses several different neurofuzzy representations which can be employed to overcome the curse of dimensionality. This Section reviews some of these relevant approaches producing alternative neurofuzzy representations which exploit structural redundancy. Several off-line iterative construction algorithms based on these representations are also described. Typically these algorithms are based on the simple algorithm (COSMOS) described in Section 12.3. This review is directed towards the development of a generic neurofuzzy construction algorithm that finds the most appropriate representation; no one modelling strategy will be appropriate for every problem.

12.4.1 Global partitioning

One of the simplest methods for trying to alleviate the curse of dimensionality is to split the model into smaller submodels (divide-and-conquer). When the final model is a sum of such smaller submodels, the model is *globally partitioned*. Global partitioning can be represented by the analysis of variance (ANOVA) decomposition and should be applied to any modelling technique. Many functions can be represented using the ANOVA decomposition which is given by :

$$f(x) = f_0 + \sum_{i=1}^{n} f_i(x_i) + \sum_{i=1}^{n} \sum_{j=i+1}^{n} f_{ij}(x_i, x_j) + \cdots + f_{1,2,\ldots,n}(x_1, \ldots, x_n)$$

where the function $f(x)$ is simply an additive decomposition of simpler subfunctions. This decomposition describes the relationships between different input variables. For many functions certain interactions are redundant, and hence their associated subfunctions can be removed from the ANOVA decomposition resulting in a less complex, more parsimonious model. When using lattice based neurofuzzy systems to represent the subfunctions of this decomposition, there can realistically be no interactions involving more than four input variables. This restriction is due to the curse of dimensionality existing in these lattice networks. Often this constraint is not fulfilled by the true function but an adequate approximation can still be obtained by the restricted ANOVA decomposition.

A globally partitioned, lattice based, neurofuzzy system is simply an additive decomposition of small lattice based neurofuzzy systems, and hence both network transparency and the use of simple linear training algorithms is retained. The weight vectors of each of the submodels can be concatenated and those training algorithms derived for conventional neurofuzzy systems can be applied. The advantage of these globally partitioned models can be demonstrated by applying it to the Powell function:

$$y = (x_1 + 10x_2)^2 + 5(x_3 - x_4)^2 + (x_2 - 2x_3)^4 + 10(x_1 - x_4)^4$$

This function is an ideal candidate for an ANOVA decomposition as it is a sum of four subfunctions. The resulting neurofuzzy system would take the form illustrated in Figure 12.6, consisting of four conventional 2-dimensional neurofuzzy models: $s_1, \ldots, s_4$. The resulting fuzzy rules of this model have antecedents only involving two input variables, making them simpler and more interpretable than the conventional rules. To illustrate the model reduction, assume each input is represented by seven fuzzy membership functions. A conventional, four-dimensional, lattice based neurofuzzy model consists of approximately 2400 different rule premises, while the ANOVA decomposed model with its four submodels consists of approximately 200. The number of rules depends on the form of the output sets, using piecewise linear B-splines would produce two rules for every individual rule premise. The ANOVA decomposition produces a significant reduction in the number of rules while still retaining model quality.

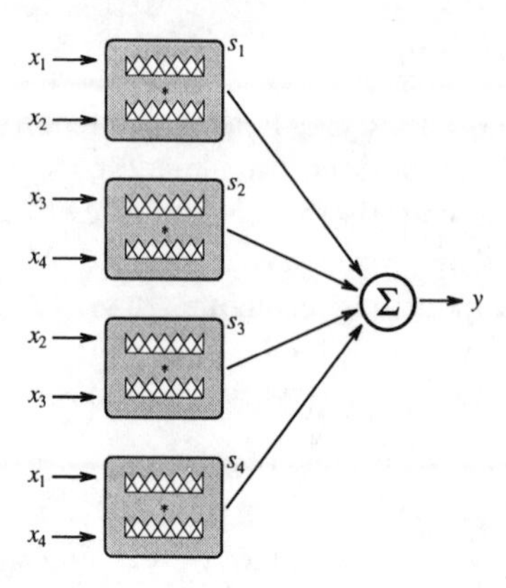

Figure 12.6 *Globally partitioned model of the Powell function*

The construction algorithm adaptive spline modelling of observational data (ASMOD) was developed by Kavli [9] to identify an ANOVA decomposition, from a representative training set. This was designed as an algorithm which iteratively constructs globally partitioned B-spline models, hence it is directly applicable to the construction of lattice based neurofuzzy models. The ASMOD algorithm has been shown to work well on various examples [5, 9, 10], and an overview of the algorithm is given here.

12.4.1.1 ASMOD: adaptive spline modelling of observational data

The ASMOD algorithm is an off-line iterative construction algorithm based on the principles of COSMOS (Section 12.3). Basically a succession of step refinements are performed on the model until an adequate approximation is obtained. There are two distinct types of candidate refinements: building and pruning. The original ASMOD algorithm just employed building refinements that fell into one of three categories: univariate addition; tensor multiplication; and knot insertion. These result in a more complex model that can reproduce the current model exactly, a distinct advantage of employing B-spline basis functions as opposed to alternatives, i.e. RBFs. Model construction is usually initiated with an empty model, but can be initialised by a set of fuzzy rules containing any *a priori* knowledge of the system. The three different construction refinement strategies are now described.

Univariate addition: To incorporate a new input variable into the model a new one-dimensional submodel is introduced. This gives the updated model the ability to additively model this new variable. Univariate submodels are only incorporated into the existing model if the new input is not in the current model. To try to make the model construction process more efficient these univariate submodels are drawn from an *external store*. This external store contains a set of univariate submodels of different fuzzy set densities for each input variable. For B-spline input sets of order

2 a typical choice of fuzzy set densities for one-dimension is shown in Figure 12.7. As a general rule it is good to encode as much *a priori* knowledge as possible into any construction algorithm and such knowledge can be used to design this external store.

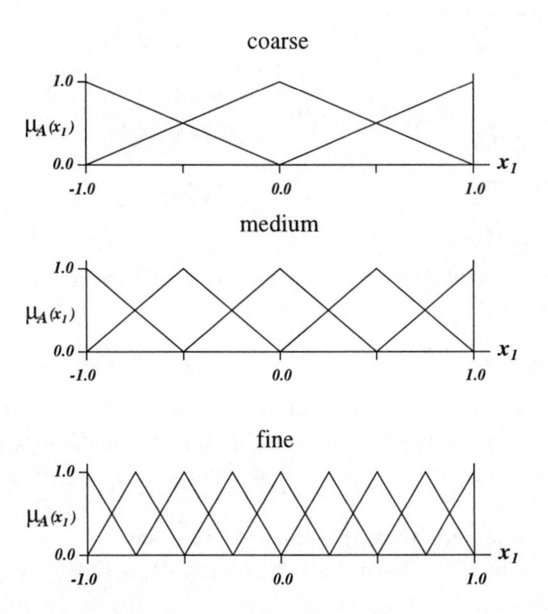

Figure 12.7 *An example of a univariate external store containing three densities of fuzzy sets: course, medium and fine*

The univariate addition refinement can be demonstrated by considering an arbitrary function of five variables where at the kth iteration the model is given by $y_k = s_1(x_1) + s_2(x_2, x_3)$, as shown in Figure 12.8. The possible univariate addition

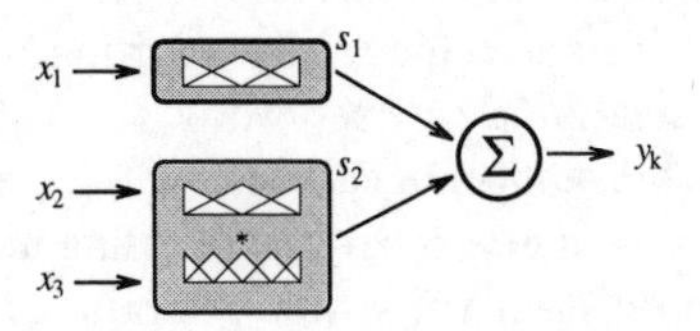

Figure 12.8 *The current model at the kth iteration, used to demonstrate a number of ASMOD refinements*

refinements are the addition of submodels depending on either x_4 or x_5. Each univariate submodel in the external store defined on these input variables is added to the current submodel to produce a new candidate model. These candidate models are then trained on the training set and the optimal one is chosen as the new model,

y_{k+1}. Assuming the external store only contains one (empty) univariate submodel for each input variable the two possible refinements for this example are shown in Figure 12.9.

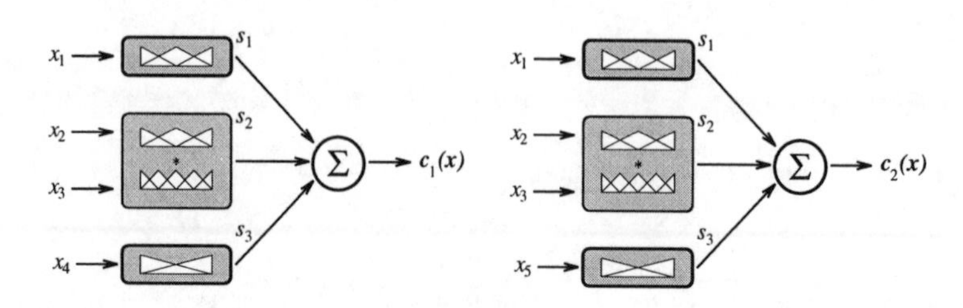

Figure 12.9 *Possible candidate models produced by the univariate addition refinement*

Tensor multiplication: A submodel of the current model is given the dependency on another input variable, by allowing tensor multiplication of an existing submodel with a univariate submodel from the external store. Tensor multiplication is only allowed if the new input variable appears in the current ASMOD model, but not in the submodel to which it is to be multiplied. The submodels are tensor multiplied with univariate submodels drawn from the external store. By having an external store containing different fuzzy set densities, the candidate models are provided with a range of possible flexibilities. As an example, once again consider the kth iteration in the construction of an arbitrary function of five variables where $y_k = s_1(x_1) + s_2(x_2, x_3)$. With an external store consisting of one univariate model for each input variable there are three possible tensor multiplication refinements as illustrated in Figure 12.10.

Knot insertion: Model flexibility of a submodel can be increased by the insertion of a new basis function. This is achieved by inserting a new knot into one of its univariate knot vectors, creating a new set of fuzzy membership functions for this input variable. To reduce the number of candidate refinements, the locations of these new knots are restricted to lie halfway between existing interior knots present in the current submodel. The insertion of a new knot results in one new basis function on the associated input vector producing a series of new multi-dimensional basis functions. If the kth iteration of the five-dimensional example is used again (Figure 12.8), there are a total of eight different knot insertion refinements. Candidate models produced from performing knot insertion on submodel s_1 are shown in Figure 12.11.

During model construction redundant model flexibility (fuzzy rules) may be introduced to certain regions due to the exclusion of modelling capabilities else-

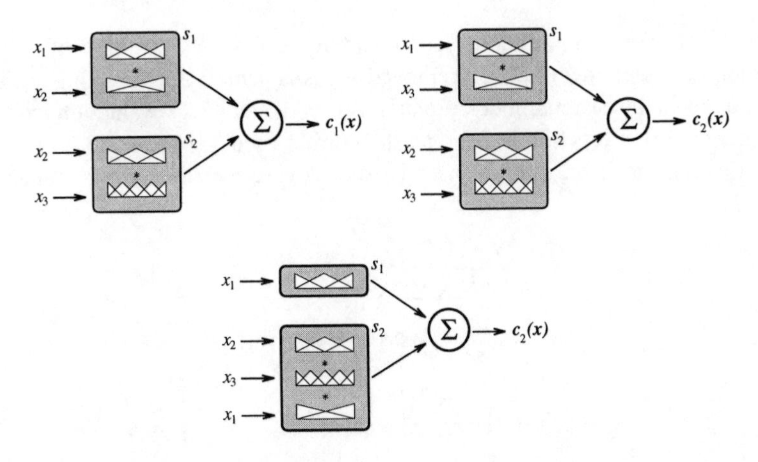

Figure 12.10 *Possible candidate models produced by the tensor multiplication refinement*

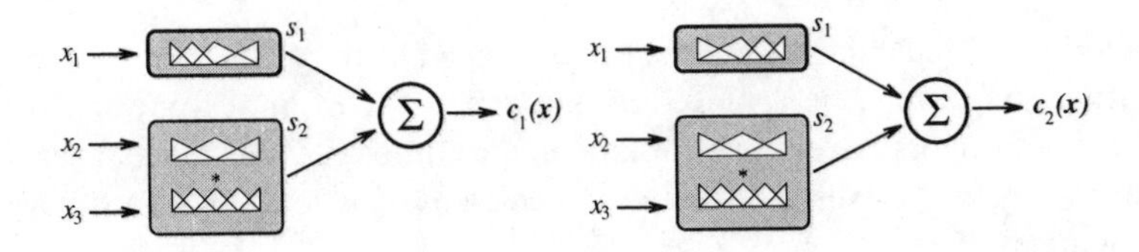

Figure 12.11 *A subset of the possible knot insertion refinements made on* y_k

where. To allow for the correction of such mistakes made during construction, pruning refinements have recently been introduced to the ASMOD algorithm [12]. Each building refinement requires an opposite pruning refinement and hence for the ASMOD algorithm there are three different pruning refinements.

Submodel deletion: Superfluous submodels in the ANOVA decomposition that have been previously identified are removed. This is achieved by deleting each of the submodels in the current model, producing a set of candidate evaluation models, each containing one less submodel than the current model. These are trained on the training set and if an improvement is found the associated submodel can be discarded.

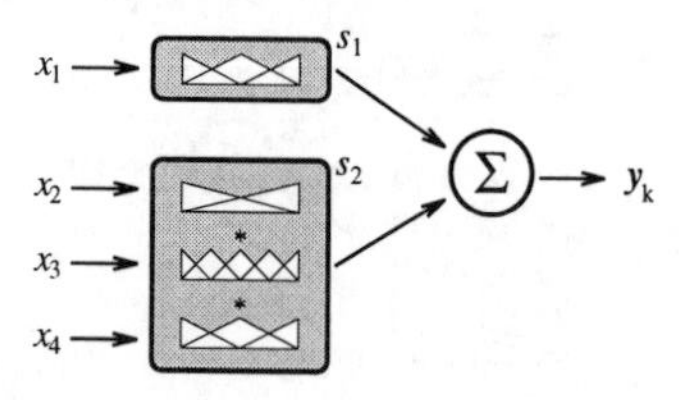

Figure 12.12 *A model used to demonstrate submodel splitting*

Submodel split: Some of the interactions of certain input variables in the current model may be redundant. This is prevented by allowing the splitting of existing submodels into two new submodels. The dimension of the two new submodels must both be at least one dimension less than the original submodel. As an example consider the ASMOD model given by $y_k = s(x_1) + s(x_1, x_2, x_3)$, shown in Figure 12.12. When considering the refinement submodel splitting there are three different candidate models as shown in Figure 12.13. If the best of these models is shown to perform better than the current model then it is replaced.

Knot deletion: Redundant flexibility can be pruned by the deletion of input membership functions, which is achieved by the deletion of a knot from one of the submodel knot vectors. Every interior knot in the entire ASMOD model is considered for knot deletion, which results in the deletion of several multi-dimensional basis functions.

The general philosophy of the current ASMOD implementation is slightly different from the COSMOS algorithm described in Section 12.3. Testing for knot insertions during model construction can prove computationally expensive. However, if the external univariate stores are sufficiently representative, it is possible to avoid knot insertions until the additive dependencies have been identified. This

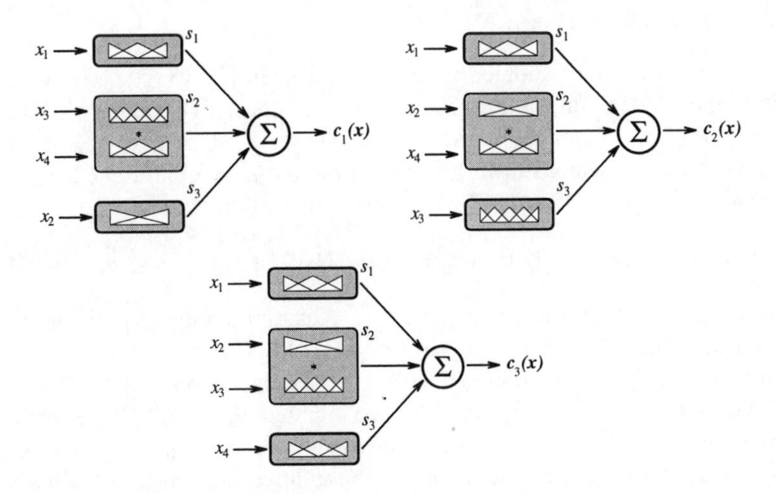

Figure 12.13 *The possible candidate models produced by performing submodel splitting of the model shown in Figure 12.12*

approach has been shown to result in a more accurate, efficient algorithm. The iterative construction algorithm is hence split into two stages:

1. The additive decomposition is first identified by using the the model refinements univariate addition and tensor multiplication. Both building and pruning refinements are carried out until no model improvement is be found;

2. When the input variable interactions of the final model have been found the specific structure of the submodels are identified. This is achieved by carrying out a series of knot insertions and deletions. Due to the historical development of the algorithm a constructive stage is followed by a pruning stage, i.e. a series of knot insertions are followed by a succession of knot deletions.

Obviously, throughout model construction a method for determining which candidate refinement (if any) performs best, should be chosen. Techniques based on minimising the prediction risk of the model are outlined in Section 12.3. Any of these methods are applicable to the ASMOD algorithm, and presently the Bayesian statistical significance measure is used, combining the mean square output error across the training set with a measure of the model's complexity. At each iteration of the algorithm the refinement that produces the largest reduction in Bayesian statistical significance measure is chosen. If a succession of optimal refinements (the failure margin) fail to improve the statistical significance of the whole model then the algorithm is terminated. The training of these ASMOD models is relatively straightforward, as the output of the complete model is linear with respect to the concatenated weight vector. Therefore training can be achieved by applying any linear optimisation technique to this concatenated weight vector.

12.4.2 Local partitioning

The lattice representation adopted by conventional neurofuzzy systems is restrictive, it fails to model local functional behaviour parsimoniously. Due to the properties of a lattice, the inclusion of new rules often leads to the production of new redundant ones. This can be demonstrated by considering the lattice partitioned fuzzy rule base shown in Figure 12.14*a*. If the new rule

IF x_1 is *almost zero* AND x_2 is *almost zero* THEN y is *positive medium* (0.7)

is added to the rule base, the new fuzzy input set distribution shown in Figure 12.14*b* is produced. This addition is equivalent to an ASMOD knot insertion in the knot vector defined on x_1 at the location 0.0, as described in Section 12.4.1.1. Due to the properties of a lattice, the inclusion of this one required rule, results in the creation of twelve redundant ones. This is a direct consequence of the curse of dimensionality, a property still existing in the submodels of the additive decomposition described in Section 12.4.1.

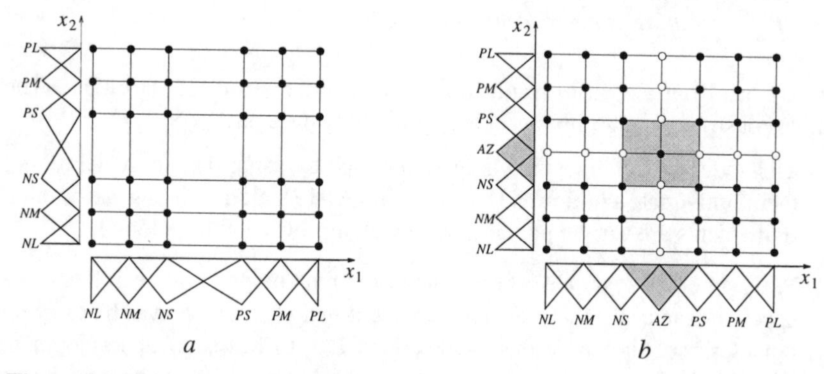

Figure 12.14 *The distribution of fuzzy input sets is illustrated before (a) and after (b) the insertion of a new fuzzy rule (the shaded region in (b)). Each ∘ in (b) represents an additional rule created as a result*

This drawback can be overcome by exploiting a more local representation, which can be achieved by abandoning the lattice representation and using a strategy that *locally partitions* the input space. In the above example local partitioning would allow the inclusion of the single rule. Popular non-lattice representations include k-d trees and quad-trees, two-dimensional examples of which can be seen in Figure 12.15. K-d trees were originally developed as an efficient data storage mechanism [16]. They result from a succession of axis-orthogonal splits made across the entire domain of an existing partition. Each split produces new regions which themselves can potentially be split. The k-d tree representation has been exploited by many researchers in both the neural network and fuzzy logic communities. These approaches include the popular MARS algorithm developed by Friedman [7] and an

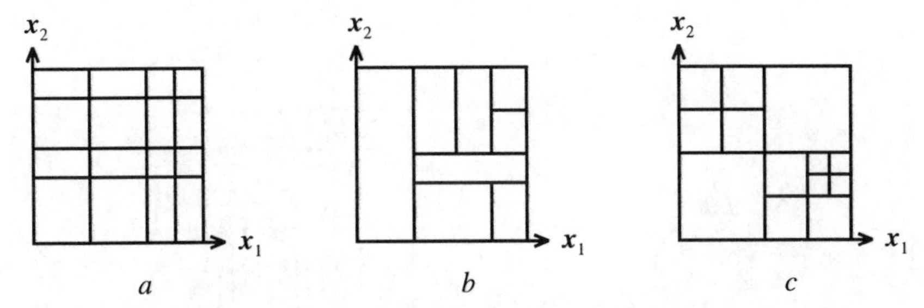

Figure 12.15 *Examples of three different input space partitions: (a) Lattice produced partition, the type of partition produced by ASMOD; (b) K-d tree partition; (c) Quad-tree partition*

extension to the ASMOD algorithm developed by Kavli, called adaptive B-spline basis function modelling of observational data (ABBMOD) [9]. This ABBMOD algorithm is directly applicable to the construction of neurofuzzy systems and is described in the following Section. Quad-trees, as employed in [19], are less flexible than k-d trees, but result in a more structured representation better suited to the extraction of rules. In the following this type of partition is extended to form a multi-resolution approach producing a hierarchy of local lattice based models of differing resolution.

12.4.2.1 ABBMOD: adaptive B-spline basis function modelling observational data

The ABBMOD algorithm is very similar to that of ASMOD, where a model is iteratively constructed. To allow for a k-d tree partition of the input space, each univariate basis function is specified by a *defining* vector, and the tensor product of these produces multivariate basis functions. This representation allows for multidimensional basis functions to be treated independently, hence allowing local model flexibility. Construction is performed by iteratively refining the current model, until an adequate model is identified. At each iteration a set of candidate refinements is identified and the refinement that produces the largest increase in statistical significance is selected. Before construction starts, an external store of empty univariate submodels are constructed, one for each input variable. Due to the nature of the algorithm an external store containing a set of univariate submodels of different fuzzy set densities cannot be used. If knowledge of the relationship to be modelled is not known, the model is initialised with the most statistically significant univariate submodel in the store. Otherwise, *a priori* knowledge can be encoded into a set of fuzzy rules (using lattice or k-d tree input space partitioning) and used to initialise the model. As required by any iterative construction algorithm the series of possible refinements contain both building and pruning steps.

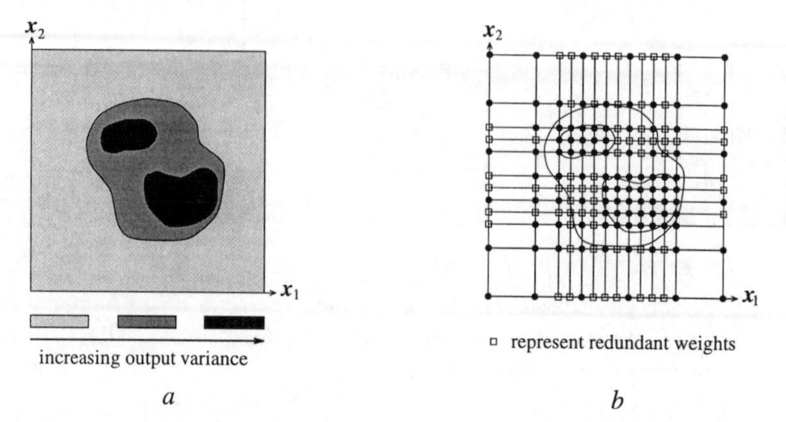

Figure 12.16 *A 2-dimensional surface consisting of regions of different output variances (a) illustrates the 2-dimensional surface, and (b) shows the inadequate partitioning of the input space produced by global partitioning, where a box represents a redundant rule premise*

12.4.2.2 Hierarchy of multi-resolution models

The use of k-d trees to locally partition the input space of neurofuzzy models can result in models with some undesirable properties. This result suggests that an alternative input space modeling strategy, exploiting local model behaviour is required. In this Section an input space partitioning strategy that is based on the multi-resolution approach of Moody [13], and the partitioning strategy of quad-trees is proposed. The resulting model is a hierarchy of lattice based neurofuzzy models of different resolutions combining the generalisation properties of coarse models with the accuracy of fine ones. Existing at the top of the hierarchy is a large coarse model, called the *base submodel*, that models the general trend of the system, aiding generalisation. At the bottom are small finely partitioned *specialised submodels* which model local behaviours. The advantage of such a model can be demonstrated by considering the two-dimensional function illustrated in Figure 12.16*a*. The input space is separated into regions of different output variance. Regions of low output variance can be modelled by coarse resolution neurofuzzy models, while regions with high output variance can only be modelled accurately by fine resolution neurofuzzy models. If a single lattice model is employed the partitioning of the input space illustrated in Figure 12.16*b* would be produced. Due to the regions of different output variance this strategy produces many redundant rules. Hierarchical local partitioning would produce a partition of the input space shown in Figure 12.17*a*, consisting of four different lattice based submodels ($s_0, \ldots, s_3$). The hierarchical structure of this model is illustrated in Figure 12.17*b*.

The hierarchy of multi-resolution models is represented by a tree type structure, which is used as an efficient mechanism for matching the input to a relevant

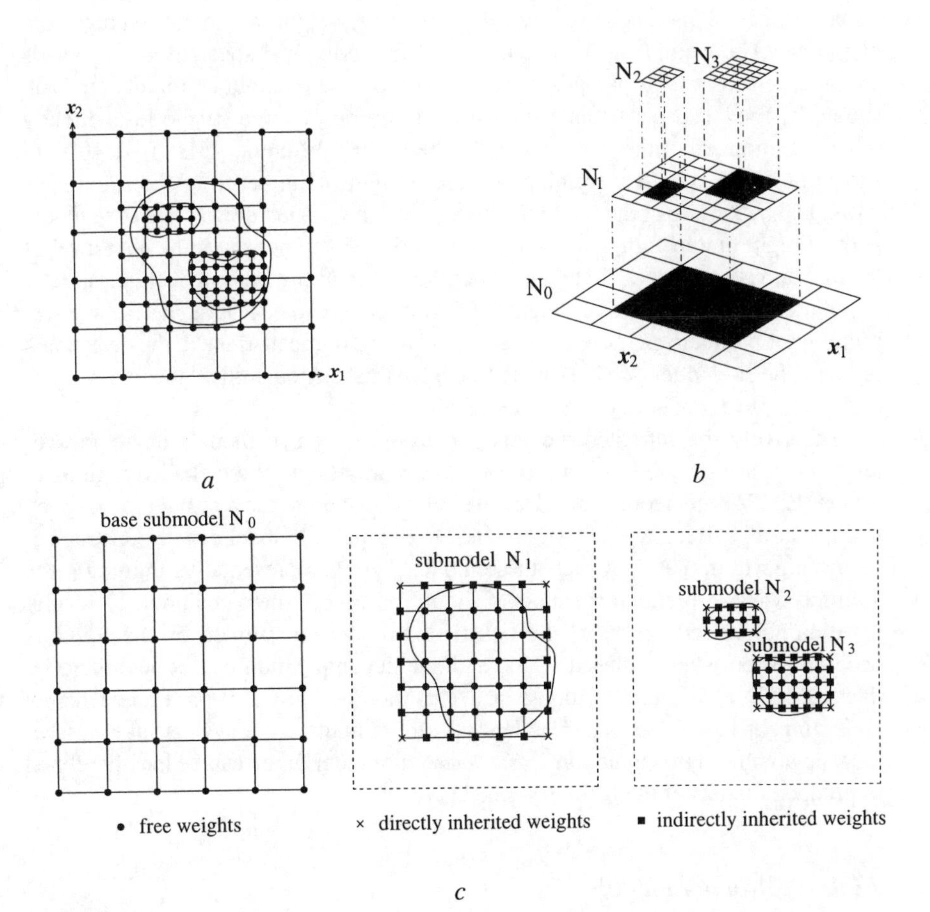

Figure 12.17 *The hierarchical model representing the 2-dimensional surface shown in figure, (a) shows the resulting partition of the input space, (b) illustrates the hierarchy of models, and (c) shows the individual submodels on which the weights are classified*

submodel. This is achieved by ascending this tree until the input is found to lie in the domain of one the submodels. The output of this submodel is the output of the overall network, as only one submodel contributes to the overall output. One of main difficulties of this type of representation is maintaining continuity in the output surface. This is achieved by weight sharing, weights are translated down the hierarchy. The weights at the boundaries of the individual specialised submodels are inherited from their parent's weight vector to ensure output continuity. In each submodel model, a particular weight falls into one of three categories: directly inherited, indirectly inherited, and free. These three different types of weights are shown in Figure 12.17*c*. Training is achieved by teaching each of the submodels individually, starting at the top of the hierarchy and working down. The base model is first taught on the complete training set. Next the submodels of the next level of the hierarchy are taught. First all the boundary weights are inherited, either directly or indirectly and then the free weights of these models can be taught on a limited size training set producing a more accurate model of the output surface in the designated region. The submodels are iteratively trained in this fashion until all the free weights of the complete model have been identified.

Effectively the inherited weights and those of coarse models on regions of the input space covered by finer models (as represented by the shaded regions in Figure 12.17*b*) are redundant. Despite this redundancy they provide a basis by which a complete fuzzy rule base for each individual submodel can be constructed.

An algorithm that constructs these models from a representative training set is required. These hierarchical models should be used, where required, to locally partition submodels of a globally partitioned model represented by the ANOVA decomposition when required. This additive decomposition can be identified by the ASMOD algorithm, resulting in a parsimonious globally partitioned model consisting of lattice based submodels. As the output of the system may possess local behaviour in the input domains of these submodels, they can be locally refined to produce models of the form described here.

12.4.3 *Product models*

The curse of dimensionality is a direct consequence of trying to model the non-linearities given by input variable interactions. Conventionally this is achieved by multivariate basis functions, but a more parsimonious way would be to use *product* networks, where the output of several low dimensional models are multiplied to produce a higher dimensional approximation. Product decomposition is given by:

$$f(x) = \prod_{u=1}^{U} f_u(X_u)$$

where X_u is a small subset of the input variables ($X_u \subset X$). Obviously this type of decomposition is only suitable for certain types of functions. As an example consider the approximation of the function:

$$f(x) = \sin(x_1) * \cos(x_2)$$

which would conventionally be represented by a two-dimensional model as shown in Figure 12.18*a*. Using product decomposition this function would be modelled by the product of two univariate submodels (s_1 and s_2) as shown in Figure 12.18*b*. This reduction in complexity can be demonstrated by considering a system with seven univariate linguistic variables on each input. The product model produces 14 fuzzy rule premises, whereas the conventional system uses 49, of which 35 are effectively redundant. This type of product decomposition is further investigated in [3].

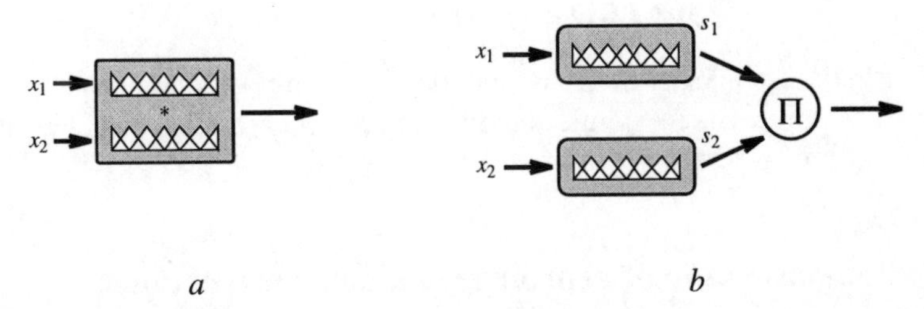

Figure 12.18 *Demonstration of a simple product network (b) shows the equivalent product decomposition of (a)*

12.4.4 Hierarchical models

All the neurofuzzy models discussed have consisted of models which directly map the input variables to the outputs. Hierarchical models consist of several submodels, where the output of one submodel forms the input to another. The models represent *deep* knowledge and can be employed to approximate functions parsimoniously, where the number of rules is a linear function of the number of input variables n. Hierarchical models, consisting of two-dimensional submodels, produce the models with the minimum number of rules. Some multi-dimensional functions possess a functional decomposition which can be exploited by such hierarchical models. Consider the following 4-dimensional equation:

$$y = 0.5(x_1 x_4)^2 \exp\left(x_3 \sin\left(x_1 + x_2\right)^3\right)$$

which can be functionally decomposed into the following form:

$$y = f(g_1(x_1, x_4), g_2(x_3, g_3(x_1, x_2)))$$

A functional decomposition of this form could be represented by the hierarchical model illustrated in Figure 12.19. Using the standard seven fuzzy membership functions defined on each input variable, roughly 200 different rule premises are formed to approximate this function. Conventionally the same level of approximation accuracy would be produced by a four-dimensional lattice based model, consisting of approximately 2400. This hierarchical representation not only provides a substantial

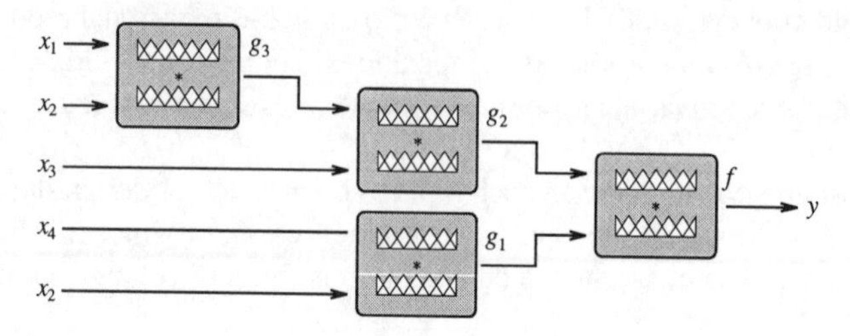

Figure 12.19 *Hierarchical decomposition*

reduction in the complexity of the system, but also results in rules conveying *deep* system knowledge. The fuzzy rules produced by the conventional model would be *shallow*, not truly representing the system.

12.5 Demonstration of neurofuzzy model construction

This Section describes initial results produced from several neurofuzzy construction algorithms. The algorithms used are based on B-spline construction algorithms such as ASMOD and ABBMOD which respectively employ *global* and *local* input space partitioning, as described in Section 12.4. As indicated there are many different neurofuzzy representations, each with their individual merits. No one model will be appropriate for every problem and hence when trying to identify a representative model, it maybe advantageous to combine different strategies. This idea of combining different strategies is demonstrated by attempting to combine the lattice and k-tree representations of ASMOD and ABBMOD to produce an improved hybrid algorithm.

To demonstrate these proposed approaches consider the 2-dimensional Gaussian function given by

$$f(x) = \exp\left(-5\left[(x_1 - 0.5)^2 + (x_2 - 0.5)^2\right]\right) + e(t)$$

where $e(t)$ are noise samples drawn from a normal distribution with a mean of zero and a variance of 0.02. This function was used to generate a uniformly distributed training set of 441 data pairs, as shown in Figure 12.20*a*. The true underlying function that the resulting model should identify is shown in Figure 12.20*b*. This surface is used as test for the generalisation property of the generated models. Three different construction algorithms where tested on this function: ASMOD; ABBMOD; and a hybrid algorithm combining the previous two approaches. The results of these construction algorithms are described in the following Sections, and are summarised in Table 12.1.

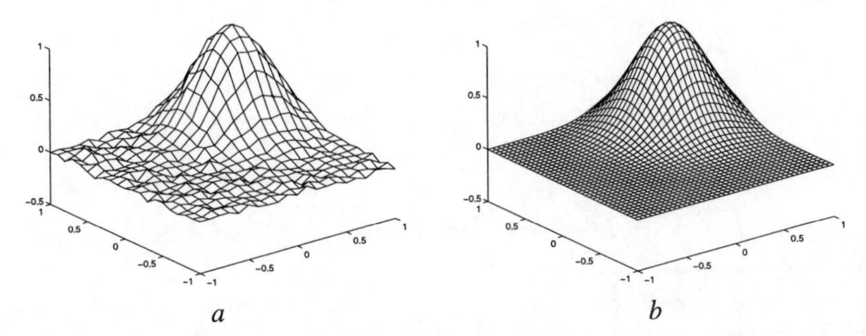

Figure 12.20 *The 2-dimensional Gaussian function, (a) shows the training set and (b) shows the true underlying surface*

Table 12.1 *Summary of the results of the different construction algorithms*

Model	Number Parameters	Bayes Stat. Sig.	MSE across training set	MSE across test set
ASMOD	35	-3083.18	0.000567	0.000274
ABBMOD	20	-3105.47	0.000663	0.000340
HYBRID	23	-3260.78	0.000448	0.000174

12.5.1 ASMOD

The ASMOD algorithm identifies the additive dependencies and the individual structure of the submodels depicted by the training set. When applied to the described training set the model shown in Figure 12.21*a* was constructed. Its construction is illustrated in Figure 12.22, showing how the Bayesian statistical significance measure and the number of parameters change during construction. The statistical significance is shown to decrease throughout construction. At the 4th iteration no construction refinement produced an increase in the statistical significance measure and hence model pruning was commenced. Model pruning significantly reduced the complexity of the final model, resulting in a more parsimonious model with a final Bayesian statistical significance of -3083.18. The response of this final lattice based model is shown in Figure 12.21*c*, which is to be compared with the true underlying function shown in Figure 12.20*b*.

An attraction of the neurofuzzy approach is the ability to extract a fuzzy rule base from the resulting network. This allows validation and ultimately gives insight into the behaviour of the process being modelled. The conventional lattice based neurofuzzy representation is inherently transparent, and to demonstrate the fuzzy rules are extracted from the model. Initially each of the input univariate basis functions must be given a fuzzy linguistic label. These must be chosen as natural

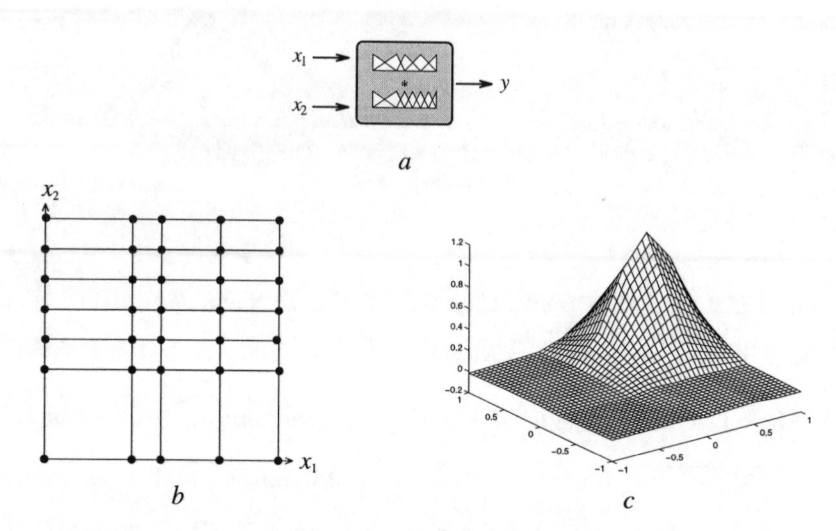

Figure 12.21 *The model produced by the ASMOD algorithm, (a) shows the model, (b) shows produced partitioning of the input space and (c) shows the model response*

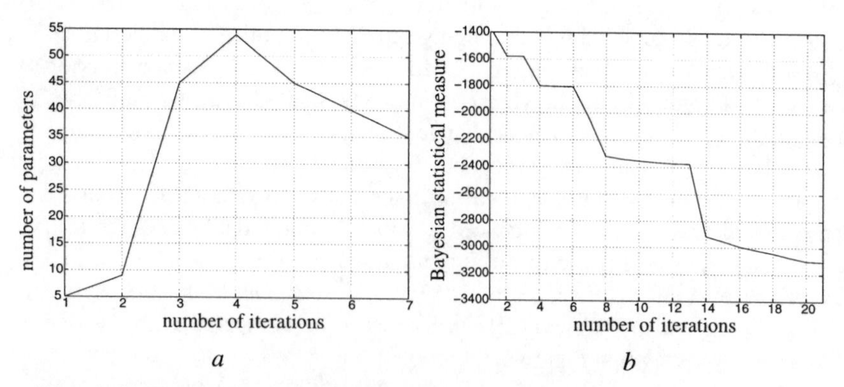

Figure 12.22 *The history of the ASMOD construction algorithm (a) shows the number of parameters (rules) changing as the model is built, (b) shows the improving Bayesian statistical significance*

descriptions of the value of the input variables and are chosen as those described in Figure 12.23*a*. The ANDing of these univariate input sets produces the antecedents of the rules. Each antecedent, has an associated consequence drawn from the fuzzy output set distribution. This fuzzy output set distribution must be chosen and for simplicity a subset of the standard seven fuzzy set distribution was used, evenly distributed in the interval $[-1.2, 1.2]$ The universe of discourse of these fuzzy output sets is determined by the domain of y, and the five sets illustrated in Figure 12.23*b* where employed. Each of the antecedents are inferred with each of the output sets to produce a rule which must have an associated rule confidence. These rule confidences are found by converting the weights of the trained neurofuzzy system. For fuzzy sets defined by B-splines this is achieved by the direct, invertible mapping

$$c_{ij} = \mu_{B^j}(w_i)$$

where μ_{B^j} is the jth fuzzy output set. Here, the vector of rule confidences c_i is found by evaluating the membership of weight w_i of all the output sets. For example, the first weight of the network is $w_i = -0.0165$ generating the rule confidence vector $c_i = (0.025, 0.975, 0.0, 0.0, 0.0)^T$, as illustrated in Figure 12.23*b*. The first two rules of the rule base would therefore be:

IF x_1 is *Coarse NL* and x_2 is *Coarse NL* THEN y is *NS* (0.025)
OR IF x_1 is *Coarse NL* and x_2 is *Coarse NL* THEN y is *AZ* (0.975)
⋮ ⋮

where NL is Negative Large, NS is Negative Small and AZ is Almost Zero.

As expected the ASMOD algorithm identifies a model that generalises well and possesses transparency producing a fuzzy rule base. The example does not test the ability of the ASMOD algorithm to globally partition the input space successfully, but ASMOD is a relatively well developed algorithm and has been well tested in the literature [5, 9, 10]. One drawback of the final model is that it contains 35 parameters, where the local behaviour of the response suggests some of the exterior ones are redundant. This drawback, described in Section 12.4.2, is a direct consequence of the lattice based input space partitioning strategy.

12.5.2 ABBMOD

The 2-dimensional Gaussian function of Figure 12.20 possesses local behaviour and to capture this a lattice neurofuzzy model produces redundant rules. Hence, it would be expected that using a representation that locally partitions the input space would be advantageous. The ABBMOD algorithm employs a k-tree representation producing local partitions. Each multivariate basis function is defined separately, and hence individual basis function refinements can be performed. The ABBMOD algorithm was applied to the training set and the model shown in Figure 12.24*a* was produced.

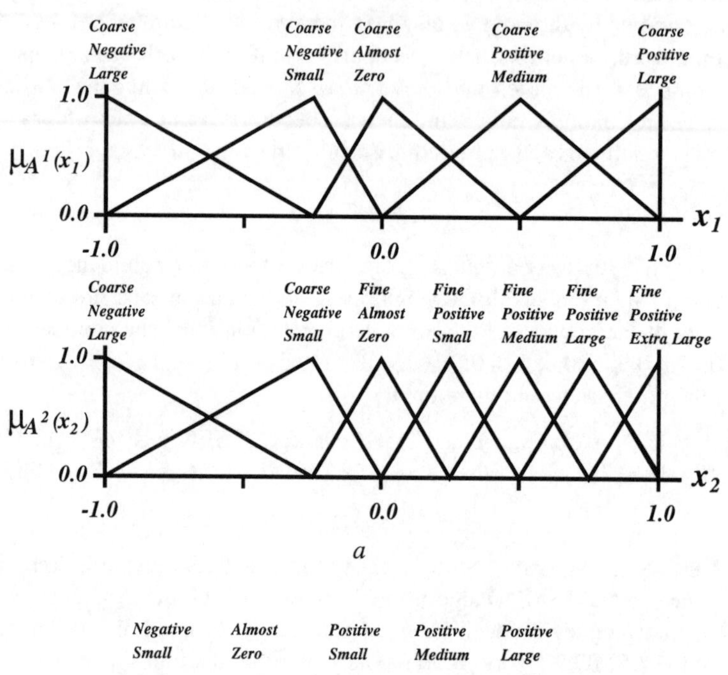

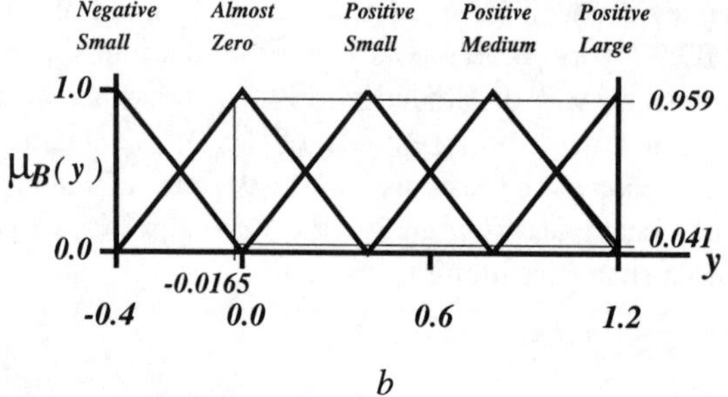

Figure 12.23 *The fuzzy sets of the neurofuzzy model (a) shows the linguistic fuzzy input sets on the input variables, and (b) shows the linguistic fuzzy output sets*

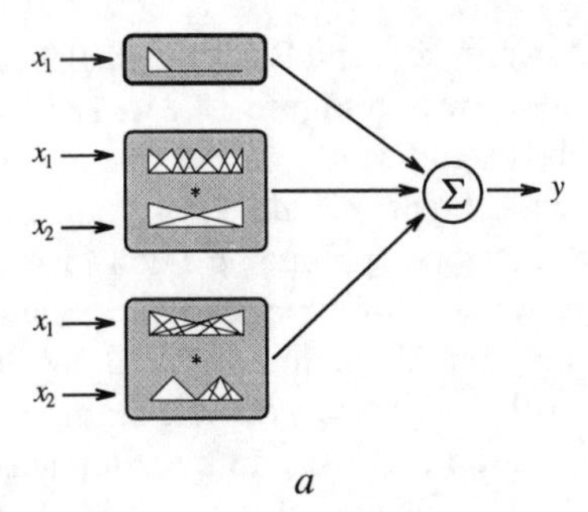

a

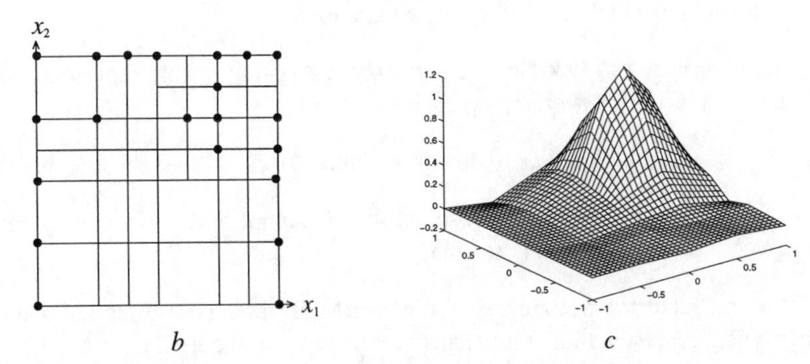

b *c*

Figure 12.24 *The model produced by the ABBMOD algorithm, (a) shows the model, (b) shows produced partitioning of the input space and (c) shows the models response*

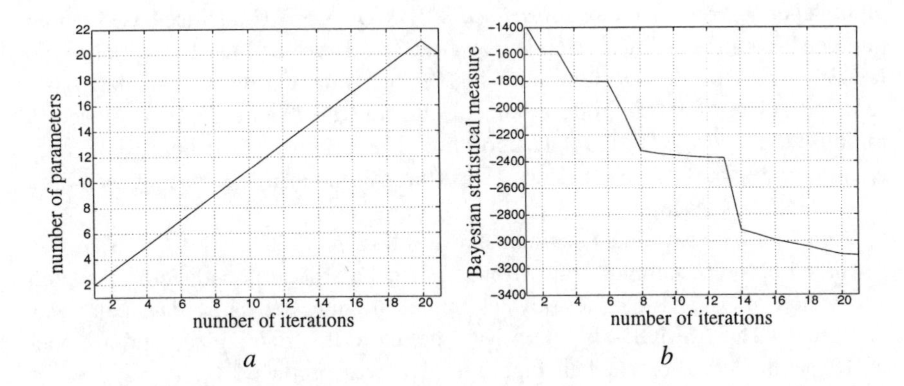

a *b*

Figure 12.25 *The history of the ABBMOD construction algorithm (a) shows the number of parameters (rules) changing as the model is built, (b) shows the improving Bayesian statistical significance*

The changing number of parameters and Bayesian statistical significance of the model during construction are shown in Figure 12.25. The statistical significance converges to -3105.47, which is significantly greater than the ASMOD model. The response of this model is shown in Figure 12.24*c*. Unfortunately, the resulting model consists of three separate tree networks, producing a non-local partition of the input space. This results in overlapping basis functions degrading the transparency of the final model and suggesting an ill-conditioned training problem. Extracting meaningful fuzzy rules from the produced network seems pointless with twelve different overlapping fuzzy sets on x_1. Despite the simplicity of this example it has highlighted some salient disadvantages of the ABBMOD algorithm, as both a modelling technique and a neurofuzzy representation:

- Construction is too flexible, leading to an non-interpretable representation consisting of lots of overlapping basis functions;
- Slow construction time due to the large number of possible step refinements;
- In the ABBMOD algorithm tensor products cannot be split, meaning that erroneous steps cannot be corrected;
- The tree structure used to represent the ABBMOD model is a computationally inefficient representation with respect to the lattice structure.

12.5.3 Hybrid approach

To try to overcome the deficiencies of the ASMOD and ABBMOD approaches, and to illustrate the advantage of combining different strategies, a hybrid approach is proposed consisting of two steps: an ASMOD step which identifies the correct input dependencies; and then a *restricted* ABBMOD step. The lattice based model produced from the ASMOD step is converted to a k-tree and used to initialise the *restricted* ABBMOD algorithm allowing the insertion and deletion of new basis functions. The ASMOD produced model, illustrated in Figure 12.21, was used to initialise the *restricted* ABBMOD algorithm. The response of the improved model is shown in Figure 12.26 but it clearly shows that this model inadequately generalises across the input space.

Consider the univariate Gaussian shown in Figure 12.27*a*. This Gaussian was perturbed by noise samples drawn from a normal distribution with zero mean and a variance of 0.02, and an evenly distributed training set of 21 data pairs was generated. The ASMOD algorithm was applied to this training set, and a model of 15 parameters illustrated in Figure 12.27*d* was produced. The response of this model across a sample interval finer than the interval of the training pairs is shown in Figure 12.27*c*. The non-smooth response shows that the model has fitted the noise, and hence possesses poor generalisation properties. The shaded basis functions in Figure 12.27*d* are the *ill-conditioned* basis functions that cause

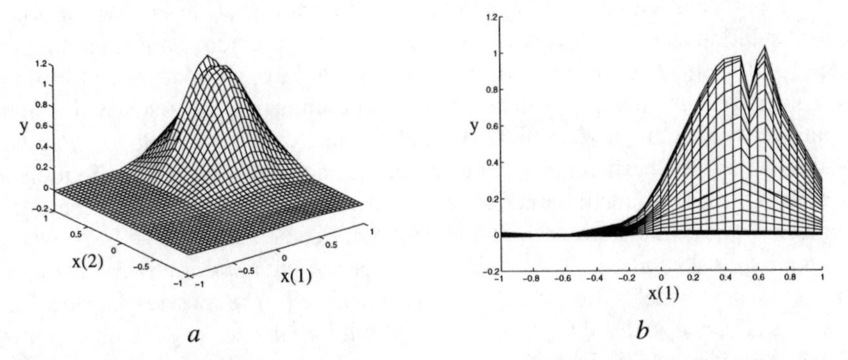

Figure 12.26 *The response of the hybrid model: (a) and (b) are different views of the same response*

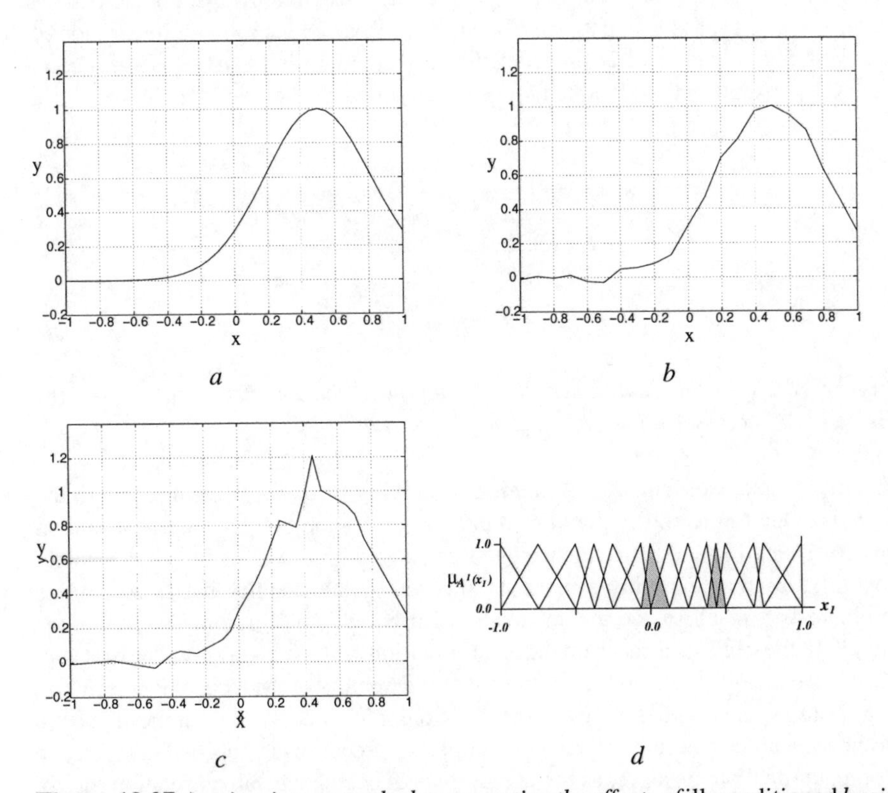

Figure 12.27 *A univariate example demonstrating the effects of* ill-conditioned *basis functions, (a) shows the true univariate Gaussian function, (b) shows the noisy training set, (c) shows the produced model response across a test set, and (d) shows the final univariate model, where the shaded basis functions represent the ill-conditioned ones*

this poor generalisation. An ill-conditioned basis function is one whose local autocorrelation matrix is ill-conditioned. Local autocorrelation matrices are formed from local submodels constructed on the support of the appropriate basis function, hence ignoring the continuity constraints often dominating the associated weight identification. In this trivial univariate example the local autocorrelation matrices of the two shaded basis functions are singular as only two data points lie in their supports. Local ill-conditioning of this kind is not picked up in the condition number of the complete model, as the continuity constraints of B-splines ensure a unique global solution. The condition number of the autocorrelation matrix of this univariate model is 109.4582, which is considered to be relatively small. This example is very trivial and the lack of generalisation is due to the extremely limited size of the training set. Despite this, it is clear that precautions must be taken to guard against ill-conditioned basis functions.

Returning to the 2-dimensional example, the lack of generalisation across the complete input space is due to one *ill-conditioned* basis function. An illustration of this is shown in Figure 12.28 on which the training points lying in the basis function support are superimposed. It can be seen that along x_2 the basis functions are not

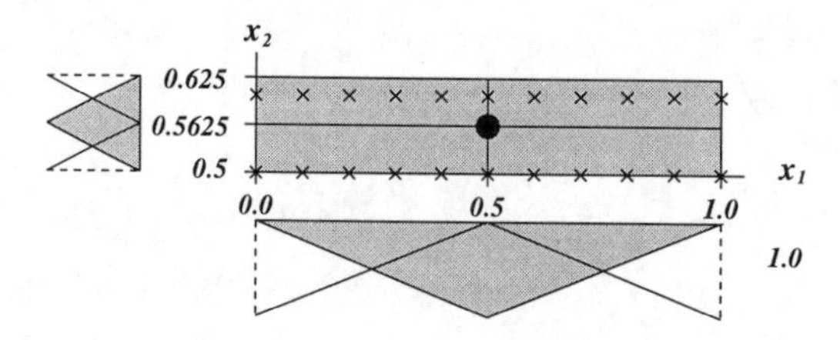

Figure 12.28 *The ill-conditioned basis function of the 2-dimensional model, the crosses represent the data points in this region*

sufficiently excited, with the 22 training pairs lying on two lines orthogonal to the x_2 axis. Due to the poorly distributed data in this region, the local autocorrelation matrix is singular and hence the weight associated with this basis function is solely identified by the continuity constraints imposed by the B-spline basis. Again this problem is a result of the unrealistically distributed, small training set, but does highlight the need for a check on the local condition of basis functions. The best way to achieve this would be to look at the condition number of the local autocorrelation matrix associated with any new basis function. If this condition number is above some threshold, then the refinement can be discarded. This approach may prove computationally expensive, as the eigenvalues of each local autocorrelation matrix have to be computed. A simple heuristic that would work in many cases would be to check the number and distribution of the training pairs in the supports of the new basis functions, but there are no guarantees that this type of heuristic would perform sufficiently well. The results of such tests can be incorporated into an active learning [15] scheme, to guide the selection of new training pairs in the region. This is only possible for systems where the acquisition of new data is relatively inexpensive. Active learning and data modelling of this type are becoming common techniques for aiding neural network learning. It should be stressed that system identification should be an iteractive procedure between the modeller and the data, as the data might not have been sufficiently gathered and also the identification procedure may perform poorly.

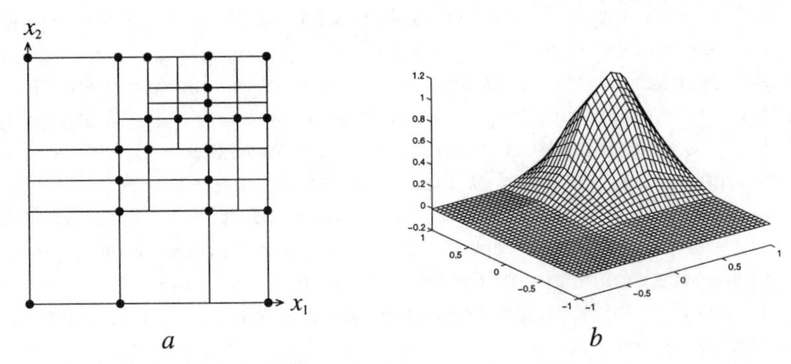

Figure 12.29 *The model produced by the halted hybrid algorithm, (a) shows the partitioning of the input space produced by this model, and (b) shows the model response*

It has been shown that the inclusion of ill-conditioned basis functions must be prevented to avoid producing models that generalise poorly. As an initial experiment construction was halted when a basis function with these properties was encountered. The hybrid construction algorithm was repeated and prematurely halted at the 25th iteration. This produced a model consisting of 23 parameters, the input space partitioning and response of which is shown in Figure 12.29. The response shows that the model successfully generalises across the input space. The final Bayesian statistical significance was -3260.78, appearing to perform better than both ASMOD and ABBMOD. As with the ABBMOD produced model, the transparency of the model still remains a problem, with lots of overlapping basis functions. As this hybrid model uses the k-d tree representation many of the disadvantages associated with the ABBMOD model still remain.

12.6 Conclusions

Neurofuzzy systems have been shown to be ideal candidates for *grey box* system modelling, combining the positive attributes of associative memory networks and fuzzy systems. Neurofuzzy systems provide a modelling technique to which conventional mathematical analysis can be applied, while still maintaining the transparency of fuzzy systems. This transparency provides a means by which *a priori* knowledge in the form of fuzzy rules can easily be incorparated into the design of the model. Also, when some form of construction algorithm is used to identify the structure of the model, the fuzzy rule model interpretation can give a qualitative insight into the modelled system's behaviour.

Unfortunately the size of conventional neurofuzzy systems is limited by the curse of dimensionality. Also, *a priori* knowledge from which an adequate model can be constructed is usually unavailable, and hence the neurofuzzy models have to be constructed by other means. These two problems have motivated the development of parsimonious neurofuzzy construction algorithms. In this chapter several off-line neurofuzzy construction algorithms that try to construct parsimonious neurofuzzy models from a representative training set have been presented. These algorithms are based on the general iterative construction algorithm, called COSMOS, as described in Section 12.3. The curse of dimensionality is a direct consequence of the lattice structure employed by conventional neurofuzzy models. To circumvent this problem, alternative neurofuzzy representations that exploit different structures to reduce the size and complexity of the resulting neurofuzzy system have been reviewed. These methods, the majority of which are based other modelling techniques, are presented in Section 12.4.

There are many different neurofuzzy representations and each one is appropriate for modelling different types of systems. They should give a more appropriate model of the system, hence producing models with generally better modelling properties e.g. better generalisation and simpler more representative rules. One of the main problems here is deciding when to apply an appropriate representation, a common problem faced by all modellers. The results of Section 12.5, despite being very preliminary, suggest that combining different approaches into a hybrid construction algorithm may be beneficial. The ultimate aim is to let the construction algorithm decide which representation to use, in which a mixture of different representations could be theoretically employed. An essential attribute when discussing neurofuzzy systems is the transparency of the final model, providing the modeller with insight into the physical behaviour of the system. The different neurofuzzy representations all have a different level of transparency, with some models conveying deep but often opaque knowledge. In applications where the transparency of the resulting model is important, less transparent representations must only be selected if a sufficient improvement in approximation quality is obtained. This strategy can be incorporated as a weight in the model performance criteria, where conventional neurofuzzy models incur no penalty. This presents another problem where an appropriate weight for each different neurofuzzy representation must be determined.

12.7 Acknowledgments

The authors are grateful for the financial support provided by GEC (Hirst Research Centre), Lucas Aerospace and the EPSRC during the preparation of this chapter.

12.8 References

[1] Bellman, R., 1961, *Adaptive Control Processes*. Princeton University Press

[2] Bishop, C, 1991, Improving the generalisation properties of radial basis function neural networks. *Neural Computation* 3: 579–588

[3] Bossley, K.M, Brown, M., and Harris, C.J., 1995, Neurofuzzy model construction for the modelling of non-linear processes. submitted to European Control Conference 1995, Rome.

[4] Brown, M. and Harris, C.J., 1994, *Neurofuzzy Adaptive Modelling and Control*. Prentice Hall, Hemel Hempstead

[5] Carlin, M., Kalvi, T. and Lillekjendlie, B., 1994, A comparision of four methods for non-linear data modelling. *Chemomtics and Intelligent Laboratory Systems* 23: 163–177

[6] Farlow, S.J., 1984, The GMDH algorithm. In *Self-Organising Methods in Modelling*, pp. 1–24, Marcel Decker, Statistics:textbooks and monographs vol. 54

[7] Friedman, J.H., 1991, Multivariate adaptive regression splines. *The Annals of Statistics* 19, no. 1: 1–141

[8] Haykin, S., 1994, *Neural Networks: A Comprehensive Foundation*. Macmillan College Publishing Company, New York

[9] Kavli, T., 1992, *Learning Principles in Dynamic Control*. Ph.D. Thesis, University of Oslo, Norway

[10] Kavli, T., 1993, ASMOD: an algorithm for Adaptive Spline Modelling of Observation data. *International Journal of Control* 58, no. 4: 947–968

[11] Mackay, D.J.C., 1991, *Bayesian Methods for Adaptive Models*. Ph.D. Thesis, California Institute of Technology, Pasadena, California

[12] Mills, D.J, Brown, M. and Harris, C.J., 1994, Training neurofuzzy systems. *IFAC Artificial Intelligence in Real-Time Control*. pp. 213–218

[13] Moody, J., 1989, Fast learning in multi-resolution hierarchies. In *Advances in Neural Information Processing Systems I*, pp. 29–39, Morgan Kaufmann

[14] Moody, J., 1994, Prediction risk and architecture selection for neural networks. In *From Statistics to Neural Networks: Theory and Pattern Recognition Applications*, edited by Cherkassky, V., Friedman, J. and Wechsler, H., Springer-Verlag

[15] Murray-Smith, R., 1994, *A local model network approach to nonlinear modelling*. Ph.D. Thesis, Department of Computer Science, University of Strathclyde

[16] Overmans, M.H. and Leeuwen, J.V., 1982, Dynamic multidimensional data structures based on quad and k-d trees. *Acta Infomatica* 17: 267–285

[17] Shewchuk, J.R., 1994, An introduction to the conjugate gradients method without the agonizing pain. Tech. Rep. CMU-CS-94-125, School of Computer Science, Carnegie Mellon University

[18] Sjöberg, J., 1995, *Nonlinear system identification with neural networks*. Ph.D. Thesis, Department of Electrical Engineering, Linköping University, S-581 83 Linköping, Sweden, (available via FTP as `joakim.isy.liu.se/pub.misc/NN.PhDsjoberg.ps.Z`)

[19] Sun, C., 1994, Rule-base structure identification in a adaptive network based inference system. *IEEE Transactions on Fuzzy Systems* 2, no. 1

[20] Wang, L., 1994, *Adaptive Fuzzy Systems and Control*. Prentice Hall, Englewood Cliffs, New Jersey

[21] Werntges, H.W., 1993, Partitions of unity improve neural function approximation. *IEEE International Conference on Neural Networks* 2: 914–918

Index